Sustainable Groundwater Development for Improved Livelihoods in Sub-Saharan Africa

This book provides a comprehensive analysis of the benefits and challenges of intensifying groundwater irrigation in Sub-Saharan Africa (SSA) for improving smallholder agrarian livelihoods. Only about 3% of the groundwater resources of Sub-Saharan Africa are used for irrigated agriculture despite the relative abundance of groundwater. The majority of the region's smallholders are highly dependent on seasonal dryland cropping, making them extremely vulnerable to uncertain weather patterns and droughts. Improved irrigation capabilities through sustainable groundwater development could unleash smallholder farming and make it a major driver of economic growth, poverty reduction, climate resilience, and improved food security. So, why is groundwater so underused? Tapping into groundwater requires a major shift in farming practices, and it has its own challenges and requirements – smallholder access to land and finance for irrigation infrastructure and equipment, gendered and equitable adoption options, supply chains, energy access, resource availability, and policy and institutional support.

The chapters in this book present a picture that is not only heterogeneous across the region but also hold some common denominators. They serve to enrich the discourse and help better understand the barriers along the pathways toward the sustainable and transformative adoption of groundwater irrigation. The scientific information provided herein would be of interest to researchers, practitioners, decision-makers, and planners with interest in the region. This book was originally published as a Special Issue of *Water International* journal.

Paul Pavelic is Senior Researcher at the International Water Management Institute (IWMI), Lao PDR and is also Associate Professor with Flinders University, Australia, by affiliation. Paul has led or participated in many interdisciplinary studies in Africa and Asia. His areas of expertise include groundwater assessment and management, conjunctive water use, urban hydrology and water reuse, agricultural water management, and managed aquifer recharge.

Karen G. Villholth is Director of Water Cycle Innovation (Pty) Ltd, South Africa. She has more than 30 years of experience in research for development. She worked for the International Water Management Institute (IWMI) for 13 years in Asia and Africa and chaired the Groundwater Solutions Initiative for Policy and Practice (GRIPP) for more than 5 years. Her areas of expertise include transboundary waters, groundwater and food security, groundwater resources assessment and integrated modelling, climate change and groundwater, groundwater-dependent ecosystems, Groundwater-Based Natural Infrastructure (GBNI), groundwater protection, 'Whole of the Water Cycle' management, and governance.

Shilp Verma is Senior Researcher, Water-Energy-Food Policies at the International Water Management Institute (IWMI), India. An alumnus of the University of Delhi, the Institute of Rural Management, Anand (IRMA) and UNESCO-IHE, Institute for Water Education, Delft, Netherlands, Shilp has more than 20 years of experience in the water, energy, agriculture, and rural livelihoods domain. In his current role, he leads the IWMI-Tata Water Policy Program in India besides providing leadership to IWMI's work on sustainable resource management and economic growth in Asia and Africa. His research covers a range of themes relevant for smallholder farming and water security in South Asia and Sub-Saharan Africa.

Routledge Special Issues on Water Policy and Governance

***Edited by:* Cecilia Tortajada** *(IJWRD) – Institute of Water Policy, Lee Kuan Yew School of Public Policy, NUS, Singapore, and* **James Nickum** *(WI) – International Water Resources Association, France*

Most of the world's water problems, and their solutions, are directly related to policies and governance, both specific to water and in general. Two of the world's leading journals in this area, the *International Journal of Water Resources Development* and *Water International* (the official journal of the International Water Resources Association), contribute to this special issues series, aimed at disseminating new knowledge on the policy and governance of water resources to a very broad and diverse readership all over the world. The series should be of direct interest to all policy makers, professionals and lay readers concerned with obtaining the latest perspectives on addressing the world's many water issues.

Source-to-Sea Management
Edited by Josh Weinberg, Qinhua Fang, Sarantuyaa Zandaryaa, Greg Leslie and James Nickum

Strengthening Cooperation over Transboundary Groundwater Resources
Edited by Gabriel Eckstein and Alice Aureli

Wicked Problems of Water Quality Governance
Edited by James E. Nickum, Raya Marina Stephan and Henning Bjornlund

Groundwater
Recent Advances in Interdisciplinary Knowledge
Edited by Raya Marina Stephan, James E. Nickum and Philippus Wester

Public Banks, Public Water
Exploring the Links in Europe
Edited by Thomas Marois and David A McDonald

Smart Water Management
Truly Intelligent or Just Another Pretty Name?
Edited by Henning Bjornlund, Stephanie Kuisma, James E. Nickum and Raya Marina Stephan

For more information about this series, please visit:
www.routledge.com/Routledge-Special-Issues-on-Water-Policy-and-Governance/book-series/WATER

Sustainable Groundwater Development for Improved Livelihoods in Sub-Saharan Africa

Edited by
Paul Pavelic, Karen G. Villholth and Shilp Verma

Routledge
Taylor & Francis Group
LONDON AND NEW YORK

IWRA

First published 2023
by Routledge
4 Park Square, Milton Park, Abingdon, Oxon, OX14 4RN

and by Routledge
605 Third Avenue, New York, NY 10158

Routledge is an imprint of the Taylor & Francis Group, an informa business

British Library Cataloguing in Publication Data
A catalogue record for this book is available from the British Library

ISBN13: 978-1-032-39191-5 (hbk)
ISBN13: 978-1-032-39193-9 (pbk)
ISBN13: 978-1-003-34878-8 (ebk)

DOI: 10.4324/9781003348788

Typeset in Minion Pro
by codeMantra

Publisher's Note
The publisher accepts responsibility for any inconsistencies that may have arisen during the conversion of this book from journal articles to book chapters, namely the inclusion of journal terminology.

Contents

Citation Information

The chapters in this book were originally published in the journal *Water International*, volume 38, issue 4 and 6. When citing this material, please use the original page numbering for each article, as follows:

Introduction

Identifying the barriers and pathways forward for expanding the use of groundwater for irrigation in Sub-Saharan Africa
Paul Pavelic, Karen G. Villholth and Shilp Verma
Water International, volume 38, issue 4 (2013) pp. 363–368

Chapter 1

Groundwater irrigation for smallholders in Sub-Saharan Africa – a synthesis of current knowledge to guide sustainable outcomes
Karen G. Villholth
Water International, volume 38, issue 4 (2013) pp. 369–391

Chapter 2

Smallholder groundwater irrigation in Sub-Saharan Africa: country-level estimates of development potential
Paul Pavelic, Karen G. Villholth, Yunqiao Shu, Lisa-Maria Rebelo and Vladimir Smakhtin
Water International, volume 38, issue 4 (2013) pp. 392–407

Chapter 3

Understanding smallholder irrigation in Sub-Saharan Africa: results of a sample survey from nine countries
Tushaar Shah, Shilp Verma and Paul Pavelic
Water International, volume 38, issue 6 (2013) pp. 809–826

Chapter 4

Do policy and institutional factors explain the low levels of smallholder groundwater use in Sub-Saharan Africa?
Srinivas Chokkakula and Mark Giordano
Water International, volume 38, issue 6 (2013) pp. 790–808

Chapter 5

Small pumps and poor farmers in Sub-Saharan Africa: an assessment of current extent of use and poverty outreach
Regassa E. Namara, Gebrehawaria Gebregziabher, Meredith Giordano and Charlotte De Fraiture
Water International, volume 38, issue 6 (2013) pp. 827–839

Chapter 6

Small pumps and the poor: a field survey in the Upper East Region of Ghana
Saa Dittoh, Joseph A. Awuni and Margaret A. Akuriba
Water International, volume 38, issue 4 (2013) pp. 449–464

Chapter 7

Improving the supply chain of motor pumps to accelerate mechanized small-scale private irrigation in Zambia
Willem Colenbrander and Barbara van Koppen
Water International, volume 38, issue 4 (2013) pp. 493–503

Chapter 8

Gender aspects of smallholder private groundwater irrigation in Ghana and Zambia
Barbara Van Koppen, Lesley Hope and Willem Colenbrander
Water International, volume 38, issue 6 (2013) pp. 840–851

Chapter 9

Integrated assessment of hydrogeology and water quality for groundwater-based irrigation development in the Raya Valley, northern Ethiopia
Tenalem Ayenew, Merhawi GebreEgziabher, Seifu Kebede and Sileshi Mamo
Water International, volume 38, issue 4 (2013) pp. 480–492

Chapter 10

Cost-benefit analysis and ideas for cost sharing of groundwater irrigation: evidence from north-eastern Ethiopia
Gebrehaweria Gebregziabher, Karen G. Villholth, Munir A. Hanjra, Muleta Yirga and Regassa E. Namara
Water International, volume 38, issue 6 (2013) pp. 852–863

Chapter 11

Predicting groundwater recharge in Ghana by estimating evapotranspiration
Fathi M. Anayah, Jagath J. Kaluarachchi, Paul Pavelic and Vladimir Smakhtin
Water International, volume 38, issue 4 (2013) pp. 408–432

Chapter 12

Groundwater potential for dry-season irrigation in north-eastern Ghana
E. Obuobie, D. Ofori, S. Agodzo and C. Okrah
Water International, volume 38, issue 4 (2013) pp. 433–448

Chapter 13

Constraints and opportunities for groundwater irrigation arising from hydrologic shifts in the Iullemmeden Basin, south-western Niger
Bio Mohamadou Torou, Guillaume Favreau, Bruno Barbier, Paul Pavelic, Mahamadou Illou and Fatoumata Sidibé
Water International, volume 38, issue 4 (2013) pp. 465–479

Notes on Contributors

Sampson Kwaku Agodzo, Agricultural Engineering Department, Kwame Nkrumah University of Science and Technology, Kumasi, Ghana.

Margaret A. Akuriba, School of Applied Economics and Management Sciences, University for Development Studies, Tamale, Ghana.

Fathi M. Anayah, College of Engineering and Technology, Palestine Technical University, Tulkarm.

Joseph A. Awuni, Department of Agricultural and Resource Economics, University for Development Studies, Tamale, Ghana.

Tenalem Ayenew, School of Earth Sciences, Addis Ababa University, Ethiopia.

Bruno Barbier, CIRAD, Ouagadougou, Burkina Faso.

Srinivas Chokkakula, Centre for Policy Research, New Delhi, India.

Willem Colenbrander, Livingstone Green Initiative, Livingstone, Zambia.

Saa Dittoh, WACWISA, University for Development Studies, Tamale, Ghana.

Guillaume Favreau, Institut de recherche pour le développement, Montpellier, France.

Charlotte De Fraiture, UNESCO-IHE, Institute of Water Education, Delft, The Netherlands.

Merhawi GebreEgziabher, Bren School of Environmental Science & Management, University of California, USA.

Gebrehaweria Gebregziabher, International Water Management Institute, Addis Ababa, Ethiopia.

Mark Giordano, School of Foreign Service, Georgetown University, Washington, D.C., USA.

Meredith Giordano, Office of Global Operations, Peace Corps, USA.

Munir A. Hanjra, International Water Management Institute, Pretoria, South Africa.

Lesley Hope, Independent Consultant, Ghana.

Mahamadou Illou, Department of Geography, University of Zinder, Niger.

Jagath J. Kaluarachchi, Utah Water Research Laboratory, Utah State University, Logan, USA.

Seifu Kebede, School of Engineering, University of KwaZulu-Natal, Durban, South Africa.

Barbara van Koppe, International Water Management Institute, Pretoria, South Africa.

Sileshi Mamo, School of Earth Sciences, Bahir Dar University, Ethiopia.

Regassa E. Namara, Sector of Water, The World Bank, USA.

Emmanuel Obuobie, Water Research Institute, Council for Scientific and Industrial Research, Ghana.

Deborah Ofori, Water Research Institute, Council for Scientific and Industrial Research, Ghana.

Collins Okrah, Water Research Institute, Council for Scientific and Industrial Research, Ghana.

Paul Pavelic, International Water Management Institute (IWMI), Vientiane, Laos.

Lisa-Maria Rebelo, International Water Management Institute, Colombo, Sri Lanka.

Fatoumata Sidibé, 2iE, Ouagadougou, Burkina Faso.

Vladimir Smakhtin, The United Nations University Institute for Water, Environment and Health (UNU-INWEH), Toronto, Canada.

Tushaar Shah, International Water Management Institute, India.

Yunqiao Shu, International Water Management Institute, Pretoria, South Africa.

Bio Mohamadou Torou, International Water Management Institute, Ouagadougou, Burkina Faso.

Shilp Verma, International Water Management Institute (IWMI), India.

Karen G. Villholth, Water Cycle Innovation (Pty) Ltd, South Africa, previously International Water Management Institute, Pretoria, South Africa.

Muleta Yirga, Department of Economics, Mekelle University, Ethiopia.

The affiliations refer to those relevant at the time of first publishing of this material. Only affiliations of the Editors and a few contributors have been updated, as relevant.

Identifying the barriers and pathways forward for expanding the use of groundwater for irrigation in Sub-Saharan Africa

Paul Pavelic, Karen G. Villholth and Shilp Verma

Sub-Saharan Africa (SSA) must significantly increase the area under irrigation for a host of pressing reasons that include addressing rural poverty, improving food security and countering droughts and famines. What remains less well known is the role that groundwater will play within the portfolio of potential irrigation strategies. Surface-water systems requiring large-scale public investments, which have captured most of the attention to date, do not always perform adequately, and are ill suited to targeting widely dispersed smallholder farming communities. Groundwater reserves are thought to be vast and broadly accessible, but estimates of their potential are still uncertain and the best development strategies are not always clear. Major challenges emerge on a number of fronts, notably inadequate knowledge of aquifer systems, the costs of drilling wells and operating pumps, and the level of technical and institutional capacity and overall governance in place to support sustainable management.

This special issue of *Water International* (along with a supplementary section in a subsequent issue this year) brings together the key findings of two major research efforts supported by the Rockefeller Foundation and the Bill & Melinda Gates Foundation and led by the International Water Management Institute (IWMI) with the support of numerous participating partners, many of whom are featured herein. The individual papers cover a number of topics, including:

- the current state of knowledge of groundwater irrigation (GWI) in SSA
- groundwater resource estimation in various regions and at different scales
- the significance of groundwater resources to smallholders and the broad-scale opportunities for livelihood enhancement that could emerge from sustainable development
- multi-country smallholder surveys on GWI and alternative water management options
- the major types of smallholder irrigators and their field-level realities and aspirations
- cross-cutting aspects, including the social and environmental impacts of intensive groundwater use

- the financial viability of GWI and cost-sharing approaches
- the gendered adoption patterns of small-scale private irrigation
- current policy and institutional context and barriers

The issue begins with a synthesis paper by Villholth, who presents updated data to support the case that the use of groundwater for irrigation by smallholder farmers is expanding more rapidly than previously thought throughout the sub-continent, and in recent years has mirrored the situation observed in India in the early stages of the Green Revolution. Importantly, most of the growth is being driven spontaneously by the farmers themselves, spurred by improved access to low-cost technologies for pumps and drilling services as well as market opportunities for their produce. There is a slow but marked shift in the emphasis of regional and international policies towards smaller-scale irrigated agriculture, along with recognition of the potential and suitability of groundwater. However, supportive national policies, strategies and enabling environments in the form of institutional arrangements and integrated approaches are needed to further unlock the potential of GWI for smallholders in SSA. While this is being done, increased focus on equitable access and environmental sustainability is required.

Smallholder irrigation, which encompasses GWI but also other sources, has thus emerged as a new priority in SSA to accelerate agricultural growth and enhance food security and the livelihoods of households in rural areas. What are the best ways for farmers, NGOs, donors and governments to make investments, recognizing the various modes of irrigation: gravity-flow irrigation from ponds and rivers, and manual or motor-pump irrigation from surface water or groundwater? Shah, Verma, and Pavelic (forthcoming) address this question by synthesizing results from field surveys of over 1500 smallholders in nine SSA countries. Those using any form of irrigation were found to be better off than those relying exclusively on rainfed farming, except in those few areas where the size of landholdings and rainfall are abundant enough to produce sufficient income. Irrigation using motor pumps leads to the best improvement in agricultural productivity. Motor-pump irrigator households tend to be male-headed and wealthier in terms of land and household assets, while manual pumps are more likely to be used by female irrigators for reasons of affordability and ease of adoption. Rainfed farmers expressed a clear preference for ownership of a well, motor pump and flexible water pipe when asked to choose from alternative modes of irrigation. The authors point out that simply making pumps more affordable in isolation, as is often suggested, would not necessarily suffice and that a more complex web of factors must be addressed conjunctively to accelerate the pace of smallholder irrigation development.

Despite the general recognition of the vastness of Africa's groundwater resources, the magnitude of that potential has not been assessed to date, even though such estimates have been derived in a number of previous studies for surface waters. Pavelic, Villholth, Shu, Rebelo, and Smakhtin apply a simple water balance to derive aggregated estimates for 13 African countries, making use of readily available data from secondary sources. The results show that a significant proportion of the rural population could potentially derive direct benefits. At the same time, the study cautions about sustainable limits and the need to balance multiple sectoral demands from a limited replenishable resource base. It highlights the need to further refine the assessment of the resource base and the contribution of groundwater needs for ecosystem services. As illustrated in the paper by Hagos, Ayenew, Mamo, and Adem (forthcoming), rapid and intensive development of groundwater supplies for urban, industrial and agricultural uses can have negative socio-economic and environmental impacts. Avoiding these requires improved local-level assessments, multi-stakeholder engagement and proactive governance arrangements at local through to regional levels, that ideally can prevent major unbalanced impacts and identify and rectify smaller problems as they emerge.

Namara, Gebregziabher, Giordano, and de Fraiture (forthcoming) address one of the key constraints to adoption identified by Villholth and Shah et al. and take inspiration from Asia, where the availability and affordability of small (and large) motorized pumps was one of the reasons behind the rapid expansion of smallholder private and communal irrigation systems. The paper presents a revealing picture of the current extent of pump-based irrigation in Ghana and Ethiopia and the socio-economic and demographic attributes of current small-pump adopters, including the variable impacts on poverty mitigation. Conclusions highlight the importance of following up pump promotion with well-targeted policy interventions and institutional support.

Policy and planning-related decision making at the national level requires reliable water resource assessments. Regional studies have not been undertaken previously in Ghana, with only limited and disparate information available to support regional planning and implementation. Anayah, Kaluarachchi, Pavelic, and Smakhtin present a new complementary approach which is suited to data-sparse settings, relies on meteorological data and overcomes uncertainties in land use and land cover. Estimates for the key water balance parameters are provided. Groundwater recharge estimates suggest that the aquifers are actively replenished and represent a significant resource that should be developed sustainably. Anayah et al. compare recharge results from previous studies and note significant differences. This lack of consistency exemplifies the lingering challenges (due to methodological as well as scale-related issues) faced in constraining recharge. It also emphasizes the need for the application of multiple methods to further reduce uncertainties.

Obuobie, Ofori, Agodzo, and Okrah investigate the opportunities and constraints of using shallow groundwater for irrigation in two watersheds of the Upper East Region of Ghana. Multiple field methods are applied, including geophysical surveying, water quality analysis, field mapping of groundwater-irrigated plots and field surveys of farming practices. In both areas, groundwater abstractions for multiple purposes are a small proportion of estimated annual replenishment, leaving an untapped potential to expand GWI. At the same time, existing irrigators face multiple challenges, including lack of appropriate technologies for drilling wells, lack of credit facilities, limited accessibility to land, and lack of access to stable markets. Most agricultural wells are dug manually, and farmers depend on traditional means for identifying suitable well sites that either lend themselves to a high risk of well failures or restrict access to the low-lying valley bottoms and along river channels, rather than the surrounding non-alluvial areas. Irrigation is mostly done manually, which adds to the drudgery of farm work, increases workload, and limits the area of cropped land. Recent changes in the policy and institutional environment at the national level seem to offer promise for future informal irrigation development.

Considering largely the same areas examined by Obuobie et al., Dittoh, Awuni, and Akuriba look more closely at the social dimensions of the rural communities and offer a complimentary perspective. The authors point to the limitations in the availability of valuable low-lying lands as a constraint to the area under cultivation. The shallow hand-dug wells, whilst making possible access to groundwater, are prone to drying up and can only be used for a few months during the dry season. Competition with water use for household domestic supplies creates tension within the communities when attempts are made to use deep wells for irrigation. Intensive use of groundwater is thought to have led to the lowering of water tables and associated loss of tree cover in some areas. The most serious constraints perceived by the farmers relate to the high cost of motor pumps and fuel and the tedium of digging and fetching water from shallow wells, rivers and streams. Expansion of irrigation should take place with mechanical methods of accessing and lifting water onto more elevated lands. Benefits would emerge from promoting the expansion of access to electricity for powering irrigation pumps and the removal of some taxes on pumps. Women's involvement in irrigation is now limited to menial tasks, while men practicing irrigation are producing crops that have traditionally been the preserve of women under rainfed conditions. Significant livelihood benefits would emerge if this "irrigation bias" against women were removed.

Torou et al. examine how regional land use changes brought about by the clearing of native vegetation in the Sahel for cultivated land can result in enhanced livelihood options through the unintended increase in groundwater resources. Income from GWI is small compared to alternative livelihood options (which themselves are limited), but probably reduces outmigration, now one of the leading alternatives. GWI involves an increasing number of small farmers, especially women, who dominate the practice but operate in male-headed households. Of the range of agricultural water management technologies available, deeper boreholes are more profitable than shallower wells as they are not prone to seasonal drying out and offer a higher degree of water control. They are less common, however, as they cost more. Access and cost are seen as greater constraints than water availability. Groundwater irrigators often lack physical capital and water their crops manually with cans and buckets. Other constraints include the limited scale of gardens brought about by the high cost of fencing, and lack of access to credit to purchase agricultural inputs. Hence, supportive measures for enhancing credit availability and agricultural services for the poor farmers could render GWI a sustainable recourse against drought and poverty in the region.

Ayenew, GebreEgziabher, Kebede, and Mamo present a detailed hydrogeological study from northern Ethiopia in a region targeted for irrigation expansion. The study reveals that there are large groundwater reserves residing within the valley-floor aquifers composed of alluvial sediment and volcanic rocks. Water quality does not pose a major constraint to agricultural water use. Ambitious plans for expanding the area under GWI in many parts of Ethiopia should be underpinned by such detailed assessments. Furthermore, subsequent monitoring and impact assessments after implementation should be emphasized, bearing in mind the lessons of negative impacts presented by the study by Hagos et al. (forthcoming). It is important to note that such analysis is rare (at best, emerging) in these countries. Nonetheless, it is critical to pursue this and to make the information available and usable early on to the respective national governments and the international development community.

Gebregziabher, Yirga, and Namara (forthcoming) firmly recognize that the transition from rainfed farming to GWI must prove to be a profitable investment for any smallholder. But economic viability is not enough; there is also the question of who should pay for the upfront investments and how to strike the appropriate balance of offering support without inducing over-dependence. These authors perform a cost-benefit analysis to assess the merits of the investments made by the government of Ethiopia over two decades in developing deep groundwater wells for smallholder irrigation in the Raya-Kobo Valley, an area where Ayenew et al. had examined the resource development potential. The analysis suggests that full cost recovery can be achieved over the expected life of the infrastructure. However, deeper questions about the opportunities for upscaling through the government's approach are raised, and alternative models are proposed that reconcile the limited upfront resources of the government administration acting as donors and the adopting farmers as recipients.

Hagos et al. (forthoming) offer an excellent example of an exception to the widely held view that groundwater development is modest at present and that opportunities for expansion abound. This case study from eastern Ethiopia shows how rapid and poorly planned development and intensive groundwater use have brought with them negative consequences that include the disappearance of a crater lake and sporadic conflict amongst water users. It also provides a methodology that combines hydrogeological and socio-economic analysis of groundwater-based smallholder irrigation in the study catchment. Whilst the system is clearly overexploited, the cumulative effect of all water users, including GWI, is the main issue. In household socio-economic terms, farmers who use groundwater for irrigation are better off than rainfed farmers. Borehole development has brought a considerable return for the users in terms of expenditure per adult equivalent. However, the paper brings to the fore the necessity of addressing, before development advances too far, the sustainable management of the resource and its many users.

Recognizing that there is very little knowledge about the way irrigation equipment markets function and benefit smallholders in SSA, Colenbrander and van Koppen examine the situation in Zambia, where smallholder GWI with motor pumps has increased considerably over the past decade. The authors have conducted interviews with NGOs, micro-finance institutions and government officials and show that smallholder farmers presently are severely inhibited in access and sustainable use of motor pumps, thereby lessening the poverty-alleviating potential of the pumps. Other weaknesses in the supply chain are revealed, such as the centralization of access to information, as well as inhibition of the flow of both goods and services by retail outlets and after-sales service centres. Recommendations are made as to how the chain can be strengthened.

Van Koppen, Hope, and Colenbrander (forthcoming) present a gender-differentiated analysis of adoption of smallholder private lift irrigation in Ghana and Zambia. Studies of this nature, focusing at the household level in SSA, are rare and could serve to inform both irrigation and gender policies. The key finding of the authors is that women are proactive irrigation adopters, in spite of the many obstacles they face in accessing productive resources. Where special investor/donor attention to gender enhances female access to production means, broader livelihood impacts are achieved. The study shows, however, that the farmers who invested privately in irrigation greatly outnumbered those benefiting from public scheme investments.

Chokkakula and Giordano (forthcoming) describe the policy and institutional landscape for smallholder GWI in SSA, with a particular focus on Ghana and Ethiopia, and raise the question: "Do policy and institutional factors explain the low levels of smallholder groundwater use in SSA?" Their analysis suggests that decision makers tend to ignore groundwater resources or are biased against its development due to a traditional focus on surface-water irrigation development. In addition, policies tend to disfavour development of the resource, partly because of the present discourse emphasizing regulation over facilitation. Some exceptions appear in Ethiopia, but plans there for groundwater development are optimistic and ambitious. Despite some progress, institutional and technical capacities are not yet commensurate with the task. The authors attempt to provide a set of measures and policy recommendations that can facilitate greater use of groundwater for smallholder irrigation.

The eight papers in this special issue, and a further six planned for a complementary section in a forthcoming issue later this year, were put together to provide a substantial and comprehensive analysis of *Intensifying Groundwater Use in Sub-Saharan Africa for Improving Smallholder Agrarian Livelihoods*. Collectively, they offer one of the most comprehensive, diverse, and up-to-date sets of studies on the topic. This is intended to provide relevant and novel scientific information and insights that would be of interest to researchers, practitioners, decision makers and planners addressing agrarian poverty, livelihood enhancement and groundwater management issues in SSA. As GWI strategies for SSA emerge and farmer-driven development is recognized as a viable alternative to state-directed irrigation in providing livelihoods, studies such as these should both inform and guide further development.

Taken together, the papers in this collection paint a picture that is not only complex and heterogeneous across regions, but also sometimes contradictory. All of these aspects serve to test the existing paradigm and enrich the discourse on a number of levels and help us to understand the array of barriers along the pathways toward the sustainable adoption of GWI – but also that, even without outside help, the farmers are going down those paths.

The manuscripts marked 'forthcoming' were reviewed externally but not finalized in time to appear in this issue.

References

Anayah, F. M., Kaluarachchi, J. J., Pavelic, P., & Smakhtin, V. (2013). Predicting groundwater recharge in Ghana by estimating evapotranspiration. *Water International*, *38*(4), 408–432. doi: 10.1080/02508060.2013.821642.

Ayenew, T., GebreEgziabher, M., Kebede, S., & Mamo, S. (2013). Integrated assessment of hydrogeology and water quality for groundwater-based irrigation development in the Raya Valley, northern Ethiopia. *Water International*, *38*(4), 480–492. doi: 10.1080/02508060.2013.821640.

Chokkakula, S., & Giordano, M. (2013). Do policy and institutional factors explain low levels of smallholder groundwater use in Sub-Saharan Africa? [tentative title]. *Water International* (forthcoming issue).

Colenbrander, W., & van Koppen, B. (2013). Improving the supply chain of motor pumps to accelerate mechanized small-scale private irrigation in Zambia. *Water International*, *38*(4), 493–503. doi: 10.1080/02508060.2013.819602.

Dittoh, S., Awuni, J. A., & Akuriba, M. A. (2013). Small pumps and the poor: A field survey in the Upper East Region of Ghana. *Water International*, *38*(4), 449–464. doi: 10.1080/02508060.2013.819454.

Gebregziabher, G., Namara, R. E., Giordano, M. A., & Yirga, M. (2013). Cost-Benefit and cost-sharing analysis of deep well groundwater irrigation: Evidence from northern Ethiopia. [tentative title]. *Water International* (forthcoming issue).

Hagos, F., Ayenew, T., Mamo, K., & Adem, S. (2013). Groundwater irrigation and its impact on sustainable livelihoods of smallholder farmers: The case of Haromaya watershed, Eastern Ethiopia. *Water International* (forthcoming issue).

Namara, R. E., Gebregziabher, G., Giordano, M. A., & de Fraiture, C. (2013). Small pumps and poor farmers in Sub Saharan Africa: An assessment of current extent of use and the poverty outreach. *Water International* (forthcoming issue).

Obuobie, E., Ofori, D., Agodzo, S. K., & Okrah, C. (2013). Groundwater potential for dry season irrigation in Northeastern Ghana. *Water International*, *38*(4), 433–448. doi: 10.1080/02508060.2013.814212.

Pavelic, P., Villholth, K. G., Shu, Y., Rebelo, L. -M., & Smakhtin, V. (2013). Smallholder groundwater irrigation in Sub-Saharan Africa: Country-level estimates of development potential. *Water International*, *38*(4), 392–407. doi: 10.1080/02508060.2013.819601.

Shah, T., Verma, S., & Pavelic, P. (2013). Understanding smallholder irrigation in Sub-Saharan Africa: Results of a sample survey from nine countries. *Water International* (forthcoming issue).

Torou, B. M., Favreau, G., Barbier, B., Pavelic, P., Illou, M., & Sidibé, F. (2013). Constraints and opportunities for groundwater irrigation arising from hydrologic shifts in the Iullemmeden Basin, south-western Niger. *Water International*, *38*(4), 465–479. doi: 10.1080/02508060.2013.817042.

van Koppen, B., Hope, L., & Colenbrander, W. (2013). A gender-differentiated analysis of adoption of small-scale private lift irrigation in Ghana and Zambia. [tentative title]. *Water International* (forthcoming issue).

Villholth, K. G. (2013). Groundwater irrigation for smallholders in Sub-Saharan Africa – a synthesis of current knowledge to guide sustainable outcomes. *Water International*, *38*(4), 369–391. doi: 10.1080/02508060.2013.821644.

Groundwater irrigation for smallholders in Sub-Saharan Africa – a synthesis of current knowledge to guide sustainable outcomes

Karen G. Villholth

Groundwater irrigation for smallholder farmers in Sub-Saharan Africa is growing in extent and importance. This growth is primarily driven spontaneously by the farmers themselves, spurred by improved access to low-cost technologies for pumps and drilling services as well as market opportunities for produce. This paper presents a review of the current status and knowledge of the prospects and constraints for sustainable and pro-poor groundwater irrigation in Sub-Saharan Africa. Further unlocking the potential of groundwater irrigation for smallholders will require better integrated approaches, simultaneously addressing groundwater-access constraints as well as enabling factors.

Introduction

Groundwater irrigation (GWI) is growing in extent and importance in arid and semi-arid areas of Sub-Saharan Africa (SSA) (Giordano, de Fraiture, Weight, & van der Bliek, 2012).[1] Groundwater responds to the demand of farmers for a reliable and flexible irrigation water supply. Also, better and more appropriate and cost-effective drilling and pump technologies, services and markets make it increasingly feasible. Giordano (2006) optimistically estimated the total groundwater-irrigated area in SSA to be 1–2 million ha. In Ethiopia, where development is presently promoted and is advancing rapidly with private as well as public funds, government statistics say that about 800,000 motor pumps were imported into the country between 2004 and 2010 (Namara, Gebregziabher, Giordano, & de Fraiture, 2013). Assuming that a quarter of these pumps went into GWI, each of them effectively irrigating 0.5 ha of cropland, this translates into about 100,000 ha of GWI land added over this period. Similar recent developments are seen in other parts of SSA. In Malawi, about half of the irrigated land (equivalent to about 20,000 ha) is serviced by lift irrigation (much of which is from groundwater), mostly using treadle pumps, with an increase observed over the last five to 10 years (Namara et al., 2013). In West Africa, it is now surmised that in Ghana, official figures for public irrigation schemes (mostly large-scale surface irrigation of 13,300 ha) have been surpassed several-fold by small-scale informal lift irrigation (approximately 190,000 ha), though official data are lacking (Namara et al., 2013).

Smallholders are poorer farmers, generally with landholdings that are smaller than 2 ha, privately owned, and under the complete control of the farmer (Abric et al., 2011).

Smallholder GWI is increasingly promoted by governments, donors and NGOs (Abric et al., 2011; CAADP, 2009; Chokkakula and Giordano, 2013; Kay, 2001). Small-scale irrigation, including GWI, is the most rapidly expanding type of irrigation in Africa (Frenken, 2005). Smallholder GWI is seen as an important vehicle to promote (especially rural) poverty alleviation, food security (at various scales, from local to global) and land and labour productivity, as well as rural employment and general economic development and adaptation to increasing climate variability (Ngigi, 2009). Such strategies partly reflect the recognition that large-scale irrigation schemes (mostly from dams) have not created the expected benefits nor sufficiently targeted or benefited the poorest farmers (Abric et al., 2011; CAADP, 2009; Naugle & Sellen, 2006). Compared to surface water and communal irrigation schemes, GWI is believed to be more attractive to farmers due to the individual modes of uptake and operation (Abric et al., 2011; Naugle & Sellen, 2006). Current evidence shows that smallholders prefer GWI, whether subsidized or not (Abric et al., 2011; Giordano et al., 2012). Where they can find the means and finances, farmers often access groundwater to enhance their livelihoods, especially by moving from subsistence agriculture towards market-oriented farming (Dittoh, Awuni, & Akuriba, 2013; Shah, Verma, & Pavelic, 2013).

While the understanding of the potential and role of GWI for poverty alleviation and food production in SSA is rapidly increasing due to various recent larger regional research projects, research on GWI in SSA more generally has remained fragmented, anecdotal and focused on certain geographic regions or topics. There is a need for a broader, integrated and interdisciplinary approach to the research to ensure context-sensitive and evidence-based recommendations for future development. Furthermore, various development models, ranging from community-based deep groundwater schemes to individualistic shallow groundwater promotion, have been implemented, but with limited understanding of the relative impacts on poverty and food security of these interventions. There is a need for more evidence-based approaches to the support for GWI in SSA.

Hence, this article draws on the manuscripts submitted for this collection to frame the current narrative on GWI for smallholders in SSA and to review current knowledge in order to support such integrated and streamlined approaches. Its specific objectives are to update the estimate of the current extent of GWI in SSA, to classify SSA countries in terms of GWI potential based on multiple criteria, to propose a simple typology for GWI in SSA, to review knowledge on the socio-economic aspects and impacts of GWI, and to highlight constraints on further GWI development. It concludes with recommendations for supplementary research and critical issues for further supporting sustainable GWI development in SSA.

Current groundwater irrigation development in Sub-Saharan Africa

As noted by Giordano (2006) and Frenken (2005), official statistics on GWI, and particularly smallholder GWI, in SSA are imperfect and often not representative of the realities on the ground. This is partly because of methodological problems associated with the assessments (due to the inherent difficulty of discriminating the source of irrigation water, i.e. surface or groundwater, in public statistical databases or in remote-sensing approaches) and partly because of rapid changes, quickly outpacing the data. Hence, in order to get a more realistic estimate of GWI area, data from recent case studies as well as other recently reported data were used to supplement existing FAO-based compiled data (Siebert et al., 2010). This analysis shows that there are countries, like Malawi and Ethiopia, that have GWI areas well in excess of the data presented in Siebert et al. (2010) (Table 1) – in these

Table 1. Extent of groundwater irrigation in Sub-Saharan Africa (1000 ha).

	Siebert et al. (2010)			This study			
Country	Total irrigated area[a]	Area irrigated by surface water	Area irrigated by groundwater	Total irrigated area	Area irrigated by surface water[b]	Area irrigated by groundwater	Source for data on 'Area irrigated by groundwater-This study' (when different from Siebert et al. (2010)
Angola	35.0	28.0	7.0	35.0	28.0	7.0	
Benin	7.1	5.9	1.3	7.1	5.9	1.3	
Botswana	0.62	0.33	0.29	0.62	0.33	0.29	
Burkina Faso	25.0	22.0	3.0	52.0	22.0	30.0	Abric et al. (2011)
Burundi	21.4	21.4	0	21.4	21.4	0	
Cameroon	25.7	24.7	1.0	25.7	24.7	1.0	
Cape Verde	1.8	1.6	0.26	1.8	1.6	0.26	
Central African Republic	0.069	0.069	0	0.069	0.069	0	
Chad	26.2	21.0	5.2	26.2	21.0	5.2	
Comoros	0.085	0.082	0.033	0.085	0.082	0.033	
(Republic of the) Congo	0.22	0.22	0	0.22	0.22	0	
Côte d'Ivoire	66.9	66.9	0	66.9	66.9	0	
Dem. Republic of the Congo	6.8	6.8	0	6.8	6.8	0	
Djibouti	0.39	0	0.39	0.39	0	0.39	
Equatorial Guinea	0	0	0	0	0	0	
Eritrea	42.6	32.6	10.1	42.6	32.6	10.1	
Ethiopia	290.7	288.1	2.6	405.3	288.1	117.2	Namara et al. (2013)
Gabon	4.5	4.5	0	4.5	4.5	0	
Gambia	1.4	1.4	0.010	1.4	1.4	0.010	

(Continued)

Table 1. (Continued).

Country	Siebert et al. (2010)			This study			Source for data on 'Area irrigated by groundwater-This study' (when different from Siebert et al. (2010)
	Total irrigated area[a]	Area irrigated by surface water	Area irrigated by groundwater	Total irrigated area	Area irrigated by surface water[b]	Area irrigated by groundwater	
Ghana	54.7	42.7	12.0	228.7	42.7	186.0	Namara et al. (2013)
Guinea	94.9	94.5	0.46	94.9	94.5	0.46	
Guinea-Bissau	22.6	17.7	4.9	22.6	17.7	4.9	
Kenya	97.0	96.0	0.97	174.5	96.0	78.5	Mumma, Lane, Kairu, Tuinhof, & Hirji (2011)
Lesotho	0.067	0.0.17	0.050	0.067	0.017	0.050	
Liberia	2.1	2.1	0.011	2.1	2.1	0.011	
Madagascar	1080.7	1080.7	0	1080.7	1080.7	0	
Malawi	54.1	54.1	0.030	74.0	54.1	19.9	Namara et al. (2013)
Mali	176.8	176.1	0.75	181.1	176.1	5.0	Abric et al. (2011)
Mauritania	22.8	20.4	2.4	22.8	20.4	2.4	
Mauritius	20.8	15.6	5.2	20.8	15.6	5.2	
Mozambique	40.1	39.8	0.22	40.1	39.8	0.22	
Namibia	7.6	5.9	1.6	7.6	5.9	1.6	
Niger	65.6	64.4	1.2	80.5	64.4	16.2	Abric et al. (2011)
Nigeria	218.8	154.8	64.0	337.8	154.8	183.0	Abric et al. (2011)
Rwanda	8.5	8.4	0.085	8.5	8.4	0.085	
São Tomé and Príncipe	9.7	9.7	0	9.7	9.7	0	

(Continued)

Table 1. (Continued).

Country	Siebert et al. (2010)			This study			Source for data on 'Area irrigated by groundwater-This study' (when different from Siebert et al. (2010)
	Total irrigated area[a]	Area irrigated by surface water	Area irrigated by groundwater	Total irrigated area	Area irrigated by surface water[b]	Area irrigated by groundwater	
Senegal	69.0	62.1	6.9	69.0	62.1	6.9	
Seychelles	0.20	0.20	0	0.20	0.20	0	
Sierra Leone	10.0	9.9	0.15	10.0	9.9	0.15	
South Africa	1498.0	1370.7	127.3	1713.3	1370.7	342.7	Pietersen, Beekman, & Holland (2011)
Sudan (N&S)	800.0	770.3	29.7	800.0	770.3	29.7	
Swaziland	44.9	44.0	0.90	44.9	44.0	0.90	
Togo	6.3	6.2	0.046	6.3	6.2	0.046	
Uganda	5.9	5.8	0.059	5.9	5.8	0.059	
United Republic of Tanzania	189.0	171.6	17.5	189.0	171.6	17.5	
Zambia	155.9	149.3	6.6	253.0	149.3	103.8	Pavelic, Giordano, Keraita, Rao, & Ramesh (2012); Mayerhofer, Shamboko-Mbale, & Mweene (2010)
Zimbabwe	123.9	109.6	14.3	180.6	109.6	71.0	Pavelic, Giordano, et al. (2012)
Total	5, 436.6	5, 108.1	328.5	6, 356.9	5, 108.1	1, 248.8	
Groundwater as fraction of total			0.06			0.20	

[a]Siebert et al. (2010) statistics consider only two categories of water sources for irrigation: surface water and groundwater.
[b]The surface water irrigated area was kept constant from Siebert et al. (2010) estimate, assuming that GWI is the most significantly growing irrigation sector presently (Giordano et al., 2012).

cases by factors of 660 and 45, respectively. Similar but smaller underestimations were found for Ghana, Kenya, Niger and Zambia (Table 1). In all countries, values from recent field investigations, where available, were always higher than official statistics. This implies that similar tendencies can be expected for the remainder of the countries in SSA, and hence our data may still be underrepresenting the reality. General underestimation is likely even though some of the figures updated include irrigation categorized as smallholder "lift irrigation", which could also include irrigation from surface-water sources (in the case of Ethiopia, Ghana and Malawi) or "smallholder irrigation" (in Burkina Faso, Mali, Niger and Nigeria). The total figure for GWI area is 1,248,800 ha, representing 20% of the total irrigated area in SSA (Table 1), significantly higher than earlier estimates of 6% and 10% (Giordano, 2006; Siebert et al., 2010). According to this analysis, the total GW-irrigated area has almost quadrupled (from 328,500 ha). Assuming, alternatively, that one-third of the increase in countries with data for "lift irrigation" and one-fourth of the increase in countries with data for "smallholder irrigation" are attributable to GWI – and applying the average rate of increase from these countries (except two outliers) to the rest of the countries without updated information – yields a total GWI acreage of 1,222,200 ha, or 19% of the irrigated area. Though independent data are required to verify increases in individual countries from the second method, the data suggest a significant transition in irrigation preferences among farmers.

The potential of groundwater irrigation in Sub-Saharan Africa

Despite recent expansion, irrigation in SSA (irrespective of source) is limited, compared to most other regions of the world. Only 3.3% of arable land (approximately 6 million ha) is irrigated, compared to 37% in Asia (Siebert et al., 2010). Hence, there is ample room for expanding areas under irrigation (Foster & Briceño-Garmendia, 2010). You et al. (2010) estimated the potential contribution from small-scale irrigation (e.g. ponds, small reservoirs, rainwater harvesting, and groundwater) in Africa to be anywhere from 0.3 to 16 million ha, based on a distributed multi-criteria analysis. They also found that the internal rate of return on investment in small-scale irrigation was significantly higher than for large-scale (dam-based) irrigation schemes (You et al., 2010). They did not estimate the specific contribution of groundwater. Only recently has dedicated research addressed this resource, mostly in terms of its physical limits. Pavelic, Smakhtin, Favreau, and Villholth (2012) developed a simple catchment water balance–based method for estimating allocation limits to GWI (translated also into GWI areas depending on crop water requirements) as residuals of annual groundwater recharge after accounting for existing demands, including domestic, livestock, industrial and environmental. They applied this approach to two catchments in Sahelian West Africa with good data to illustrate the considerable scope for further GWI development in these areas (Pavelic, Smakhtin, et al., 2012). A subsequent application at the national scale for 13 countries in SSA illustrated that all countries have variable but significant potential for GWI expansion, in total an area of 13 million ha, potentially serving 26 million additional smallholder households (Pavelic, Villholth, Smakhtin, Shu, & Rebelo, 2013). Hence, groundwater may not restrict development in these countries, at least in the short term.[2] This is encouraging and warrants further investigations into best options and locations, while not neglecting local conditions and impacts that may differ from the national results.

To further refine assessments, mapping of relative development potential at distributed sub-national scales has been attempted in Ethiopia (Awulachew, Erkossa, & Namara, 2010; Santini, Peiser, & Faurès, 2012) and Ghana (Evans, Giordano, & Clayton, 2012; Forkuor,

Pavelic, Asare, & Obuobie, 2013; Gumma & Pavelic, 2012). The three Ghana studies adopt a hydrogeological, a hydrogeophysical, and a biophysical suitability approach, respectively. Even granting that the objective of Forkuor et al. (2013) was broader, on the potential for adoption of small motor pumps (including for surface-water pumping), the three studies arrive at inconsistent and sometimes diverging results, illustrating the complexities of data availability, interpretation, mapping and verification of results. Still, it highlights the need to combine and integrate socio-economic approaches, such as market options, infrastructure and livelihoods options, into the more typical hydrogeological and geophysical studies.

Table 2 gives key figures relevant for GWI development in 14 selected SSA countries. These figures relate to economic growth, poverty (here "hunger index" [von Grebmer et al., 2012] as a proxy for the need for increased food productivity), current irrigation development, groundwater resources availability and accessibility, cost of development, and a couple of infrastructure parameters that reflect the degree of rural development ("Share of population without access to electricity" and "Proportion of good rural road networks"). A first observation is that all countries except South Africa have low shares of irrigated land.

For a simple multiple objective analysis, the countries can be grouped into four overall groups with respect to GWI development potential and scope. The first are relatively arid countries with low resource endowments, where only localized GWI development may make sense (Kenya and South Africa). The slightly more GW-affluent countries (Mali, Niger and Tanzania), though semi-arid, have zones of various extents where (further) GWI may be feasible and desirable, considering poverty levels. The next group contains more promising countries: Burkina Faso, Ethiopia, Ghana, Malawi, Mozambique, Nigeria and Zambia. These are low-income countries (perhaps excepting Ghana and Nigeria) with rapid current GWI development yet with untapped physical potential. In accordance, some of these countries exhibit low and falling prices of GWI development (Xenarios & Pavelic, 2011). In the last category are Rwanda and Uganda, countries with relatively low aridity (NB: high aridity index), signifying less need (under present population and climate conditions) for irrigation and hence little documented development of GWI (Table 1). Most of the remaining countries in SSA probably fall into one of these four categories. Table 2 further shows that the 14 countries vary significantly in electrification rates and proportion of good rural networks, factors that could significantly influence GWI uptake, making simple generalizations or recommendations based on constraints inappropriate. As an example, Ghana generally has both good roads and relatively good electrification. Mozambique is quite the opposite. Burkina Faso has good roads but poor electrification, while nearby Nigeria has good electrification but poor roads. Constraints to GWI development will be discussed later in this paper.

Groundwater irrigation typology

While the focus in this paper is on GWI for smallholders, groundwater is accessed through diverse schemes, varying by scale, funding source, ownership of land and resources, type and depth of groundwater utilized, crops grown, degree of market orientation and systems used to extract the groundwater. A simple typology that encompasses the most prominent forms of GWI in SSA is suggested in Table 3, which distinguishes between two overall parameters: depth of the groundwater utilized and funding source. According to this classification, which may not be exclusive or exhaustive, smallholder schemes fall under Types 2, 3 and 4. Type 1 represents larger-scale (>100 ha), mechanized, often export-oriented crop

Table 2. Key figures related to groundwater irrigation in selected countries in SSA.

Country	Country-level GDP[a] (US$/capita)	Hunger index[b] (% decrease, 1990 – 2012)	Aridity index[c]	Non-irrigated arable land[d], Mha	% of arable land irrigated (SW+GW)[d]	GW irrigated land[e], 1000 ha	GW renewable resource[d], km^3/yr, ($m^3/yr/cap.$)	Share of land w. sufficient aquifer yield[f], %	Cost of drilling[g]: Per well, US$ (depth of wells)	Cost of drilling[g]: Per m, US$	Share of pop. w/o electricity access[h]	Proportion of good rural road network[i], %
Burkina Faso	663	17.2 (26.8)	0.38	5.9	0.42	30.0	9.5 (543)	100	10,000–20,000 (40–60m)	166–500	90	63
Ethiopia	359	28.7 (32.0)	0.45	13.7	2.1	117.2	20.0 (231)	95	8,000–26,700 (60–100 m)	136–258	85	33
Ghana	1528	8.9 (58.4)	0.73	4.3	1.2	186.0	26.3 (1030)	79	6000 (60 m)	100	46	51
Kenya	850	19.3 (6.8)	0.34	5.3	1.8	78.5	3.5 (82)	96	8,400–21,000 (700–300 m)	70–120	85	49
Malawi	350	16.7 (42.8)	0.69	3.5	1.5	19.9	2.5 (157)	100	5000 (50 m)	100	91	41
Mali	668	16.2 (41.7)	0.15	6.2	2.8	5.0	20.0 (1226)	94	14,260–36,800 (30–58m)	320–641	83	–
Mozambique	582	23.3 (34.4)	0.58	5.0	0.79	0.22	17.0 (695)	94	4,900–7,900 (31–55m)	136–166	88	18

(*Continued*)

Table 2. (Continued).

Country	Country-level GDP[a] (US$/capita)	Hunger index[b] (% decrease, 1990 – 2012)	Aridity index[c]	Non-irrigated arable land[d], Mha	% of arable land irrigated (SW+GW)[d]	GW irrigated land[e], 1000 ha	GW renewable resource[d], km^3/yr, (m^3/yr/cap.)	Share of land w. sufficient aquifer yield[f], %	Cost of drilling[g]		Share of pop. w/o electricity access[h]	Proportion of good rural road network[i], %
									Per well, US$ (depth of wells)	Per m, US$		
Niger	399	22.3 (38.7)	0.06	14.9	0.44	16.2	2.5 (150)	92	–	157–503	91	31
Nigeria	1490	15.7 (34.9)	0.66	33.7	0.64	183.0	87.0 (522)	52	3,300–13,000 (50 m)	66–260	53	14
Rwanda	605	19.7 (30.1)	0.77	1.3	0.65	0.085	7.0 (621)	100	–	–	95	0
South Africa	8066	5.8 (15.9)	0.33	12.9	10.4	342.7	4.8 (95)	92	2,618–4,761 (50–200 m)	13–95	25	31
Tanzania	553	19.3 (16.8)	0.61	9.8	1.9	17.5	30.0 (630)	64	5,000–13,500 (50–100m)	100–193	89	31
Uganda	477	16.1 (13.9)	0.70	6.6	0.09	0.059	29.0 (814)	35	11,000	183	91	0
Zambia	1413	23.3 (6.0)	0.59	3.2	4.7	103.8	47.0 (3385)	93	2,600–55,000 (30–120m)	86–458	81	16

[a]GDP: Gross Domestic Product. Wikidepia (2013).
[b]von Grebmer et al. (2012).
[c]Global Potential Evapotranspiration (Global-PET) and Aridity Index (Global-Aridity). CGIAR-CSI (2009). AI < 0.65: Arid/semi-arid areas with high aridity.
[d]FAO (2013).
[e]This study, see Table 1.
[f]Assuming an aquifer (well) yield of 0.5 l/s as a lower threshold for feasible smallholder irrigation (Pavelic et al., 2013; MacDonald, Bonsor, Dochartaigh, & Taylor, 2012).
[g]Pavelic, Giordano, et al. (2012a). These figures pertain to deep wells (most likely motorized drilling). Shallow manual well drilling will be cheaper (Abric et al., 2011).
[h]UNDP-WHO (2009).
[i]Foster and Briceño-Garmendia (2010).

Table 3. Typology of GWI systems.

			Depth of wells[a]	
			Deep	Shallow
Funding source	Private[b]		1. Commercial, larger-cale, mechanized, export-oriented	2. Informal, small-scale, farmer-driven
		Examples:	*Flower farms in Ethiopia, Center-pivot grain farms in Zambia*	*Vegetable growing schemes in Northeast Ghana*
	Public[c]		3. Deep systems, subsidized	4. Shallow systems, subsidized
		Examples:	*Public schemes in Raya-Kobo in Ethiopia*	*Fadama systems in Nigeria*

[a]Somewhat arbitrarily, the following distinction has been made: Deep wells are >20 m deep (typically drilled wells), shallow wells <20 m deep (typically dug wells).
[b]Investment derived from the farmers themselves.
[c]Investment is derived from external sources, like the public sector, NGOs, international donors, etc.

production, such as flower farms in Ethiopia (Pavelic, Giordano, Keraita, Rao, & Ramesh, 2012), grain producers in the outskirts of Lusaka, Zambia (Mayerhofer, Shamboko-Mbale, & Mweene, 2010), and similar systems in Zimbabwe and South Africa (Masiyandima & Giordano, 2007; Pietersen, Beekman, & Holland, 2011). Type 2 is characterized by private development of shallow groundwater by the farmers themselves, usually individually or in small groups, typically using rudimentary wells and extraction means, like ropes and buckets or human- or animal-operated mechanical pumps. With low capital investments, these systems, exemplified by the floodplain and valley-bottom systems of the Upper East Region of Ghana (Dittoh et al., 2013; Namara et al., 2011), are financed by the farmers themselves. Type 3 is deep-well public systems supported by government, donors or non-governmental organizations (NGOs) for groups of farmers, as seen in the Raya-Kobo Valley in Ethiopia (Ayenew, GebreEgziabher, Kebede, & Mamo, 2013; Gebregziabher, Yirga, & Namara, 2013) and Maunganidze in Zimbabwe (IFAD, 2007). Finally, Type 4 is exemplified by shallow well smallholder schemes subsidized with irrigation structures and input from the public sector, donors and NGOs, such as the *fadama*[3] systems in Nigeria, supported since the 1990s mostly by the World Bank (Abric et al., 2011; Nkonya, Phillip, Mogues, Pender, & Kato, 2010). Key figures for some smallholder schemes are given in Table 4. These will be further discussed in the remainder of this paper.

Agronomic and socio-economic aspects and impacts of GWI

Numerous studies indicate that farmers favour GWI, as the resource is always available and they have autonomy over its control. This leads to lowered risk in investments for other inputs (seeds, fertilizers and energy), which in turn leads to intensification, diversification,

Table 4. Data from irrigation cases, including GWI, in SSA, with focus on smallholder systems.

Country, region	Irrigation started	Plot size per HH[a] (ha)	GW depth, well type	Combined w. rainfed?	Lifting device, irrigation device	Crops, GWI season	GWI practitioners	Land tenure	GWI type (cf. Table 3)	GWI dev. stage[b]	Farming w. highest productivity[c]	External financial support	Reference
Ethiopia Raya-Kobo Valley	>1995	~ 0.25	Deep, wells (60-170 m)	Yes	Elec. pumps, furrow/ buckets/ sprin-kler	Onion/ tomato/ pepper Dry season	Smallholder farmers	Lease/ share cropping/ own	3	2	Motor pump irrigation from GW	Partly gov't/ NGOs	Gebregziabher et al. (2013); Ayenew et al. (2013); Abate (2006)
Ghana Upper East Region	1890's	0.01–0.21	Shallow, dugouts/ HDWs[d]	Yes	Rope and buckets	Tomato/ onion/ pepper Dry season	Mostly women and youth	Lease	2	1	Bucket irrigators from GW	Limited	Obuobie, Ofori, Agodzo, & Okrah (2013); Namara et al. (2013); Dittoh et al. (2013)
Niger Niamay Capital District	>1990	0.13	Deep + shallow, various	Yes	Hand/ foot pumps, buckets	Onion/ tomato/ cabbage Dry season	2/3 women	Individual/ collec-tive	2	1	GW irrigation	Partly gov't/ NGOs	Torou et al., 2013
Nigeria Northern Nigeria	>1993	0.5–1.0	Shallow, bore-holes	Yes	Motor pumps, flooding	Onion, cabbage, pepper, tomatoes All year	Smallholder farmers	Individual/ lease	4	2-3	GW irrigation	World Bank/ Gov't	Abric et al. (2011); Nkonya et al. (2010); Dabi (no year)

[a]HH: Household; [b]GWI development stage acc. to Deb Roy and Shah (2003). 1 signifies early stages of groundwater development while 4 indicates mature and over-development of groundwater; [c]Gross revenue per ha; [d]HDW: Hand-dug well.

higher and more stable outputs, and higher productivity and net incomes (per area under cultivation and per labour input) as compared to rainfed or surface-water systems (for SSA, see Shah et al., 2013; Dittoh et al., 2013).

GWI is almost exclusively utilized in high-value horticulture production for market (mostly vegetables and to a smaller degree fruits, flowers, herbs, etc.), and rarely for staple and non-food crops. This reflects the higher production costs, including labour, of GWI, increasing market demands for horticulture crops, higher suitability of smaller irrigated and intensively managed plots, and greater willingness of farmers to risk pursuing higher-value crops. Contrary to common arguments for increasing irrigation in SSA, smallholder GWI does not appear to be converting rainfed staple crops (i.e. grain) to being irrigated, even on a supplementary basis (Shah et al., 2013). What is observed generally is that groundwater supports dry-season irrigation on separate (often much smaller) dedicated plots, for garden-style crop production. Often, rainfed cropping and the irrigated gardens coexist and combine to serve the livelihood strategies of the farmer households (Dittoh et al., 2013) by addressing their needs for a more diversified and nutritious diet. Often, women take a primary role in irrigated gardening while men focus on rainfed (Namara et al., 2013; Shah et al., 2013; van Koppen, Hope, & Colenbrander, 2013). Livestock, often partly dependent on groundwater for their drinking water and fodder production, also forms an important source of income, food and wealth security for many of the smallholder systems observed, depending on factors such as climate, population density, traditional production systems, land tenure, soils and culture.

In the Upper East Region of Ghana, Dittoh et al. (2013) found that farmers using groundwater by manual means obtained at least 20% higher net revenues per area irrigated than those using pumps or surface water (Table 4). Similarly, Kamwamba (2012) found that value added per area for GWI (pump or manual) was at least twice that of other irrigation systems in Malawi. It also provided the largest value added per labourer. Kamwamba-Mtethiwa, Namara, de Fraiture, Mangisoni, and Owusu (2012) found that treadle-pump users using shallow groundwater or surface water in Malawi were better off than non-adopters. Ajayi and Nwalieji (2010) found a significant increase in the number of farmers in higher income brackets upon adoption of vegetable production as a result of the *fadama* interventions in Nigeria. Abate (2006) found similar but less well-documented results in the deep-well irrigation schemes in the Kobo Valley in Ethiopia. However, for various reasons, crop productivity from GWI (Abric et al., 2011) still lags behind international benchmark standards FAOSTAT (2013).

GWI may not unilaterally alleviate poverty for the poorest segment of farmers. While Namara et al. (2011) found that small farmers in Ghana significantly improved their livelihoods through access to GWI – and local rural societies were stabilized through reduced distress migration due to increased livelihoods and employment opportunities – in a later study, they found that direct GWI benefits in other areas mostly accrued to the wealthier farmers (Namara et al., 2013). Although treadle pumps support irrigation livelihoods, they do not significantly alleviate labour for lifting water (Kamwamba, 2012), nor are they commonly available to the poorest or to women (Adeoti, Barry, Namara, Kamara, & Titiati, 2007; Kamwamba-Mtethiwa et al., 2012). Most studies focusing on distributional impacts of irrigation development, including GWI, find that farmers adopting motor-pump irrigation are better off than exclusively rainfed farmers or adopters of other types of irrigation (Kamwamba, 2012; Namara et al., 2013; Shah et al., 2013). However, such studies fail to account for *ex ante* differences between groups, and hence they are inconclusive as to whether the uptake of GWI alleviates poverty or simply maintains or aggravates existing wealth disparity (Kamwamba-Mtethiwa et al., 2012).

Women farmers, either as sole heads of households or as married companions to their husbands, participate in and derive benefit from groundwater irrigation (van Koppen et al., 2013). Limited results suggest that GWI may suit well the roles and capabilities of women. However, while preliminary findings suggest that GWI is also benefiting women, they also suggest that women are disadvantaged due to the lack of secure land tenure, as traditionally they have no right of land inheritance (van Koppen et al., 2013; Torou et al., 2013). Hence they are limited in obtaining loans (Wahaj, Hartl, Lubbock, Cleveringa, & Nepveu, 2007) and thereby acquiring irrigation equipment. This may partly explain why female-headed households have smaller plots irrigated by groundwater and tend to use manual methods for lifting water (Kamwamba, 2012; Namara et al., 2013; Shah et al., 2013; van Koppen et al., 2013). Also, female farmers may have less labour available than men, due to the generally smaller size of female-headed households (Naugle & Sellen, 2006; van Koppen et al., 2013). Traditional "women's" chores such as fetching water and firewood further reduce the time that can be devoted to other activities (Coulter, Kebede, & Zeleke, 2010). Hence, gender-related factors may cause GWI to underperform its expected potentials. These findings point to the need to consider gender and distributional aspects in order to understand statistics of GWI adoption rates and poverty impacts, and to derive appropriate and acceptable recommendations for further support in order to increase equality in access to water, land, credit and technology. Preliminary research indicates that where women farmers are favoured in grant and credit schemes, positive outcomes are achieved for the overall communities (van Koppen et al., 2013).

Constraints to further groundwater irrigation development

In analyzing the constraints facing poor farmers in increasing productivity from groundwater-irrigated crop production in SSA, one may distinguish between direct factors that are specific to accessing groundwater and ones that are indirect and linked to common constraints faced by farmers, such as knowledge and markets for purchasing inputs and selling outputs. Villholth, Ganeshamoorthy, Rundblad, and Knudsen (2013) proposed such a generalized framework, essentially building on a value-chain concept, in which GWI is seen as an input to individual food-crop value chains as well as the outcome of its own value chain.

Direct constraints refer to accessing groundwater by means of wells and lifting devices. The techniques used depend on groundwater and hydrogeological conditions. For shallow, high-yielding aquifers, very simple technologies can be applied, such as manual digging and human lifting of water. In this case, the primary constraint is the availability of labour. Deeper drilling and various types of mechanized pumps may be more common as they become available or necessary due to groundwater-level decline. In this phase or in areas with deeper groundwater, a major constraint for small farmers is access to mechanized technologies.

Pumps

Although no data exist on the distribution and prevalence of the various types of pumps at the regional level, preliminary research indicates that most smallholder GWI in SSA at present is done with manual lifting and small pumps using diesel or petrol (Abric et al., 2011). In Malawi, motorized pumping is used on 2800 ha, while treadle pumps (driven by human foot power) contribute to 13,000 ha (Namara et al., 2013). In a larger survey in two of the nine regions in Ethiopia, Namara et al. (2013) found that 31% of farmers used water-lifting devices. Of these, 84% used buckets, while 16% used motor pumps. The major advantage of motorized pumps is their considerably greater capacity to lift water, making it possible to expand irrigated surface areas. Their major disadvantages include higher capital and recurrent costs (for fuel and maintenance) (Perry, 1997). Namara et al. (2013) found diesel to be the energy source most used for motor pumps in Ethiopia, followed by petrol and then electricity, reflecting the unit cost and availability of these sources in the rural areas. Pump prices range from USD60 to USD360 for human-powered pumps and from USD80 to more than USD2000 for motorized pumps (3–5 hp) (Abric et al., 2011). Pumps are increasingly imported from South-East and East Asia (China and India), displacing Europe and Japan (Perry, 1997), which are higher-cost sources (Abric et al., 2011; Colenbrander & van Koppen, 2013).

Pump rental markets have emerged among smallholders (Namara et al., 2013), but groundwater sales by well owners to non-owners, as reported in parts of India (Mukherji, 2004), have not been documented in SSA, presumably because of lower population and irrigation intensity.

The pump supply chains are underdeveloped and unsustaianable in a variety of ways. The provision of sales and services is concentrated in urban areas (Colenbrander & van Koppen, 2013); spare-part availability is unreliable; quality control is limited; and supply diminishes or breaks down altogether after supportive external project interventions end (Abric et al., 2011). Large travel distances present farmers with high transaction costs in acquiring pumps (Takeshima, Adeoti, & Salau, 2009).

Wells

Well construction can be classified into three types: manual digging (superficial water only), manual drilling (mostly shallow wells, less than about 20 m and through unconsolidated materials) and motorized drilling (mostly deep drilling and through harder formations), with costs increasing from the rudimentary to the more advanced technologies. Abric et al. (2011) show prices for low-cost shallow manual drilling in West Africa as approximately one-tenth of prices given for deep wells (Table 2). Hence, manual drilling has been promoted and adopted widely in West Africa as a suitable approach for smallholder irrigation (Abric et al., 2011). UNICEF and International Development Enterprises have produced maps of appropriate areas for manual drilling (IDE, 2013; UNICEF, Practica, & and Enterprise Works, 2011).

Credit and capital

Capital is the largest constraint faced by poor farmers in SSA in accessing new technology. Farmers can draw from their own savings, borrow from family, use remittances or sales of farm proceeds, go to private money lenders, obtain loans from formal credit institutions, or benefit from different subsidy or donation models. Subsidies have been broadly used for pumps, wells, fertilizers, energy and other inputs (e.g. Abric et al., 2011), theoretically increasing the chance of favouring the poorest. Unfortunately, they often do not (Wiggins & Leturque, 2010). Also, subsidies tend to distort the market and to generate a culture of dependency (Naugle & Sellen, 2006). They also undermine policies of water and energy demand management (Rosegrant & Perez, 1997). Treadle pumps tend not to be used as efficiently when subsidized as they are when purchased with farmers' own funds (Abric et al., 2011). Where subsidies are used, they need to be "smart": targeted to those who need them; limited in time; and designed to enhance rather than supplant commercial production and distribution (Wiggins & Leturque, 2010). Furthermore, ensuring a degree of consistency and transparency in subsidy schemes across various initiatives and programmes is critical to their success (Abate, 2006; Abric et al., 2011).

Direct grants or subsidies for GWI development have been applied in Ethiopia (Gebregziabher et al., 2013), Zimbabwe (deep wells) (IFAD, 2007) and Nigeria (shallow wells) (Abric et al., 2011), among others. Though the public investments are viable over the service life of the projects, outsourcing, especially for deep well interventions, is costly and not economically sustainable for the public sector (Gebregziabher et al., 2013). It is estimated that the total investment costs of the *fadama* systems in Nigeria have been USD1650/ha (Abric et al., 2011), just less than a suggested optimal upper limit for capital investment in small-scale irrigation of USD2000/ha (Foster & Briceño-Garmendia, 2010), while capital costs (including power transmission lines and concrete distribution canals) of USD10,940/ha for deep-well systems were found in Zimbabwe (IFAD, 2007), and USD4900/ha (including electric power and powerhouse and installation for drip and sprinkler systems) in Ethiopia (Gebregziabher et al., 2013). Gebregziabher et al. (2013) propose a cost-sharing mechanism through long-term pay-back to the government from the smallholder farmers to sustain financial viability. No experience with such recovery schemes yet exists.

Loans from formal credit institutions are often out of reach of farmers, both physically, as these facilities are not present in the rural areas, and financially, due to high interest rates and farmers' lack of collateral (Chokkakula and Giordano, 2013). There is a large potential for micro-finance schemes that specifically target rural households for irrigation investments (Chokkakula and Giordano, 2013; Colenbrander & van Koppen, 2013); yet little attention has been paid to supporting GWI. Present impediments include lack of diffusion of micro-finance to rural areas and lack of knowledge of GWI technologies by lenders (Colenbrander & van Koppen, 2013). Ideally, micro-finance would be designed to break the persistent land-tenure constraints that disadvantage small farmers, especially women, in getting access to credit. Also, tax and duty exemptions on imported pumps for smallholders could improve their availability (Colenbrander & van Koppen, 2013; Giordano et al., 2012)

Energy

Energy for mechanized groundwater pumping typically comes from fossil fuels or electricity. Electrical pumping has lower running costs and is cleaner (Pavelic, Giordano et al., 2012), but it is constrained by low rural electrification rates, which average about 12% in SSA (Table 2) (Energy for Development and Poverty Reduction, 2013). Naturally, groundwater use and irrigation tend to increase with rural electrification, as seen in South Africa, Kenya and Ethiopia (Pavelic, Giordano et al., 2012). Similarly, mechanized GWI was significantly promoted by public fuel subsidies in Nigeria, lowering the cost of fuel to one-half to one-third that of neighbouring countries (Abric et al., 2011). Promoting rural electrification, combined with rational supply and pricing policies, is an efficient way of promoting GWI development. While individual metering of use entails excessive transaction costs, alternative approaches of flat-rate tariffs for individuals or groups of users, combined with restriction and separation of power for irrigation (Pietersen et al., 2011) would show promise in the SSA context. The current prospects for using alternative energy sources like biofuels and solar power for irrigation are limited but should be further explored (Abric et al., 2011; Burney, Woltering, Burke, Naylor, & Pasternak, 2010).

Land tenure

Insecure land tenure is consistently mentioned as a constraint to the further development of small-scale irrigation (CAADP, 2009; Namara et al., 2011; Shah et al., 2013). But this issue is not easy to address, as tenure is highly complex and diverse across the settings, with local arrangements often bound to traditional and informal regulations. Land tenure rights are often highly skewed towards the wealthier farmers. Where much of the land is rented or under communal tenure, smallholder insecurity hampers irrigation expansion (IFAD, 2007). Recent large land acquisitions by foreign investors tend to support large-scale farming, and so are also affecting smallholders' access to cultivated land or water (Allan, Keulertz, Sojamo, & Warner, 2012).

Markets for produce

To be profitable, GWI must be used for cash crops. Cash crops need markets. Those in turn necessitate a demand for the crops, proximity to outlets, reliable road infrastructure, and options for selling the crops at favourable prices. Generally, increasing urbanization and diversified diets, general economic development, and increasing means for and interest in processed food have increased the demand for vegetables and fruit in SSA (Perry, 1997). Limited road networks, especially in the rural areas (Foster and Briceño-Garmendia (2010) (Table 2), restrict accessibility to markets, except for increasing peri-urban agriculture (FAO, 2011). Because they lack storage facilities and telecommunication capabilities, smallholder farmers often find themselves trapped in inferior negotiating positions and selling at low prices to middlemen at times of oversupply (Chokkakula and Giordano, 2013; Villholth, Ganeshamoorthy et al., 2013). The proliferation of mobile phones is quickly remedying the telecommunication and information barriers in many areas (How we made it in Africa, 2012).

Groundwater resources: availability, assessment and impacts

Groundwater resources are generally plentiful in SSA (MacDonald et al., 2012) and still developed to a smaller extent, traditionally, and most critically for rural domestic water supply (Carter & Beven, 2008). However, their sustainable exploitation is limited by replenishment rates (even more than by storage capacity), extractability in some regions, and groundwater's role in the hydrological cycle and as a provider of environmental services (Colvin, Le Maitre, Saayman, & Hughes, 2007). Hence, with the ongoing rapid expansion of groundwater use for irrigated agriculture and emerging evidence of resource degradation in groundwater-dependent agricultural areas (Hagos, Ayenew, Mamo, & Ade, 2013; Ngigi, 2009; Pietersen et al., 2011), it is imperative that environmental sustainability issues are critically examined and accounted for in the planning, development and management of GWI.

Much smallholder GWI in SSA, especially when fully farmer-driven, occurs on an opportunistic, unregulated basis, with little regard to the resource endowment. Even in larger government and donor-driven development, either little prior comprehensive assessment of the groundwater resource potential is carried out (Abric et al., 2011) or significant uncertainty in estimates prevails (Gebregziabher et al., 2013). In a few cases, an emerging broader conceptual, qualitative and quantitative understanding of the groundwater systems is accompanying development, for example in Ghana (Obuobie, Ofori, Agodzo, & Okrah, 2013), Niger (Favreau et al., 2009) and Ethiopia (Ayenew et al., 2013). Hence, dedicated and integrated studies are required to comprehend the circulation patterns of groundwater at various scales as well as the possible impacts of development and factors such as climate change and land use modifications that affect the resource base (Taylor et al., 2010; Villholth Tøttrup, Stendel, & Maherry, 2013). While development of the knowledge base and of the resource run in parallel for GWI, the knowledge generated does not necessarily feed into planning and is mostly unrelated to investigations of socio-economic aspects, missing the wider issues of integrated sustainability (Silliman, Hamlin, Crane, & Boukari, 2008).

The negative impacts of groundwater development on agriculture include depletion and degradation of the resource as well as potential social conflicts. In arid and semi-arid areas, salt accumulated in upper soils due to evapotranspiration on previously uncultivated land may generate transitory salt leaching and soil and groundwater contamination upon clearing and irrigation (Favreau, Seidel, Leduc, Marlin, & Mariotti, 2004). The major issue in most cases, however, relates to overdrafting of groundwater, leading to continuous lowering of the groundwater level and increasing costs of exploitation (Abric et al., 2011), as well as potential effects on groundwater-dependent surface water bodies (Hagos et al., 2013). Conflicts with non-farmers and with those drawing their drinking water from the same resource, especially during drought, are likely (Coulter et al., 2010, Hagos et al., 2013).

Often, the source of the water used for irrigation is unclear. In the primary phases of development and for rudimentary systems, groundwater is often drawn from shallow aquifers, many of which (as with the *fadama* systems) are in direct connection with surface-water systems. Hence, there is a need to recognize the complementarity of these resources, their interconnectedness, and the scope for planned conjunctive use, to increase water use efficiency, mitigate undesirable fluctuations in water supply, and control shallow water-table levels and soil salinity (Evans & Evans, 2012, Foster & van Steenbergen, 2011). This may to some extent already be ongoing in an unplanned manner, as farmers with access to motorized pumps find themselves enabled to access multiple water sources in an opportunistic manner. The importance of such access transcends groundwater, empowering farmers to extend their domain of water control. However, issues of well siting (Carter, Chilton, Danert, & Olschewski, 2010; MacDonald, Davies, & Dochartaigh, 2001), drilling technologies (Abric et al., 2011) and maintenance of wells and pumps (Furey & Danert, 2012) are particular to GWI in SSA and still cause concerns for overall sustainability.

Policy and institutional aspects of groundwater irrigation

International and regional development policies increasingly favour smallholder GWI (Abric et al., 2011; CAADP, 2009), but these are not reflected in national policies and institutions. Ethiopia does have significant, almost unrealistically optimistic, plans to expand groundwater development, including for irrigation (Ministry of Water Resources, 2011), but, as in many countries, institutional capacities are insufficient and roles are unclear. Often, GWI falls between the mandates of the agricultural and WASH (water, sanitation and hygiene) sectors. In addition, irrigation planning continues to favour large-scale public surface water schemes, or (e.g. in Ethiopia) community-based government-driven groundwater development (CAADP, 2009; Chokkakula and Giordano, 2013). Farmer-driven smallholder GWI receives little recognition or support. Chokkakula and Giordano (2013) argue that an excessive focus on regulation under the guise of integrated water resources management (IWRM) is at fault. Whatever the current paradigm, the real culprit is more likely inertia in irrigation development strategies and planning, reflecting vested professional and bureaucratic interests. Smallholder GWI development will require policy makers to recognize that the state can be a facilitator of individual initiative as well as a regulator.

Discussion and conclusions

Sub-Saharan Africa is finally catching up with GWI development after lagging behind other developing economies (Shah, Burke, & Villholth, 2007). The estimated expansion of 248,800 ha to 1,177,000 ha over a 7-year period corresponds to growth rates in India in the beginning of the 1960s (Narayanamoorthy, 2010). Concurrently, a wealth of knowledge has been generated over the last decade through the dedicated research reviewed in this paper. Groundwater irrigation is very popular among farmers, and may presently be expanding at a higher rate than other types of irrigation (Frenken, 2005). Hence, smallholder GWI may well be a suitable and socio-economically acceptable alternative and supplement to large-scale surface water irrigation. However, a number of issues remain to be resolved:

- Getting a better understanding of the extent and depth of the benefits and the socio-economic context and variable forms under which smallholder GWI exists across SSA
- Understanding the residual potential for GWI as a resource while avoiding negative environmental impacts from increased use
- Developing better ways to mainstream GWI development through well-targeted, integrated, informed and feasible interventions
- Building the human and institutional capacity to manage groundwater and GWI at various levels

Presently, the high economic barriers faced by smallholder farmers imply that (1) the poorest farmers are unlikely to initially adopt and directly benefit from GWI;[4] and therefore that (2) increased attention must be devoted to credit and financing schemes that lower this barrier, especially for women and marginalized groups, who constitute a significant proportion of farmers.

Further promotion of smallholder GWI requires the consideration of various models of support, as indicated in Table 3, while taking into account resource conditions and financing options. There is a trade-off between the shallow, informal systems (Types 2 and 4) and the deep, more formal systems (Types 1 and 3). The deeper systems tend to provide more secure and perennial water availability, but require larger initial investments, higher maintenance and stricter farmer organization to deal with water sharing (the latter particularly for Type 3). The public sector also tends to have better control of implementation, farmer training, and monitoring of deep systems, whilst farmer-driven informal systems tend to be spontaneous, autonomous and still largely unrecognized by the public sector.

Whether support comes in the form of direct investment in deep wells or indirect subsidies to farmers for shallow ones, the reviewed research uniformly affirms the fundamental nature of supplemental support to enabling factors, such as energy, credits or subsidies for agricultural inputs, infrastructure for transport and markets, and extension services on cropping, irrigation, marketing and groundwater management (Lipton, Litchfield, & Faurès, 2003; Nkonya et al., 2010). One notable success story was an EU-funded project in Zimbabwe, where farmers risked investing in inputs to obtain higher yields and achieved a 265% increase in farm income due to the provision of an assured market by NGO-established grower associations and of reliable groundwater supplies (IFAD, 2007).

Ensuring community engagement, preferably empowerment, and local management of land and water resources should not be neglected. Despite GWI being primarily a private enterprise that benefits farmers individually, long-term viability often hinges on joint beneficiary action in areas such as management of land and water, marketing of produce, and negotiation for grants and policy change. Hence, building on existing organizations such as farmer groups or farmers' cooperatives may be more feasible for irrigation and groundwater management than the conventionally prescribed water user associations (Ghazouani, Molle, & Rap, 2012). Further, basic farmer education repeatedly emerges as influential for greater adoption and better outcomes of GWI (Burney & Naylor, 2012).

In sum, governments have to realize their dual role as facilitator as well as regulator in the development of GWI in SSA and this across relevant sectors. Understanding the groundwater resources and its users and managing them, require up-front attention and integration into policy and decision making. This becomes the more critical as larger shares of groundwater are being appropriated for agriculture.

Acknowledgments

The data and material in this article derive primarily, but not exclusively, from outputs of two recently terminated projects led by the International Water Management Institute: Groundwater in Sub-Saharan Africa: Implications for Food Security and Livelihoods (2009–2012, funded by the Rockefeller Foundation) and Agricultural Water Management Solutions (2009–2012, funded by the Bill & Melinda Gates Foundation).

Notes

1. Sub-Saharan Africa is defined to include all of Africa with the exception of the North African countries of Algeria, Egypt, Libya, Morocco and Tunisia.
2. Population increases, welfare increases and climate change impacts have not been accounted for.
3. *Fadama* refers to the low-lying areas in Nigeria, where groundwater is shallow and in direct connection with surface-water systems that are flooded or waterlogged part of the year (Tarhule & Woo, 1997).
4. However, general agricultural development may still have an overall positive poverty alleviation impact, through for instance reduced food prices and increased labour options for the poorest (Lipton et al., 2003).

References

Abate, A. (2006). *Small holder farmers' experience on pressurized irrigation systems in Kobo Valley* (pp. 18–21). Proceedings of MoARD/MoWR/USAID/IWMI Symposium: Best Practices and Technologies for Small Scale Agricultural Water Management in Ethiopia.

Abric, S., Sonou, M., Augeard, B., Onimus, F., Durlin, D., Soumaila, A., & Gadelle, F. (2011). *Lessons learned in the development of smallholder private irrigation for high- value crops in West Africa*. Washington, DC: World Bank.

Adeoti, A., Barry, B., Namara, R. E., Kamara, A., & Titiati, A. (2007). *Treadle pump irrigation and poverty in Ghana* (IWMI Research Report No. 117). Colombo, Sri Lanka: International Water Management Institute.

Ajayi, A. R., & Nwalieji, A. H. (2010). Impact of the Anambra State Fadama Project Phase –1 on the socio-economic life of the rural farmers. *Journal of Human Ecology*, *29*(2), 129–139.

Allan, T., Keulertz, M., Sojamo, S., & Warner, J. (Eds.). (2012). *Handbook of land and water grabs: Foreign direct investment and food and water security*. London: Routledge.

Awulachew, S. B., Erkossa, T., & Namara, R. (2010). *Irrigation potential in Ethiopia – constraints and opportunities for enhancing the system*. IWMI, International Water Management Institute.

Ayenew, T., GebreEgziabher, M., Kebede, S., & Mamo, S. (2013). Integrated assessment of hydrogeology and water quality for groundwater irrigation development in the Raya Valley, Northern Ethiopia. *Water International*, *38*(4), 480–492. doi: 10.1080/02508060.2013.821640

Burney, J. A., & Naylor, R. L. (2012). Smallholder irrigation as a poverty alleviation tool in Sub-Saharan Africa. *World Development*, *40*(1), 110–123. doi:10.1016/j.worlddev.2011.05.007

Burney, J., Woltering, L., Burke, M., Naylor, R., & Pasternak, D. (2010). Solar-powered drip irrigation enhances food security in the Sudano–Sahel. *PNAS*, *107*(5), 1848–1853.

Carter, R. C., & Beven, J. E. (2008). Groundwater development for poverty alleviation in Sub-Saharan Africa. In S. Adelana & A. MacDonald (Eds.), *Applied groundwater studies in Africa. IAH selected papers on hydrogeology* (Vol. 13, pp. 25–42). London: Taylor & Francis.

Carter, R., Chilton, J., Danert, K., & Olschewski, A. (2010). *Siting of drilled water wells – A guide for project managers*. Rural Water Supply Network Field Note 2010–5. June 2012.

CGIAR-CSI (2009). Global Potential Evapotranspiration (Global-PET) and Aridity Index (Global-Aridity). Retrieved from: http://csi.cgiar.org/aridity. Consultative Group on International Agricultural Research - Consortium for Spatial Information. Accessed 26 July 2013.

Chokkakula and Giordano (2013). Improving groundwater irrigation access to smallholders in Sub-Saharan Africa: Policy and institutional realities and challenges. *Water International* (forthcoming issue).

Colenbrander, W., & van Koppen, B. (2013). Improving the supply chain of motor pumps to accelerate the mechanization of small private irrigation in Zambia. *Water International*, *38*(4), 493–503. doi: 10.1080/02508060.2013.819602

Colvin, C., Le Maitre, D., Saayman, I., & Hughes, S. (2007). *An introduction to aquifer dependent ecosystems in South Africa*. WRC Report TT 301/07, Water Research Commission, Pretoria.

Comprehensive Africa Agriculture Development Programme (CAADP). (2009). *Sustainable land and water management*. The CAADP Pillar I Framework. Tool for use by Countries in Mainstreaming and Upscaling of Sustainable Land and Water Management in Africa's Agriculture and Rural Development Agenda.

Coulter, L., Kebede, S., & Zeleke, B. (2010). *Water and livelihoods in a highland to lowland transect in Eastern Ethiopia*. Water Economy Baseline Report. Working Paper 16 of RIPPLE (Research-inspired Policy and Practice Learning in Ethiopia and the Nile region).

Dabi, D. D. (no year). *Asessing the vulnerability of fadama (floodplain) agriculture to climate variability and change – A case study of semi-arid Northern Nigeria*. Unpublished paper. Retrieved from https://docs.google.com/viewer?a=v&q=cache:8a4RbRILJ_sJ:www.start.org/Program/advanced_institute3_web/p3_documents_folder/Dabi_Final%2520Report.doc±Dabi,±D.D±irrigation±fadama±nigeria&hl=en&gl=za&pid=bl&srcid=ADGEEShHPISowbU3YAIPbB6DLNT3yBNls327h_swjt2hxGugBLqkL0xAGMEbvkLzDD8HIwOEBIicSDgiOhBoojopwXCNOAQwiENANBsXLt2DsWWdoZYMcs-Dir_Hdr0wkCjqfqX7Lx3r&sig=AHIEtbQvC3WyvPBDSRIl8diZyUn1NeHVJA

Deb Roy, A., & Shah, T. (2003). Socio-ecology of groundwater irrigation in India. In R. Llamas & E. Custodio (Eds.), *Intensive use of groundwater: Challenges and opportunities* (pp. 307–335). The Netherlands: Balkema Publishers.

Dittoh, S., Awuni, J. A., & Akuriba, M. A. (2013). Small pumps and the poor: A field survey in the Upper East Region of Ghana. *Water International*, *38*(4), 449–464. doi: 10.1080/02508060.2013.819454

Energy for Development and Poverty Reduction (2013). http://www.energyfordevelopment.com/2010/04/rural-electrification-by-numbers.html. Accessed 26 July 2013.

Evans, W. R., & Evans, R. (2012). *Conjunctive use and management of groundwater and surface water*. Thematic Paper 2. Groundwater Governance: A Global Framework for Country Action. GEF ID 3726.

Evans, A. E. V., Giordano, M., & Clayton, T. (Eds.). (2012). *Investing in agricultural water management to benefit smallholder farmers in Ghana*. AgWater Solutions Project Country Synthesis Report. Colombo, Sri Lanka: International Water Management Institute (IWMI), (IWMI Working Paper 147), 37pp. doi:10.5337/2012.209

FAO (2011). *The Place of Urban and Peri-urban Agriculture (UPA) in National Food Security Programmes*. Rome: FAO.

FAO (2013). AQUASTAT database, Food and Agriculture Organization of the United Nations (FAO). Retrieved July 6 2013, from: http://www.fao.org/nr/water/aquastat/data/query/index.html?lang=en.

FAOSTAT (2013). http://faostat.fao.org/site/567/default.aspx#ancor. Accessed 26 July 2013.

Favreau, G., Cappelaere, B., Massuel, S., Leblanc, M., Boucher, M., Boulain, N., & Leduc, C. (2009). Land clearing, climate variability, and water resources increase in semiarid southwest Niger: A review. *Water Resources Research*, *45*. doi:10.1029/2007WR006785

Favreau, G., Seidel, J. L., Leduc, C., Marlin, C., & Mariotti, A. (2004). *Land clearance as a cause of groundwater nitrate increase in semiarid Niger* (pp. 105–106). International Workshop on the Application of Isotope Techniques in Hydrological and Environmental Studies, 6–8 September 2004. Paris, France: UNESCO.

Forkuor, G., Pavelic, P., Asare, E., & Obuobie, E. (2013). Modelling potential areas of groundwater development for agriculture in northern Ghana using GIS/RS. *Hydrological Sciences Journal*, *58*(2), 437–451. doi:10.1080/02626667.2012.754101

Foster, V., & Briceño-Garmendia, C. (Eds.). (2010). *Africa's infrastructure – A time for transformation*. A Co-publication of the Agence Française de Développement and the World Bank.

Foster, S., & van Steenbergen, F. (2011). Conjunctive groundwater use: A 'lost opportunity' for water management in the developing world? *Hydrogeology Journal*, *19*, 959–962. doi10.1007/s10040-011-0734–1

Frenken, K. (2005). *Irrigation in Africa in figures*. AQUASTAT Survey – 2005. FAO Water Reports, No. 29. Food and Agriculture Organization of the United Nations.

Furey, S. G., & Danert, K. (2012). *Sustainable Groundwater Development: Use, protect and enhance*. Rural Water Supply Network. Publication No 2012–1.

Gebregziabher, G., Yirga, M., & Namara, R. E. (2013). Cost-benefit analysis and ideas for cost-sharing of groundwater irrigation: evidence from Northern Ethiopia. *Water International* (forthcoming issue).

Ghazouani, W., Molle, F., & Rap, E. (2012). *Water Users Associations in the NEN region IFAD interventions and overall dynamics*. Draft submitted to IFAD.

Giordano, M. (2006). Agricultural groundwater use and rural livelihoods in sub-Saharan Africa: A first-cut assessment. *Hydrogeology Journal*, *14*, 310–318. doi:10.1007/s10040-005-0479–9

Giordano, M., de Fraiture, C., Weight, E., & van der Bliek, J. (2012). *Water for wealth and food security*. Supporting Farmer-Driven Investments in Agricultural Water Management. Synthesis Report of the AgWater Solutions Project.

Gumma, M. K., & Pavelic, P. (2012). Mapping of groundwater potential zones across Ghana using remote sensing, geographic information systems, and spatial modelling. *Environmental Monitoring and* Assessment, *185*(4): 3561–79. doi:10.1007/s10661-012-2810–y

Hagos, F., Ayenew, T., Mamo, K., & Ade, S. (2013). Groundwater use for irrigation and its impact on the livelihoods of smallholder farmers: The case of Haromaya watershed, Eastern Ethiopia. *Water International* (forthcoming issue).

How we made it in Africa (2012). http://www.howwemadeitinafrica.com/four-ways-how-technology-can-boost-african-farming/22908/. Accessed 26 July 2013.

IDE (2013). http://ethiopia.ideorg.org/OurStrategies/WellDrilling.aspx. Accessed 26 July 2013.

International Fund for Agricultural Development (IFAD). (2007). *Agricultural water development for poverty reduction in Eastern and Southern Africa*. Main Report, Zero Draft, Rome, 101pp.

Kamwamba, G. H. (2012). *Socio-economic survey for small pumps and agrarian livelihoods in Malawi* (Draft unpublished paper). Colombo: International Water Management Institute (IWMI).

Kamwamba-Mtethiwa, J., Namara, R., de Fraiture, C., Mangisoni, J., & Owusu, E. (2012). Treadle pump irrigation in Malawi: Adoption, gender and benefits. *Irrigation and Drainage*, *61*(5), 583–595. doi:10.1002/ird.1665

Kay, M. (2001). *Smallholder irrigation technology: Prospects for Sub- Saharan Africa*. Knowledge Synthesis Report Number 3. Rome: FAO/IPTRID.

Lipton, M., Litchfield, J., & Faurès, J.-M. (2003). The effects of irrigation on poverty: A framework for analysis. *Journal of Water Policy*, *5*, 413–427.

MacDonald, A., Bonsor, H. C., Dochartaigh, B. É. Ó., & Taylor, R. G. (2012). Quantitative maps of groundwater resources in Africa. *Environmental Research Letters*, *7*. doi: 10.1088/1748–9326/7/2/024009

MacDonald, A. M., Davies, J., & Dochartaigh, B. É. Ó. (2001). *Simple methods for assessing groundwater resources in low permeability areas of Africa*. British Geological Survey Commissioned Report CR/01/168N. BGS Keyworth UK.

Masiyandima, M., & Giordano, M. (2007). Sub-Saharan Africa: Opportunistic exploitation. In M. Giordano & K. G. Villholth (Eds.), *The agricultural groundwater revolution: Opportunities and threats to development* (pp. 79–99). Wallingford, UK: CABI, in ass. w. IWMI.

Mayerhofer, C., Shamboko-Mbale, B., & Mweene, R. (2010). *Development of a Groundwater Information & Management Program for the Lusaka Groundwater Systems*. Survey on Commercial Farming and Major Industries – Land Use, Groundwater Abstraction & Potential Pollution Sources. BGR and Republic of Zambia.

Ministry of Water Resources. (2011, February). *Ethiopia: Strategic framework for managed groundwater development*. GWMATE, World Bank.

Mukherji, A. (2004). Groundwater markets in Ganga-Meghna-Brahmaputra basin – Theory and evidence. *Economic and Political Weekly*, July 31, 2004. pp. 3514–3520.

Mumma, A., Lane, M., Kairu, E., Tuinhof, A., & Hirji, R. (2011, June). *Kenya Groundwater Governance*. Case Study. World Bank Water Papers.

Namara, R. E., Awuni, J. A., Barry, B., Giordano, M., Hope, L., Owusu, E. S., & Forkuor, G. (2011). *Smallholder shallow groundwater irrigation development in the upper east region of Ghana*. IWMI Research Report No. 143. International Water Management Institute: Colombo, Sri Lanka, 35pp. doi:10.5337/2011.214

Namara, R. E., Gebregziabher, G., Giordano, M. A., & de Fraiture, C. (2013). Small pumps and poor farmers in Sub Saharan Africa: An assessment of current extent of use and the poverty outreach. *Water International* (forthcoming issue).

Narayanamoorthy, A. (2010). India's groundwater irrigation boom: Can it be sustained? *Water Policy*, *12*(4), 543–563. doi:10.2166/wp.2010.042

Naugle, J., & Sellen, D. (2006). *Investing in smallholder irrigation*. Agricultural & Rural Development Notes. World Bank.

Ngigi, S. N. (2009). Climate change adaptation strategies. Water resources management options for smallholder farming systems in Sub-Saharan Africa. New York: The MDG Centre for East and Southern Africa, The Earth Institute at Columbia University.

Nkonya, E., Phillip, D., Mogues, T., Pender, J., & Kato, E. (2010). *From the ground up. Impacts of a Pro-Poor Community-Driven Development Project in Nigeria*. IFPRI, 77pp. doi: 10.2499/978089629179

Obuobie, E., Ofori, D., Agodzo, S. K., & Okrah, C. (2013). Groundwater potential for dry season irrigation in North-eastern Ghana. *Water International*, *38*(4), 433–448. doi: 10.1080/02508060.2013.814212

Pavelic, P., Giordano, M., Keraita, B., Rao, T., & Ramesh, V. (2012). *Groundwater availability and use in Sub-Saharan Africa. A review of 15 countries*. Colombo, Sri Lanka: International Water Management Institute (IWMI), 276pp. doi:10.5337/2012.213.

Pavelic, P., Smakhtin, V., Favreau, G., & Villholth, K. G. (2012). Water-balance approach for assessing potential for smallholder groundwater irrigation in Sub-Saharan Africa. *Water SA*, *38*(3), 399–406. International Conference on Groundwater Special Edition 2012. doi.org/10.4314/wsa.v38i3.18

Pavelic, P., Villholth, K. G., Smakhtin, V., Shu, Y., & Rebelo, L.-M. (2013). Smallholder groundwater irrigation in Sub-Saharan Africa: Country-level estimates of development potential. *Water International*, *38*(4), 392–407. doi: 10.1080/02508060.2013.819601

Perry, E. (1997). Low-cost irrigation technologies for food security in sub-Saharan Africa. In: *FAO: Irrigation technology transfer in support of food security*. (FAO Water Reports – 14). Proceedings of a Subregional Workshop Harare, Zimbabwe, 14–17 April 1997.

Pietersen, K., Beekman, H. E., & Holland, M. (2011). *South African Groundwater Governance*. Case Study. Report prepared for the World Bank in partnership with the South African Department of Water Affairs and the Water Research Commission. June 2011. WRC Report No. KV 273/11.

Rosegrant, M. W., & Perez, N. D. (1997). *Water resources development in Africa: A review and synthesis of issues, potentials, and strategies for the future*. EPTD Discussion Paper No. 28.

Santini, G., Peiser, L., & Faurès, J.-M. (2012). *Assessing the potential for poverty reduction through investments in agricultural water management*. A Methodology for Country Level Analysis. FAOWATER.

Shah, T., Burke, J. J., & Villholth, K. G. (2007). Groundwater: a global assessment of scale and significance. In: D. Molden (Ed.), *Water for food, water for life* (pp. 395–423). Comprehensive Assessment of Water Management in Agriculture Synthesis Report. Earthscan.

Shah, T., Verma, S., & Pavelic, P. (2013). Irrigation against agrarian poverty in small-holder agriculture – A nine country survey of 1550 farmers in East and West Africa. *Water International* (forthcoming issue).

Siebert, S., Burke, J., Faures, J. M., Frenken, K., Hoogeveen, J., Döll, P., & Portmann, F. T. (2010). Groundwater use for irrigation – A global inventory. *Hydrology and Earth System Sciences Discuss*, *7*, 3977–4021. doi:10.5194/hess-14-1863-2010

Silliman, S. E., Hamlin, C., Crane, P. E., & Boukari, M. (2008). International collaborations and incorporating the social sciences in research in hydrology and hydrologic engineering. *Journal of Hydrologic Engineering*, *13*(1): 13–19. doi:10.1061/(ASCE)1084-0699(2008) 13

Takeshima, H., Adeoti, A., & Salau, S. (2009). *Transaction costs and investment in irrigation pumps: Evidence from Nigeria*. Nigeria Strategy Support Program. Brief No. 7.

Tarhule, A., & Woo, M. (1997). Characteristics and use of shallow wells in a stream fadama. *Applied Geography*, *17*(I), 29–42.

Taylor, R., Longuevergne, L., Harding, R., Todd, M., Hewtison, B., Lall, U., . . . Shamsudduha, M. (2010). *Groundwater and global hydrological change – Current challenges and new insight*. Xth Kovacs Colloquium held in Paris, 2–3 July 2010, Paris, France.

Torou, B. M., Favreau, G., Barbier, B., Pavelic, P., Illou, M., & Sidibé, F. (2013). Constraints and opportunities for groundwater irrigation arising from hydrologic shifts in the Iullemeden Basin, south-western Niger. *Water International*, *38*(4), 465–479. doi: 10.1080/02508060.2013.817042

UNDP-WHO (2009). *The energy access situation in developing countries: A review focusing on the least developed countries and Sub-Saharan Africa*. New York: UNDP.

UNICEF, Practica, and Enterprise Works. (2011). *Maps of areas appropriate for manual drilling in Africa. Benin, Central African Republic, Chad, Cote d'Ivoire, Liberia, Madagascar, Mali, Mauritania, Niger, Senegal, Sierra Leone, Togo*. Retrieved from http://www.unicef.org/wash/index_54332.html

van Koppen, B., Hope, L., & Colenbrander, W. (2013). Gender aspects of small-scale private irrigation in Africa. *Water International* (forthcoming issue).

Villholth, K. G., Ganeshamoorthy, J., Rundblad, C. M., & Knudsen, T. S. (2013). Smallholder groundwater irrigation in sub-Saharan Africa: An interdisciplinary framework applied to the Usangu plains, Tanzania. *Hydrogeology Journal*, *21*(4): 863–885. doi:10.1007/s10040-013-1016-x

Villholth, K. G., Tøttrup, C., Stendel, M., & Maherry, A. (2013). Integrated mapping of groundwater drought risk in the Southern African Development Community (SADC) region. *Hydrogeology Journal*. doi:10.1007/s10040-013-0968-1

von Grebmer, K., Ringler, C., Rosegrant, M. W., Olofinbiyi, T., Wiesmann, D., Fritschel, H., . . . Rahall, J. (2012). *Global Hunger Index. The Challenge of Hunger: Ensuring Sustainable Food Security under Land, Water, and Energy Stresses*. Deutsche Welthungerhilfe, IFPRI and Concern World Wide. Bonn/Washington, DC/Dublin, October 2012.

Wahaj, R., Hartl, M., Lubbock, A., Cleveringa, R., & Nepveu, A. (2007). *Gender and water. Securing water for improved rural livelihoods: The multiple-uses system approach*. IFAD, 28pp.

Wiggins, S., & Leturque, H. (2010). *Raising agricultural productivity in Africa – Options for action, and the role of subsidies*. Africa Progress Panel Policy Brief.

Wikipedia (2013). List of African countries by GDP (nominal). Retrieved July 26 2013 from: http://en.wikipedia.org/wiki/List_of_African_countries_by_GDP_(nominal).

Xenarios, S., & Pavelic, P. (2011). *Factors affecting costs of groundwater development in Sub-Saharan Africa*. 6th Rural Water Supply Network (RWSN) Forum, Kampala, Uganda, 29 November–01 December 2011.

You, L., Ringler, C., Nelson, G., Wood-Sichra, U., Robertson, R., Wood, S., . . . Sun, Y. (2010). *What Is the Irrigation Potential for Africa*? A Combined Biophysical and Socioeconomic Approach. IFPRI Discussion Paper 00993. June 2010.

Smallholder groundwater irrigation in Sub-Saharan Africa: country-level estimates of development potential

Paul Pavelic, Karen G. Villholth, Yunqiao Shu, Lisa-Maria Rebelo and Vladimir Smakhtin

The abundance of groundwater resources of Sub-Saharan Africa is generally well recognized, but quantitative estimates of their potential for irrigation development are lacking. This study derives estimates using a simple and generic water balance approach and data from secondary sources for 13 countries. Even with conservative assumptions and accounting for water demands from other sectors, including the environment, a 120-fold increase (by 13.5 million hectares) in the area under groundwater irrigation is possible for the countries considered. This expansion could improve the livelihoods of approximately 40% of the present-day rural population.

Introduction

Groundwater is recognized as a valuable source of clean and reliable drinking water for Sub-Saharan Africa (SSA) (United Nations Economic Commission for Africa [UNECA], 2011). Increasingly, it is also seen as a largely untapped resource for agricultural development that could dramatically improve food security and livelihoods by protecting against poor and highly variable wet season rainfall and by enabling more productive use of land during the dry season (Kay, 2001; Shah, Burke, & Villholth, 2007). Groundwater irrigation offers a positive supplement or alternative to rainfed farming and is central to achieving some of the major Millennium Development Goals and to supporting economic growth in SSA.

Mounting evidence for increasing climate variability and climate change in SSA (Intergovernmental Panel on Climate Change [IPCC], 2007) implies significant challenges for maintaining traditional rainfed agricultural production systems (MacDonald, Calow, MacDonald, Darling, & Dochartaigh, 2009). Greater rainfall variability significantly affects surface-water resources, implying increased need for water storage and irrigation. Whilst groundwater systems are not entirely insulated from climate-related impacts on recharge and utilization (Villholth, Tøttrup, Stendel, & Maherry, 2013), the response is typically far more tempered than with surface-water systems (UNECA, 2011).

Just 4% of the cultivated area in SSA is equipped for irrigation, so strategies to increase this area are a priority (You et al., 2010). The use of groundwater to support and enhance agricultural development throughout SSA has been steady increasing but has historically been very low compared with many other regions in the world (Giordano, 2006). As far back as three decades ago, there was judged to be significant potential to expand the limited amount of groundwater irrigation (World Bank, 1981), a situation that has largely remained unchanged to the present. A demonstrative sample of 13 SSA countries reveals that groundwater is drawn upon to irrigate from just 10 ha to less than 100,000 ha in each one, or less than 0.5% of the arable land (Figure 1). Only around 10% of the area under full or partial irrigation control in SSA is reportedly supplied from groundwater (FAO, 2005), although this figure is likely to be an underestimate because informal groundwater use by private, individual small- and micro-scale irrigators is rarely captured in official survey data. This is supported by the work of Villholth (2013) and Giordano (2006). The latter showed that FAO (AQUASTAT) figures from nine SSA countries underestimate actual groundwater use by 44% on average. In any case, the areas under irrigation from groundwater are small. The reasons for this include limited awareness among development partners and rural communities, lack of economic opportunities and incentives for farmers, and limited support services for drilling, pump sales and repairs, and rural energy access (Allaire, 2009; Tuinhoff, Foster, van Steenbergen, Talbi, & Wishart, 2011; Villholth, 2013).

In more recent years there has been no waning in the number of strategies being suggested to address food security through irrigation in SSA. For example, the World Water Council vision for 2025 includes increasing water supply for agriculture by 15–20% from investments in new large and small dams and from groundwater exploitation (International Commission on Irrigation and Drainage [ICID], 2001). More ambitious is the report prepared in 2005 by the Commission for Africa, which called for a doubling of the area under irrigation by 2015 (Commission for Africa, 2010), although by 2010 it was clear that the actual growth in the intervening five years had been just 0.9%. The Comprehensive Africa Agriculture Development Programme (CAADP) under the New Partnership for Africa's Development (NEPAD) has adopted land and water management as the first of its four pillars for priority investment, and in so doing aims to significantly expand the area under

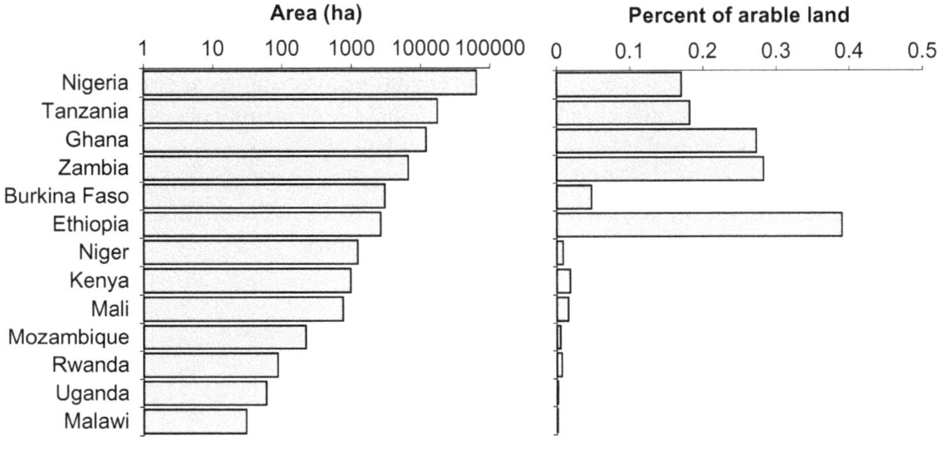

Figure 1. Estimates of groundwater use for irrigation, expressed in terms of actual irrigated area and as a percentage of arable land for 13 Sub-Saharan African countries. Sources: Siebert et al. (2010) and FAOSTAT (http://faostat.fao.org/).

reliable water control. Groundwater features in its portfolio of strategies (CAADP, 2009). In response, there has been enhanced interest in groundwater irrigation by governments in many SSA countries and by a wide range of international donors and investors (e.g. Utz, 2008; World Bank, 2011). The emerging evidence is of variable but often limited efforts from governments, whilst development progresses in certain areas where farmers (individually or in small groups) get access to the resource and production means to make a viable livelihood from groundwater irrigation (Giordano, de Fraiture, Weight, & van der Bliek, 2012).

One of the core arguments historically put forward for curbed expansion of groundwater use for agriculture in SSA is that aquifers are low in transmissivity and hence well/borehole yields are inadequate to support agricultural development at scales larger than minor garden irrigation, particularly in the weathered crystalline basement rocks that extend over about 40% of the African land mass (Wright, 1992; Chilton & Foster, 1995). It has also been argued that groundwater use for irrigation will have unacceptable detrimental impacts on wetlands and other groundwater-dependent ecosystems or impact on access to the groundwater for domestic supplies (Adams, 1993; Giordano & Villholth, 2007; MacDonald et al., 2009).

Despite the recognized potential of groundwater to contribute towards poverty alleviation and food security in countries such as India (Mukherji, Villholth, Sharma, & Wang, 2009) and in parts of SSA (Namara, Horowitz, Nyamadi, & Barry, 2011), the potential in SSA is not clear. To what extent can smallholder irrigation be introduced on a sustainable basis, given the physical limits of the resource? Are the opportunities evenly distributed amongst countries, or are some more promising than others? Simple approaches are needed that offer initial guidance as to what level of irrigation development, in terms of acreage and water volumes, can be sustained and how much of the landscape may be hydrogeologically suited to agricultural groundwater development. Previous assessments of the irrigation potential of Africa (Svendsen, Ewing, & Msangi, 2008; You et al., 2010) have focused on surface water resources rather than groundwater. In this paper, we address this lapse and consider groundwater issues from a water balance perspective that takes the country as the basic unit.

The 13 countries listed in Figure 1 are the main focus of the Alliance for a Green Revolution in Africa (http://www.agra.org) and were selected for analysis. They cover two broad geographic domains, with one group of five countries extending over much of West Africa and another group of eight countries over much of East and South-East Africa. Collectively they represent a range of climatic conditions, varying from arid to wet semi-arid tropics, and major aquifer typologies that include a predominance of Precambrian basement (i.e. the weathered crystalline rocks described above) as well as unconsolidated sediments and volcanic rocks (Table 1). Aquifers shared across national borders underlie 42% of the surface of Africa (Altchenko & Villholth, 2013), indicating the importance of co-management of these resources when irrigation intensifies.

Methodology

General principles

Comprehensive approaches to water allocation across the use sectors require considerable efforts in terms of information collection, technical resources and administrative support, and therefore should be made progressively and with careful planning (United Nations, 2000). To be commensurate with the existing state of knowledge and development, the

Table 1. Climatic and hydrogeologic characteristics at the country level.

	Rainfall (mm/y)[a]	Potential evaporation (mm/y)[a]	Climatic region[b]	Major aquifer type[c]
Burkina Faso	748	2000	SAT-D	PCB
Ethiopia	848	1500	SAT-D	VR
Ghana	1187	1500	SAT-W	PCB
Kenya	630	2000	SAT-D	UCS
Malawi	1181	1500	SAT-W	PCB
Mali	282	2000	SA	UCS
Mozambique	1032	1500	SAT-W	PCB
Niger	151	2200	A	UCS
Nigeria	1150	1700	SAT-W	PCB
Rwanda	1212	1500	SAT-W	PCB
Tanzania	1071	1700	SAT-W	PCB
Uganda	1180	1500	SAT-W	PCB
Zambia	1020	1500	SAT-W	PCB

[a] AQUASTAT (http://www.fao.org/nr/water/aquastat/dbase/index.stm); details on the calculation procedure for potential evaporation are given in FAO (2003).
[b] A = arid (rainfall <250 mm/y); SA = semi-arid (250–500 mm/y); SAT-D = dry semi-arid tropics (500–1000 mm/y); SAT-W = wet semi-arid tropics (>1000 mm/y).
[c] PCB = Precambrian basement; UCS = unconsolidated sediments; VR = volcanic rocks (derived from MacDonald et al. (2010)).

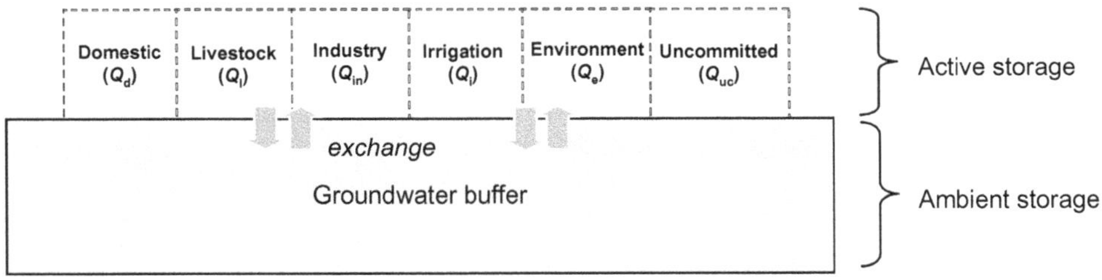

Figure 2. Total groundwater storage partitioned into ambient and active storages, with the latter indicative of both replenishment and allocation to the five primary water use sectors plus uncommitted reserves. Source: Pavelic et al. (2012a).

straightforward and generic groundwater balance–based methodology presented by Pavelic et al. (2012a) is used here to identify the opportunities for increasing groundwater irrigation supplies and to help understand how cropping choices and intensities influence the potential areal extent of irrigation. The approach accounts for existing anthropogenic uses of groundwater and for the water needed to maintain environmental objectives. Allocations of groundwater to further expand irrigation must therefore draw upon the remaining uncommitted reserves or be transferred from other uses (Figure 2). The approach does not consider groundwater quality limitations because there are insufficient data to draw upon for the assessment (MacDonald, Dochartaigh, Bonsor, Davies, & Key, 2010).

Application in focal countries

Estimating the upper limits for groundwater irrigation potential relies on an a priori estimation of present water supply and demand components on an annualized time basis. For this evaluation, data were assembled from multiple sources, and various assumptions and approximations were made in accordance with the quality of the information available, as detailed below. Most of the data were derived from the AQUASTAT and FAOSTAT databases of the Food and Agriculture Organization (FAO) of the United Nations, supplemented by various other sources (e.g. African Statistical Yearbook, 2010; Siebert et al., 2010). The baseline data assembled for this analysis are given in Table 2.

Country-level recharge data, used as an estimate of the quantum of available and annually renewable groundwater resource for the various demands, were taken from AQUASTAT and generated from rainfall infiltration estimations within the country as part of whole-of-water-cycle calculations (FAO, 2003; Margat, Frenken, & Faurès, 2005).

The quantum of groundwater abstracted for domestic water supplies at the national level has not been reported for the countries considered here. Recognizing, however, that groundwater is the source of drinking water for 75% of Africa's population (UNECA & AU & AfDB, 2000), it was assumed throughout that this percentage of the total population are supplied with groundwater at a rate of 50L per capita per day, in line with the minimum clean-water requirement for basic human activities (Gleick, 1996). Whilst this per capita consumption rate is likely to overestimate (by a factor of between two and three) the actual conditions in countries where manual fetching of water by households is common (e.g. about 20 L per capita per day in Benin; Kamagaté et al., 2007), it is applied for the purpose of ensuring the conservative overestimation of demand and accounting for future increases in domestic demand to some extent.

Groundwater use associated with livestock watering is dependent on climate. Figures were taken from total animal numbers and universal unit figures on the per capita water use for each type of livestock (both taken from the FAOSTAT database that included country-level data for cattle, pigs, sheep, goats and poultry). To again ensure that the amount was not being underestimated, all use was assumed to be sourced from groundwater. In practice, groundwater may support livestock less than surface water, as for example in the case of Tanzania, where it is estimated to account for only about 20% of total use (Dr. J.J. Kashaigili, Sokoine University of Agriculture, Tanzania, pers. comm. 17 December 2011).

Research on the link between industry and groundwater has tended to focus on industrialization's impacts on urban groundwater quality (Adelana, Fantong, Nedaw, & Duah, 2011). Industrial water abstraction estimates according to source are unreported at the national level. At the continent scale, however, levels of industrial water use are estimated to be less than half of the municipal withdrawals (FAO, 2005). For the Pangani River basin in Tanzania, about 88% of groundwater extracted is used for irrigation, 4% for industrial use and 8% for domestic use (Mato, 2002). In some countries (e.g. Burkina Faso and Uganda) it is known to be minimal, whereas in others (e.g. Ethiopia) use is relatively high and growing steadily (Pavelic, Giordano, Keraita, Ramesh, & Rao, 2012b). In view of the limited quantitative information, the industrial demand was unilaterally set at 50% of the domestic demand.

Estimation of the environmental requirements (i.e. Q_e in Figure 2) is important in two regards. First, in most countries the sum of all anthropogenic groundwater uses (domestic, livestock and industry) amounts to less than 10% of recharge (Table 2), so the environmental component represents a potentially large, unknown yet important determinant of the uncommitted component (Q_{uc}). Second, some of the most obvious prospective locations to develop smallholder groundwater irrigation are along the major river floodplains and swamps, which often are linked to and dependent on groundwater. This accentuates the importance of understanding the surface water and groundwater interactions. Presently, estimation of Q_e is a challenging task given the methodological difficulties associated with its quantification and impacts (Seward, Xu, & Brendonck, 2006). Despite these challenges,

Table 2. Country-level base data used for this analysis.

Country	Country area (km²)[a]	Arable land area (km²)[a]	Human population (millions)[b]	Rural households (millions)[c]	Mean recharge (mm/y)[d]	Domestic use (mm/y)[e]	Livestock use (mm/y)[f]	Industrial use (mm/y)[g]	GW-irrigated area – actual (ha)[h]	GW irrigation use (mm/y)[h]	SW-irrigated area – actual (ha)[h]	Non-irrigated arable land (10^6 ha)	Environmental GW requirement (mm/y) At 30% of recharge	At 50% of recharge	At 70% of recharge
Burkina Faso	274,220	59,000	15.76	1.91	34.6	0.87	1.1	0.44	3,000	0.068	22,000	5.9	10.4	17.3	24.3
Ethiopia	1,104,300	139,480	82.83	14.09	18.1	1.07	1.0	0.54	2,611	0.010	288,118	13.7	5.4	9.1	12.7
Ghana	238,540	44,000	23.84	2.29	110.3	1.47	0.38	0.73	12,000	0.154	42,710	4.3	33.1	55.1	77.2
Kenya	580,370	54,000	39.80	5.95	6.0	1.01	0.62	0.50	970	0.010	96,041	5.3	1.8	3.0	4.2
Malawi	118,480	36,000	15.26	2.80	21.1	1.83	0.52	0.92	30	0.002	54,104	3.5	6.3	10.6	14.8
Mali	1,240,190	63,610	13.01	1.94	16.1	0.18	0.23	0.090	750	0.005	176,093	6.2	4.8	8.1	11.3
Mozambique	799,380	50,500	22.89	3.36	21.3	0.42	0.083	0.21	217	0.001	39,846	5.0	6.4	10.6	14.9
Niger	1,267,000	149,400	15.29	2.00	2.0	0.18	0.24	0.090	1221	0.009	64,394	14.9	0.6	1.0	1.4
Nigeria	923,770	340,000	154.73	17.19	94.2	2.47	1.1	1.23	64,000	0.379	154,840	33.8	28.3	47.1	65.9
Rwanda	26,340	13,000	10.00	1.66	265.8	5.86	1.6	2.93	85	0.007	8,415	1.3	79.7	132.9	186.0
Tanzania	947,300	100,000	43.74	6.89	31.7	0.69	0.43	0.34	17,465	0.105	171,582	9.8	9.5	15.8	22.2
Uganda	241,040	66,000	32.71	5.56	120.3	2.02	0.89	1.01	59	0.001	5,841	6.6	36.1	60.2	84.2
Zambia	752,610	33,500	12.94	1.47	62.4	0.25	0.085	0.13	6,646	0.057	149,266	3.2	18.7	31.2	43.7

Note: GW = groundwater; SW = surface water.
[a]AQUASTAT (http://www.fao.org/nr/water/aquastat/data/query/index.html?lang=en). Last accessed on 30 January 2013.
[b]Taken from Wikipedia,http://en.wikipedia.org/wiki/List_of_African_countries_by_population (primary sources provided within). Last accessed on 30 January 2013.
[c]Rural population figures derived from Word Bank estimates for 2010 (http://databank.worldbank.org/data/views/reports/tableview.aspx); average household size for nine countries taken fromhttp://www.nationsencyclopedia.com/WorldStats/AICD-noninfrastructure-household-size.html (average values used for the other four countries). Last accessed on 20 May 2013.
[d]AQUASTAT (http://www.fao.org/nr/water/aquastat/dbase/index.stm); principles on the calculation procedure for recharge are given by Margat et al. (2005) and FAO (2003).
[e]For domestic use, it is assumed that 75% of the population are served by groundwater at 50L per day.
[f]From FAOSTAT database, assuming consumption of 40 L per day for cattle, 30 for pigs, 20 for sheep and goats, and 0.2 for poultry.
[g]Industrial water use is assumed to be equal to 50% of the domestic use.
[h]Siebert et al. (2010).

there is widespread agreement that groundwater systems often play a vital role in supporting ecosystem functions in both terrestrial and aquatic landscapes that may include environmental flows to rivers, wetlands, estuaries and coastal systems (Le Maitre, Scott, & Colvin, 1999).

There has been no identification to date of the level of groundwater dependence of these aquatic and terrestrial ecosystems, or even zones of high environmental value supported directly or indirectly by groundwater, in most parts of SSA (Maherry et al., 2011). However, some degree of guidance can be drawn from other regions. In Australia, where the allocation of groundwater to the environment has been given considerable attention over the last two decades, entitlements of groundwater for so-called environmental services, ranging from 30% to 90% of annual recharge, have been deduced across the different states and territories (Murray, Zeppel, Hose, & Eamus, 2003). The range of values reported relates as much to the different methodologies used as to the contrasting biophysical requirements in each region. Groundwater level trends, though not easily convertible into subsurface flow equivalents, also provide an indirect indicator, given that increasing abstraction potentially reduces potentiometric gradients and lateral inflows to receiving water bodies. Using national standardized protocols, over many years of application and evaluation of large datasets across diverse hydrogeological terrain, Indian researchers have deduced that signs of over-stress in terms of long-term falling groundwater levels do not occur unless around 70% of the annual average recharge rate is abstracted (Chatterjee & Purohit, 2009). Values in the 13 study countries with AQUASTAT data on surface water–groundwater exchange (i.e. that part of the renewable water resources which is common to both surface water and groundwater; FAO (2003)) range greatly, from 0 to 80%, with an average of 21%.

The current state of knowledge on SSA is grossly inadequate to arrive at defensible figures on environmental groundwater requirements. A first-pass, across-the-board figure for the allocation to the environment of 50% of annual average recharge was applied as the base case. Variations ranging from 30% to 70% of recharge were also tested. These values are commensurate with the Australian and Indian experiences mentioned above. These estimates were compared with estimates of the total water demand for all wetlands and lakes, calculated from the product of country areal estimates (drawn from the Global Lakes and Wetlands Database assembled by Lehner and Döll (2004)) and net potential evaporation figures (from AQUASTAT). Wetlands and lakes cover all open water bodies 1 km^2 or larger, are based on a single snapshot in time, and include numerous classes (freshwater marsh and floodplains, swamp and flooded forest, coastal fresh/brackish/saline wetlands, bog, and intermittent wetland/lakes). Recognizing that these figures are grossly conservative since they neglect contributions to the demand side from rivers and seawater, the intention here is to identify the countries with the greatest sensitivity that may most warrant further attention.

In addition, aquifer productivity, expressed in terms of well yield, is used here as an indicator of the suitability of the aquifers for irrigation development. The Africa-wide map reported by MacDonald et al. (2010) was used to determine the proportion of area within each yield class for each of the 13 countries. A well yield of 0.5 L/s was assumed as the lower threshold for aquifer productivity that could sustain small-scale garden irrigation (Foster et al., 2006). A 0.5 L/s well operated for 8 hours per day equates to an abstraction rate of 14.4 m^3 per day and can meet a 500 mm crop water demand within a 100-day crop cycle over an area of 0.3 ha.

Results and discussion

Uncommitted reserves

In relative terms, there is appreciable potential for enhancing groundwater irrigation in each country, as judged from the balance of the uncommitted groundwater reserves. For the 50% base case, the reserve levels range from around 40% to 50% of recharge for the majority of countries to less than 40% for the 4 remaining countries (Figure 3). Presently, the lowest potential overall is for Kenya, at less than 20%, due to the higher consumptive demand as a result of high population and livestock numbers. In no case does current irrigation use of groundwater constitute more than 0.5% of recharge (Table 2).

In the arid and semi-arid countries (i.e. Niger and Mali), where irrigation water is needed the most, there is still potential. Even for more humid countries, where groundwater is replenished the most, there are still long months with minimal or no rainfall where an irrigation demand exists. Rwanda, the wettest country overall, still experiences seasonal water deficits, and irrigation is seen as being underdeveloped and having a major role for food security (African Development Bank/Organization for Economic Co-operation and Development [AfDB/OECD], 2007). For the remaining countries the situation lies between these end-member conditions.

New irrigated areas

Table 3, which gives the percentage of arable land potentially irrigable with groundwater, shows that one country has high potential (Zambia), seven have moderate potential (Ghana, Mali, Mozambique, Nigeria, Rwanda, Tanzania and Uganda), and five have low potential (Burkina Faso, Ethiopia, Kenya, Malawi and Niger). The results are presented for three different levels of irrigation-water demand (that amount that is over and above the effective rainfall available for plants): 100, 500 and 1000 mm/y. The lowest value used here might account for minor or supplemental irrigation to meet crop water needs during a dry spell within the wet season; the intermediate value represents dry season irrigation of a

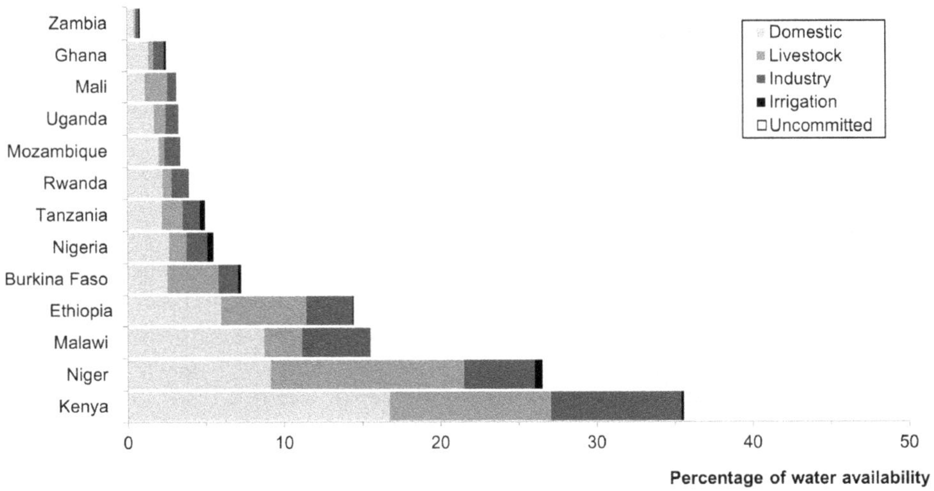

Figure 3. Sector-wise percentage water demand components relative to recharge for each county, indicating the residual (uncommitted) component, Q_{uc} (for the base case for Q_e). Note that Q_e is not indicated because it universally constitutes 50% of groundwater availability in the base case.

Table 3. Groundwater-irrigable area (as a percentage of the current non-irrigated arable land).

	Percentage irrigable area for irrigation water demand of:								
	100 mm/y			500 mm/y			1000 mm/y		
	Percentage of recharge for environmental GW requirement								
	30	50	70	30	50	70	30	50	70
Burkina Faso	102	69	37	20	13	7.4	10	6.9	3.7
Ethiopia	81	52	23	16	11	4.6	8.1	5.2	2.3
Ghana	410	288	167	82	58	33	41	29	17
Kenya	23	9.5	0	4.5	1.9	0	2.3	1.0	0
Malawi	38	24	11	7.7	4.9	2.0	3.9	2.4	1.0
Mali	216	152	87	43	30	17	22	15	8.7
Mozambique	226	158	90	45	32	18	23	16	9.0
Niger	7.3	3.9	0.59	1.5	0.79	0.12	0.73	0.39	0.059
Nigeria	166	115	63	33	23	13	17	11	6.3
Rwanda	358	250	141	72	50	28	36	25	14
Tanzania	199	138	77	40	29	15	20	14	7.7
Uganda	293	206	118	59	41	24	29	21	12
Zambia	1018	723	429	204	145	86	102	72	43

moderately water-demanding crop; and the highest value applies to a more water-intensive crop in the dry season. In this way, physical boundary conditions can be established on the cultivated area that may be used for irrigation. For the least intensive case of 100 mm/y and the base-case environmental requirement, between 3.9% and 100% of the non-irrigated area per country can be put under irrigation, whereas the areas are reduced to between 0.4% and 72% for the most intensive case. The rankings between countries are independent of the value of irrigation intensity considered. For eight of the countries under low-level irrigation, there is sufficient rainfall recharge to meet the crop water requirement if all the presently non-irrigated arable land is utilized for groundwater irrigation, whereas the situation changes to just one country for the moderate-irrigation scenario and no countries for the high-irrigation scenario. This significant dependence of possible irrigated area on unit water requirements illustrates the importance of optimizing groundwater use and increasing water and land use efficiency to achieve long-term resource sustainability and reduced environmental impacts.

Under the highest water-demand conditions described above, a total of 13.5×10^6 ha of new irrigated area could be established across the 13 countries studied (Table 4). This offers a dramatic 123-fold increase over the current irrigated areas fed by groundwater in those same countries. With farm sizes averaging around 0.5 ha in SSA (International Assessment of Agricultural Knowledge, Science and Technology for Development [IAASTD], 2009), this translates to 27 million households potentially receiving direct livelihood benefits, or around 40% of the estimated 67 million rural households within the 13 countries (Table 2). With lower irrigation demand, the figures are proportionately larger.

Uncommitted irrigation potential is present in all countries investigated, but it is not uniformly distributed. Naturally, some countries have greater prospects for expanded groundwater use than others. Hence, potential can be anticipated as well in SSA countries not evaluated here. Groundwater resources are broadly distributed, and areas with potentially suitable aquifers do not appear to be a major limiting factor with respect to achieving this expansion target. However, for the least productive aquifers, which require intensive

Table 4. Summary of new areas under irrigation and the potential number of households affected. The ranges for new irrigated area reflect the three ranges of Q_e used.

	New areas under irrigation (10^6 ha) for irrigation water demand of:			Min. and max. number of households affected (millions)
	100 mm/y	500 mm/y	1000 mm/y	
Burkina Faso	4.1 ± 1.9	0.81 ± 0.38	0.41 ± 0.19	0.43 – 12
Ethiopia	7.1 ± 4.0	1.4 ± 0.8	0.71 ± 0.40	0.62 – 22
Ghana	12.5 ± 5.3	2.5 ± 1.1	1.2 ± 0.5	1.4 – 36
Kenya	0.50 ± 0.22	0.104 ± 0.04	0.05 ± 0.02	0 – 2.0
Malawi	0.86 ± 0.50	0.17 ± 0.10	0.086 ± 0.050	0.072 – 2.7
Mali	9.4 ± 4.0	1.9 ± 0.8	0.94 ± 0.40	1.1 – 27
Mozambique	7.9 ± 3.4	1.6 ± 0.7	0.79 ± 0.34	0.91 – 23
Niger	0.59 ± 0.50	0.12 ± 0.10	0.06 ± 0.05	0.017 – 2.2
Nigeria	38.7 ± 17.4	7.7 ± 3.5	3.9 ± 1.7	4.3 – 112
Rwanda	3.2 ± 1.4	0.65 ± 0.28	0.32 ± 0.14	0.37 – 9.3
Tanzania	13.5 ± 6.0	2.7 ± 1.2	1.4 ± 0.6	1.5 – 39
Uganda	13.6 ± 5.8	2.7 ± 1.2	1.4 ± 0.6	1.6 – 39
Zambia	23.1 ± 9.4	4.6 ± 1.9	2.3 ± 0.9	2.7 – 65
Total	**135.1±60.1**	**27.0±12.0**	**13.5±6.0**	**15.0–390**

pumping, the most cost-effective and beneficial uses appear to be for supplemental, less water-demanding use or for small-scale garden use.

When the uncommitted water flux, expressed in absolute terms, is plotted against the proportion of each country where the aquifers are deemed sufficient to sustain small-scale irrigation (defined earlier), the resultant distribution tends to be far more scattered in terms of the flux than that of aquifer suitability (Figure 4). The plot-space can be interpreted such that countries near the top-right corner have the highest landscape suitability and highest potential to provide additional water for irrigation; that suitability reaches a minimum at the bottom-left corner where that potential is the lowest. Several clusters can be observed:

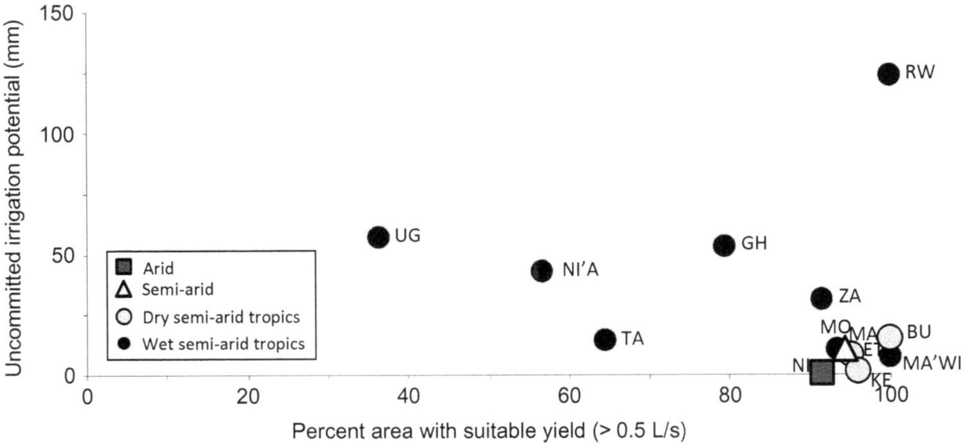

Figure 4. Relation between the uncommitted water availability (irrigation potential, base case) and the proportion of area within each country sufficient to meet small-scale irrigation. Countries are shown according their average climate as given in Table 1.

a single country with high irrigation potential and high aquifer suitability (i.e. Rwanda); countries with high irrigation potential but variable aquifer suitability (Uganda, Nigeria, Ghana and Zambia); and countries with high-to-medium aquifer suitability and relatively limited irrigation potential (Burkina Faso, Malawi, Tanzania, Niger, Mozambique, Mali, Ethiopia and Kenya). All of the countries have at least one-third of their territory suitable for smallholder irrigation, and all except one (Uganda) have at least half of their area suitable.

Estimation of uncertainty and development constraints

The groundwater-recharge data are critical in the assessment as they determine the ceiling for groundwater development potential used in the applied analysis. While official data from FAOSTAT have been used, it is realized that large uncertainty may be associated with recharge estimation (Wada et al., 2010). FAOSTAT data may be considered conservative because they account for only diffuse recharge via rainfall, not focused recharge. More rigorous estimations and comparisons are required, particularly in the face of climate change (Taylor, Koussis, & Tindimugaya, 2009) or broad-scale land use change (Torou et al., 2013), which may significantly influence recharge. This may be at the global scale (Wada et al., 2010) or the catchment scale using distributed model data (Massuel et al., 2011).

The environmental groundwater requirements are potentially a significant part of the water balance and are highly uncertain, as suggested by the broad range of values used here, ranging from 30% to 70% of recharge. This is also suggested by the data presented in Table 5, which aggregates at the national level the areal extent of potential groundwater-dependent ecosystems (lakes and wetlands) and their potential discharge fluxes (through evapotranspiration) and compares this to the corresponding recharge flux. All of the countries except four (Burkina Faso, Ghana, Nigeria, and Rwanda) have potentially more discharge than recharge, and all countries apart from two (Ghana and Rwanda) exceed the baseline environmental requirement (i.e. 50% of recharge). In all cases, the baseflow

Table 5. Calculated potential discharge (evapotranspiration) fluxes from wetlands and lakes in relation to recharge.

Country	Net potential evapotranspiration (mm/y)[a]	Lake and wetland area (km^2)[b]	Calculated discharge flux (10^6 m^3/y)	Recharge rate (10^6 m^3/y)	Ratio of discharge to recharge
Burkina Faso	1,252	5,223	6,539	9,500	0.7
Ethiopia	652	35,740	23,302	20,000	1.2
Ghana	313	7,275	2,277	26,300	0.1
Kenya	1,370	107,100	146,727	3,500	41.9
Malawi	319	36,534	11,654	2,500	4.7
Mali	1,718	68,229	117,217	20,000	5.9
Mozambique	468	77,695	36,361	17,000	2.1
Niger	2,049	81,057	166,086	2,500	66.4
Nigeria	550	130,010	71,506	87,000	0.8
Rwanda	288	5,208	1,500	7,000	0.2
Tanzania	629	185,418	116,628	30,000	3.9
Uganda	320	101,228	32,393	29,000	1.1
Zambia	480	117,862	56,574	47,000	1.2

[a]Derived from rainfall and potential evaporation data given in Table 1.
[b]From Lehner and Döll (2004).

contribution to the discharge of the major rivers has been neglected. This supports the need for greater attention to be given to the environmental groundwater allocation at the national or groundwater/river-basin level, to the extent and importance of groundwater-dependent ecosystems and to the subsurface flow processes that support these systems, which in turn support human activities in agriculture, fisheries, recreation and tourism. This is especially the case for countries with a high discharge-to-recharge ratio, such as Niger and Kenya (Table 5). Where groundwater is developed close to rivers and other surface-water bodies potentially dependent on groundwater, it is recommended to use the higher rate for environmental requirements and undertake additional local-level assessments.

Little is known about the role of groundwater in supporting ecosystems in SSA (Maherry et al., 2011). Hagos et al. (2013) analyze the impacts arising from groundwater overexploitation in parts of eastern Ethiopia, where overuse for domestic, industrial and irrigation activities has led to impacts, causing a crater lake to vanish, and human conflicts. This shows that not every location in SSA has vastly underutilized irrigation potential. Localized or larger-scale impacts may be, or become, critical, and whilst addressing underdevelopment it is critically important to foster early associated mechanisms for management of the resource at local, national and transboundary levels. Monitoring systems are also needed, together with better understanding of the nature, distribution, and interaction of aquifer systems and the role of groundwater storage and groundwater discharge in sustaining terrestrial and aquatic ecosystems, and assessments of the impact of increased distributed or localized pumping on the groundwater systems and already existing reliant populations.

The aggregation applied here for broad-scale assessment purposes neglects much of the hydrogeological complexity and variability present at the sub-national and local scales. Recharge may vary significantly within countries, especially larger ones covering a number of climate zones (e.g. Ethiopia, Kenya and Niger), implying that local-scale assessments of potential are required for more detailed analysis and policy development. Groundwater development prospect maps, currently available at the continental scale and in some cases at the country, sub-country and basin scales (UNECA, 2011; Gumma & Pavelic, 2012), provide invaluable tools to better understand the resource and to provide guidance on where to focus efforts in more detailed local-scale assessments. Efforts must also be directed towards overcoming barriers related to other factors relevant for groundwater irrigation development, such as soil conditions, infrastructure, access to markets, drillers, pumps, energy, credit options and institutional arrangements (Pavelic et al., 2012b; Villholth, 2013).

Conclusions

From an examination of data from various secondary sources, and building upon a simple water balance framework recently developed by the authors, 13 countries in SSA were evaluated to estimate the potential of groundwater irrigation in terms of acreage for various crop water demands, considering three environmental water demand scenarios. The assessment elucidated which countries are the best candidates for expanded groundwater use, and what are the likely macro-level socio-economic benefits. The conservative results reveal that 13.5×10^6 ha of groundwater irrigation development could be established across the countries examined, which could potentially improve the livelihoods of 27 million rural households. These estimates recognize that the potential may not be evenly distributed within the countries from a water resource availability and accessibility point of view.

This study serves to go beyond qualitative statements by highlighting the actual potential for groundwater development that might help to induce investment and/or changes in policy regimes that are hindering development. The results presented in this study confirm to governments, donors and development partners that large areas exist for agricultural groundwater development throughout SSA. In addition, the approach, while at present more indicative than precise, due to uncertainties in determination of components of the groundwater balance, provides a tool for further refinement of the assessments as better methodologies for variables such as recharge and environmental requirements become available. Emerging smaller-scale mapping and resource-analysis tools, together with socio-economic assessments, provide means to target the efforts and to reduce the risks associated with investments.

Notwithstanding the large development potential, the study cautions about the limits to sustainability and the consequent need to balance multiple sectoral demands on a limited replenishable resource base. Rapid development of groundwater supplies can have unintended negative socio-economic and environmental impacts; this requires improved local-level assessments and improved proactive governance arrangements at local through to regional levels in order to prevent major problems and rectify smaller ones as they emerge.

Acknowledgements

This work contributes to a research project financially supported by the Rockefeller Foundation through project number 2008-AGR-305, Groundwater in Sub-Saharan Africa: Implications for Food Security and Livelihoods, and administered through the CGIAR Research Program on Climate Change, Agriculture and Food Security (CCAFS). Comments provided by Dr Mark Giordano (IWMI) and the journal reviewer helped to improve the manuscript. Ms. Aruna Jyothi (IWMI) is kindly thanked for performing GIS-based calculations.

References

Adams, W. M. (1993). Indigenous use of wetlands and sustainable development in West Africa. *Geographical Journal*, 159(2), 209–218.

Adelana, S. M. A., Fantong, W. Y., Nedaw, D., & Duah, A. (2011). Groundwater and Health: Meeting unmet needs in sub-Saharan Africa. In J. Anthony & A. Jones (Eds.), *Sustaining Groundwater Resources: A Critical Element in the Global Water Crisis*, International Year of Planet Earth (pp. 21–33). doi: 10.1007/978-90-481-3426-7-2

African Development Bank/Organization for Economic Co-operation and Development (AfDB/OECD). (2007). African Economic Outlook 2007. Country Notes: Rwanda, p. 14. Retrieved from: http://www.oecd.org/dataoecd/27/19/38562991.pdf

African Statistical Yearbook. (2010). African Statistical Yearbook. African Development Bank Group, African Union and Economic Commission for Africa, ISSN 1561–2805, pp. 399.

Allaire, M. (2009). Drought Mitigation in Semi-Arid Africa: The Potential of Small-Scale Groundwater Irrigation. Resources for the Future. International Water Management Institute. p. 12. Retrieved from: http://sustainus.org/docs/citsci/winners/2009/Maura%20Allaire.pdf

Altchenko, Y., & Villholth, K. G. (2013). Transboundary aquifer mapping and management in Africa: A harmonised approach. *Hydrogeology Journal*. doi: 10.1007/s10040-013-1002-3

CAADP (Comprehensive Africa Agriculture Development Programme). (2009). The CAADP Pillar I Framework. Retrieved from: http://www.caadp.net/library-pillar1-documents.php

Chatterjee, R., & Purohit, R. R. (2009). Estimation of replenishable groundwater resources of India and their status of utilization. *Current Science*, 96(12), 1581–1591.

Chilton, P. J., & Foster, S. S. D. (1995). Hydrogeological characterisation and water-supply potential of basement aquifers in tropical Africa. *Hydrogeology Journal*, 3(1), 36–49. DOI: 10.1007/s100400050061

Commission for Africa. (2010). Still our common interest. The Commission for Africa. Retrieved from: http://www.commissionforafrica.info/
FAO (Food and Agriculture Organization). (2003). Review of World Water Resources by Country. FAO Water Report 23, Food and Agriculture Organization of the United Nations, Rome, Italy, ISBN 92-5-104899-1.
FAO (Food and Agriculture Organization). (2005). Irrigation in Africa in figures: AQUASTAT Survey – 2005. FAO Water Report 29 (with CD ROM). Food and Agriculture Organization of the United Nations, Rome, Italy.
Foster, S., Tuinhof, A., Garduño, H., Kemper, K., Koundouri, P., & Nanni, M. (2006). Groundwater Resource Development in Minor Aquifers: Management Strategy for Village and Small Town Water Supply. GW-MATE Briefing Note Series Note 13.
Giordano, M. (2006). Agricultural groundwater use and rural livelihoods in sub-Saharan Africa: A first-cut assessment. *Hydrogeology Journal*, 14(3), 310–318.
Giordano, M., de Fraiture, C., Weight, E., & van der Bliek, J. (2012). Water for wealth and food security. Supporting farmer-driven investments in agricultural water management. Synthesis Report of the AgWater Solutions Project, pp. 38.
Giordano, M.. & Villholth, K. G.. (Eds.). (2007). *The agricultural groundwater revolution: Opportunities and threats to development*. Comprehensive Assessment of Water Management in Agriculture Series 3. Wallingford: IWMI and CAB International.
Gleick, P. H. (1996). Basic water requirements for human activities: Meeting basic needs. *Water International*, 21, 83–92.
Gumma, M. K., & Pavelic, P. (2012). Mapping of groundwater potential zones across Ghana using remote sensing, geographic information systems and spatial modelling. *Environmental Monitoring and Assessment*. DOI 10.1007/s10661-012-2810-y
Hagos, F., Ayenew, T., Mamo, K., & Adem, S. (2013). Groundwater irrigation and its impact on livelihoods of smallholder farmers: The case of Haromaya Watershed, Eastern Ethiopia. *Water International* (forthcoming issue).
International Assessment of Agricultural Knowledge, Science and Technology for Development (IAASTD). (2009). In B. D. McIntyre et al. (Eds.), *Sub-Saharan Africa (SSA) report* (p. 153). Island Press, .
International Commission on Irrigation and Drainage (ICID). (2001). Strategy for Implementation of ICID's Concerns Emanating from the Sector Vision on 'Water for Food and Rural Development'. International Commission on Irrigation and Drainage.
Intergovernmental Panel on Climate Change (IPCC). (2007). Summary for Policymakers. In Climate Change 2007: The Physical Science Basis. Contribution of Working Group I to the Fourth Assessment Report of the Intergovernmental Panel on Climate Change. S. Solomon, D. Qin, M. Manning, Z. Chen, M. Marquis, K. B. Averyt, M. Tignor, & H. L. Miller (Eds.), Cambridge, UK and New York: Cambridge University Press.
Kamagaté, B., Séguis, L., Favreau, G., Seidel, J. -L., Descloitres, M., & Affaton, P. (2007). Hydrological processes and water balance of a tropical crystalline bedrock catchment in Benin (Donga, upper Ouémé River). *Comptes Rendus Geoscience*, 339, 418–429.
Kay, M. (2001). Smallholder irrigation technology: prospects for sub-Saharan Africa. IPTRID, FAO, Rome. Knowledge Synthesis Report No. 3.
Le Maitre, D. C., Scott, D. F., & Colvin, C. (1999). A review of information on interactions between vegetation and groundwater. *Water SA*, 25, 137–152.
Lehner, B., & Döll, P. (2004). Development and validation of a global database of lakes, reservoirs and wetlands. *Journal of Hydrology*, 296, 1–22.
MacDonald, A. M., Calow, R. C., MacDonald, D. M. J., Darling, W. G., & Dochartaigh, B. É. Ó. (2009). What impact will climate change have on rural groundwater supplies in Africa?. *Hydrological Science Journal*, 54(4), 690–703.
MacDonald, A. M., Dochartaigh, B. É. Ó., Bonsor, H. C., Davies, J., & Key, R. (2010). Developing quantitative aquifer maps for Africa. British Geological Survey Internal Report, IR/10/103. p. 34.
Maherry, A., Colvin, C., Roberts, W., May, F., Masinyandima, M., Le Maitre, D., & Mwenge Kahinga, J. M. (2011). SADC Groundwater Dependent Ecosystems. Final Component Report. SADC Groundwater and Drought Management Project Component C003A: Groundwater Dependent Ecosystems (GDEs): Occurrence, vulnerability, value and protection. p. 163.

Margat, J., Frenken, K., & Faurès, J. -M. (2005). Key water resources statistics in AQUASTAT: FAO's Global Information System on Water and Agriculture. IWG-Env, International Work Session on Water Statistics, Vienna, June 20–22, 2005, p. 13.

Massuel, S., Cappelaere, B., Favreau, G., Leduc, C., Lebel, T., & Vischel, T. (2011). Integrated surface water-groundwater modelling in the context of increasing water reserves of a regional Sahelian aquifer. *Hydrological Science Journal*, 56(7), 1242–1264. doi: 10.1080/02626667.2011.609171

Mato, R. R. A. M. (2002). *Groundwater pollution in urban Dar es Salaam, Tanzania: Assessing vulnerability and protection priorities* (PhD Thesis). Eindhoven University of Technology, The Netherlands.

Mukherji, A.. Villholth, K. G.. Sharma B. R.. & Wang, J. (Eds.). (2009). *Groundwater governance in the Indo-Gangetic and Yellow River Basins: Realities and challenges. Series: IAH – Selected papers on hydrogeology*, Vol. 15. CRC Press, Taylor & Francis Group. pp. 325 ISBN 978-0-415-46580–9.

Murray, B. R., Zeppel, M. J. B., Hose, G. C., & Eamus, D. (2003). Groundwater-dependent ecosystems in Australia: It's more than just water for rivers. *Ecological Management & Restoration*, 4(2), 110–113.

Namara, R. E., Horowitz, L., Nyamadi, B., & Barry, B. (2011). Irrigation development in Ghana: Past experiences, emerging opportunities, and future directions. Ghana Strategy Support Program (GSSP). GSSP Working Paper No. 0026.

Pavelic, P.. Giordano, M.. Keraita, B.. Ramesh, V.. & Rao, T.. (Eds.). (2012b). Groundwater availability and use in Sub-Saharan Africa: A review of 15 countries. Colombo, Sri Lanka: International Water Management Institute (IWMI). 276 p. doi: 10.5337/2012.213.

Pavelic, P., Smakhtin, V., Favreau, G., & Villholth, K. G. (2012a). Water balance approach for assessing potential for smallholder groundwater irrigation in sub-Saharan Africa. *Water SA*, 38(3), 399–405.

Seward, P., Xu, Y., & Brendonck, L. (2006). Sustainable groundwater use, the capture principle, and adaptive management. *Water SA*, 32(4), 473–482.

Shah, T., Burke, J., & Villholth, K. (2007). Groundwater: A global assessment of scale and significance. In D. Molden (Ed.), *Water for food, water for life: A comprehensive assessment of water management in agriculture*. London, UK: Earthscan; Colombo, Sri Lanka.

Siebert, S., Burke, J., Faures, J. M., Frenken, K., Hoogeveen, J., Döll, P., & Portmann, F. T. (2010). Groundwater use for irrigation – a global inventory. *Hydrology and Earth System Sciences Discussions*, 7, 3977–4021.

Svendsen, M., Ewing, M., & Msangi, S. (2008). Watermarks: Indicators of Irrigation Sector Performance in Sub-Saharan Africa. Africa Infrastructure Country Diagnostic, Summary of Background Paper 4.

Taylor, R. G., Koussis, A. D., & Tindimugaya, C. (2009). Groundwater and climate in Africa: A review. *Hydrological Science Journal*, 54(4), 655–664. DOI: 10.1623/hysj.54.4.655

Torou, B. M., Favreau, G., Barbier, B., Pavelic, P., Illou, M., & Sidibé, F. (2013). Constraints and opportunities for groundwater irrigation arising from hydrologic shifts in the Iullemmeden Basin, south-western Niger. *Water International*, *38*(4), 465–479. doi: 10.1080/02508060.2013.817042.

Tuinhoff, A., Foster, S., van Steenbergen, F., Talbi, A., & Wishart, M. (2011). Appropriate Groundwater Management Policy for Sub-Saharan Africa in Face of Demographic Pressure and Climatic Variability. GW-MATE, Strategic Overview Series, No. 5, p. 40.

United Nations Economic Commission for Africa. (2011). *Management of groundwater in Africa including transboundary aquifers: Implications for meeting MDGs, livelihood goals and climate change adaptation*. United Nations Economic Commission for Africa, African Climate Policy Centre, Working Paper 6, November 2011.

United Nations Economic Commission for Africa, AU & AfDB. (2000). The Africa Water Vision 2025: Equitable and Sustainable Use of Water for Socioeconomic Development. Retrieved from: http://www.uneca.org/awich/african%20water%20vision%202025.pdf

United Nations. (2000). *Principles and practices of water allocation among water-use sectors*. Economic and Social Commission for Asia and the Pacific, Water Resources Series No. 80, United Nations, New York, ISBN: 92-1-120002-4.

Utz, R. J. (2008). *Sustaining and sharing economic growth in Tanzania*. World Bank. p. 342. DOI: 10.1596/978-0-8213-7195-4.

Villholth, K. G. (2013). Groundwater irrigation for smallholders in Sub-Saharan Africa – A synthesis of current knowledge to guide sustainable outcomes. *Water International*, *38*(4), 369–391. doi: 10.1080/02508060.2013.821644.

Villholth, K. G., Tøttrup, C., Stendel, M., & Maherry, A. (2013). Integrated mapping of groundwater drought risk in the SADC region of Africa. *Hydrogeology Journal* (Accepted).

Wada, Y., van Beek, L. P. H., van Kempen, C. M., Reckman, J. W. T. M., Vasak, S., & Bierkens, M. F. P. (2010). Global depletion of groundwater resources. *Geophysical Research Letters*, 37, L20402. DOI: 10.1029/2010GL044571

World Bank. (1981). *Accelerated development in sub-Saharan Africa: An agenda for action*. Washington: World Bank.

World Bank. (2011). *Ethiopia: Strategic framework for managed groundwater development*. Federal Democratic Republic of Ethiopia, Ministry of Water Resources and GW-MATE. pp. 58.

Wright, E. P. (1992). The hydrogeology of crystalline basement aquifers in Africa. *Geological Society, London, Special Publications*, 66, 1–27.

You, L., Ringler, C., Nelson, G., Wood-Sichra, U., Robertson, R., Wood, S., Guo, Z., Zhu, T., & Sun, Y. (2010). *What is the irrigation potential for Africa? A combined biophysical and socioeconomic approach*. IFPRI Discussion Paper 00993.

Understanding smallholder irrigation in Sub-Saharan Africa: results of a sample survey from nine countries

Tushaar Shah, Shilp Verma, and Paul Pavelic

Smallholder irrigation is emerging as a development priority in Sub-Saharan Africa. Based on a survey of 1554 smallholders from nine countries, this paper compares rainfed farming with gravity-flow, manual-lift and motor-pump irrigation. Motor-pump-irrigation farmers reported the highest net value added per acre and per family worker, with gravity-flow and manual-irrigation farmers earning marginally more than rainfed-only farmers. In addition to making affordable pumps more readily available, improving the availability of working capital, enhancing security of tenure and ensuring the availability of affordable fuel are all likely to accelerate smallholder irrigation development in Sub-Saharan Africa.

Introduction

Accelerating agricultural growth has been a high priority for Sub-Saharan Africa (SSA) (FARA, 2003). Agriculture contributes an average of 30% of GDP and 67% of employment in the region (*ibid.*). Some 80% of SSA's people live in rural areas, and 70% depend on agriculture for their livelihoods (*ibid.*; Merrey & Sally, 2008). The region has 204 million malnourished people, a third of whom are children (Merrey & Sally, 2008, p. 515). Research has shown that in the countries of SSA, a given rate of GDP growth due to agricultural growth reduces poverty 11 times more effectively than does an identical rate of GDP growth due to non-agricultural growth (FAO, 2012). The population of the region is expected to increase at a rate of close to 3% annually – to over one billion in 2025 (FARA, 2003). Climate change is predicted to increase the variability of precipitation in SSA, increasing the risks associated with rainfed farming and posing new challenges of drought mitigation as well as adaptation. At the turn of the millennium, FARA (2003) estimated that SSA's agricultural economy would have to grow at a rate of 6% per year to provide for the growing population, achieve food security, attack rural poverty and speed up economic growth in a sustainable manner (*ibid.*). Irrigated land contributes roughly 25% to the value of agricultural output in SSA while occupying only 3.5% of the region's cultivated land (Foster & Briceño-Garmendia, 2009). As a result, irrigation is increasingly considered essential for accelerating agricultural growth in SSA as well as for stabilizing smallholder livelihoods. Given the limited irrigation development in SSA and the abundance of land and water resources there is scope for expanding smallholder irrigation if the existing constraints can be overcome.

Africa's irrigation history has a large colonial imprint. Colonial powers designed and managed irrigation systems such as the Office du Niger in Mali and the Gezira Scheme in Sudan as globally integrated agri-businesses, imposing mono-crop regimes, providing all inputs and marketing all outputs, and reducing the African smallholder to the status of a serf (Shah, 2009; Shah et al. 2002). With the end of colonialism, this model of irrigation management declined as emphasis shifted towards the "social and not so much the productive character of irrigation" (Ertsen, 2006). Decline in operation and maintenance, lack of institutional reform, bureaucratic sloth and inefficiency, poor service delivery, lack of demand for irrigation – all of these led to severe deterioration in the performance of large irrigation systems in SSA (Alam, 1991; ICID, 2010:165). Moreover, at around USD 10,000/ha, centrally planned large-scale irrigation projects turned out costlier to construct in SSA than elsewhere in the developing world (ICID, 2010; Lankford, 2009). As a result, international financial institutions became increasingly reluctant to invest in large irrigation projects in SSA. Many colonial powers constructed small irrigation systems to benefit smallholders, but these too were managed in a centralized, authoritarian manner, in an "estate mode". This tradition continued after the colonial era, but the performance of these small systems – in terms of productivity, equity and sustainability – was no better than the large ones (Barnett, 1984; Carter, 1989; Perret, 2002).

Since the 1980s, however, vibrant pockets of informal smallholder irrigation have emerged throughout SSA, supported sometimes by NGOs and donors, using manual or motorized pumps to lift small amounts of water from above or below ground to irrigate gardens as well as field crops (Allaire, 2009; Eguavoen et al. 2011; Lankford, 2009). Brown and Nooter (1992) found these booming in Niger, Nigeria, Mali, Mauritania, Chad, Senegal, Burkina Faso and Cameroon towards the end of the 1980s. "Smallholder farming" in this discourse is taken to mean "family farming" as distinct from operating a commercial farm, as described by Toulmin and Gueye (2005). Behind this groundswell in small-scale, informal smallholder irrigation is mostly private initiative by individuals or small groups of smallholders. These smallholder irrigation schemes depend for their water on shallow wells, ponds, streams, rivers and other sources; they often involve lifting by manual or motor pump and conveyance of water through open channels or pipes (Adams & Carter, 1987; Carter & Howsam, 1994; Eguavoen et al. 2011; Lankford, 2009). Common to these are several distinct features: a new entrepreneurial model of irrigation organization in which the smallholder is the decision maker rather than only a labourer; the use of familiar and affordable technology; and institutional arrangements that promote farmer management, either in groups or individually. Critical to their vibrancy is not the size of land irrigated, but the ownership-management model. As Vaishnav (1994:524) noted:

> Small-scale irrigation is not large-scale formal irrigation made small. It is perhaps in the management element that the key difference lies. In a small system there are no tiers of management as in large-scale schemes. The farmer alone decides when to irrigate and how much water to apply, when to start and stop pumps or other appliances, and generally runs the entire scheme with the help of the family or local community members.

Brown and Nooter (1992:4) observed: "Almost all successful irrigation occurs on plots owned and managed by individuals with individual water supply control." Researchers have noted several advantages of smallholder irrigation systems. Allaire (2009), for example, found that small-scale irrigation, especially based on groundwater, arguably has the multiple advantages of ubiquity, immediacy, low cost, drought resistance, farmer management, spatial equity, on-demand water availability and conveyance efficiency. However, implicit in the notion of "success" is financial and ecological sustainability, demand-driven irrigation service and beneficial financial impact on the smallholder households.

Grass-roots-level information and analyses to guide thinking and discussion on the irrigation development strategy most appropriate for SSA's smallholder agriculture remain lacking. The goal of this multi-country study is to make an assessment, through in-depth field research, of the livelihood and food-security implications of four modes of agricultural water management in smallholder farming systems of Sub-Saharan Africa, viz., (1) rainfed farming; (2) gravity-flow irrigation; (3) small-manual-pump irrigation; and (4) small-motor-pump irrigation.

Approach of the study

Given the growing evidence of the increasing emergence and importance of smallholder irrigation in SSA, an important question to ask is how pervasive it is. This question can be answered by a census of such systems, like, for example, the Minor Irrigation Census that the government of India conducts every five years. However, there are other questions about smallholder irrigation in SSA that can be explored using small-sample surveys. This study sets out the results of such a survey of smallholder irrigation, undertaken in nine SSA countries (Burkina Faso, Ethiopia, Ghana, Kenya, Malawi, Mali, Niger, Nigeria and Tanzania) during 2011 (Figure 1). The overarching assumptions in the design of the survey were that smallholders desire irrigation to improve on-farm water control, and that the degree of such water control depends on the mode of irrigation (defined by the water source and the technology of water mobilization and conveyance). Farmers were assigned to a different water-control class (WCC) depending upon whether they engaged exclusively in rainfed farming (WCC1) or had some of their farmland under gravity-flow irrigation (WCC2), manual-lift irrigation (WCC3), or motor-pump irrigation (WCC4) (see Table 1).

A common questionnaire was used by research partners in all nine countries to canvas information from at least 25 smallholders purposively selected from each WCC in each country. The objectives were two. The first was to carry out nine country-level studies, combining and analyzing qualitative and survey data for each; and the second was to pool the data for all nine countries and compare SSA smallholders under the four WCCs. The present paper is an attempt to meet the second objective, with the first objective being met by the country-level studies listed in Table 2. As Table 2 shows, a total of 1554 smallholders were surveyed, with a sample of at least 350 farmers under each WCC, drawn from across nine countries of East and West Africa.[1]

Some of the key questions the survey was designed to address are: What is the profile of smallholders under different WCCs? Do gender, wealth, education, age, landholding size or land tenure explain why farmers opt for different modes of irrigation? Is adoption of a certain mode of irrigation associated with the ownership of household assets and wealth? Do modes of irrigation influence crop choices? Do they affect the economic behaviour and input choices made by smallholders? What are the household-level impacts of different modes of irrigation in terms of value added per acre as well as per worker? What are the constraints that smallholders face in accessing irrigation? What do they perceive as the most desirable mode of irrigation? These questions are important given the widely shared

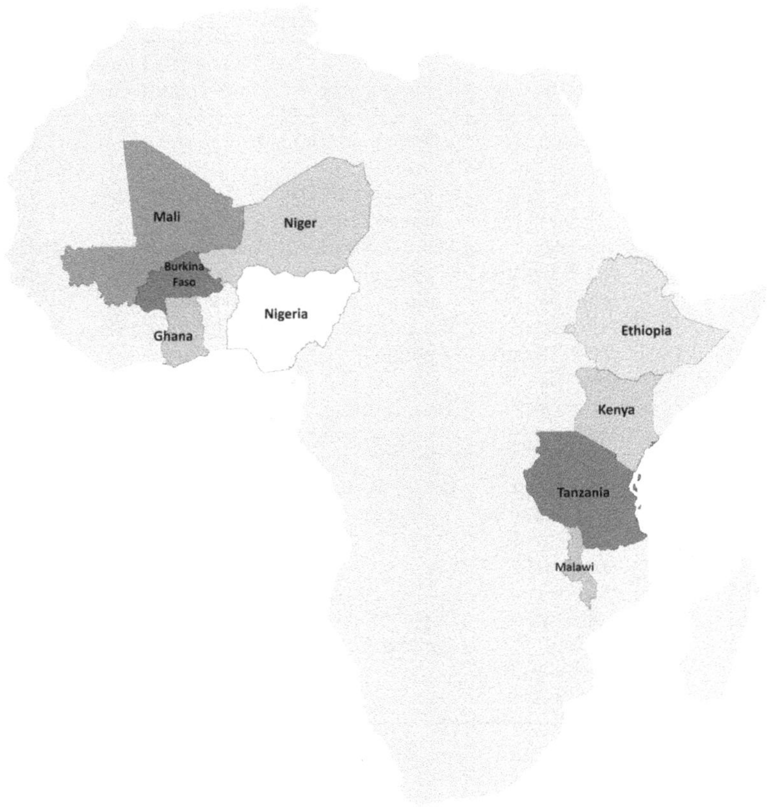

Figure 1. Countries covered in the survey.

interest among donors, NGOs and policy makers in understanding what might be the most appropriate strategies to accelerate smallholder irrigation in the SSA region.

Irrigation profile of SSA's smallholders

What explains the prevalence of rainfed farming in SSA?

Given that access to irrigation – with low or high water control – expands a smallholder's opportunity set, what are the characteristics of households that seek some means of irrigation and those that choose to pursue rainfed-only farming? One thought might be that rainfed farmers remain rainfed because they have not had an irrigation option, because of lack of access to either water, or capital, or extension and technical support. Another comes from the Asian experience, where irrigation is actively sought by smallholders to be able to intensify farming on increasingly fragmented land-holdings. Following from this, it may be asked whether rainfed farmers in SSA are those with a large enough land holding to be able to make an adequate living from extensive farming of a single rainfed crop.

There is some evidence to support these ideas in the data. For the combined sample of 1554 farmers, rainfed farmers not only have marginally larger land holdings, they also cultivate larger areas compared to irrigators (Figure 2). However, a closer look at country-level data suggests that this is not true uniformly. In Ghana, Malawi, Kenya and Ethiopia, rainfed farmers have the smallest land holdings, while in Burkina Faso, Niger, Mali and Nigeria,

Table 1. Categorization of smallholder irrigation modes into four water-control classes.

Water-control class	Water sources	Water mobilization	Water transport and application	Level of water control
WCC1 Rainfed farming	Precipitation	Effective rainfall, *in situ* moisture conservation	n/a	Low
WCC2 Gravity-flow irrigation	River, lake, canal, stream, reservoir	Dam, diversion structure, large pump scheme	Lined canals, earthen field channels	Moderate but seasonal
WCC3 Manual/Manual pump irrigation	River, lake, canal, stream, reservoir	Bucket, manual pump	Earthen channels, overland flexible pipes	Moderate but seasonal
	Groundwater (shallow wells)	Treadle pump, manual pump, bucket		High but seasonal
WCC4 Motorized-pump irrigation	Surface water (river, lake, canal, small reservoir)	Diesel or electric pump	Earthen channels, buried pipelines, overland flexible pipes	High but seasonal
	Groundwater (wells, boreholes, tubewells)	Diesel or electric pump		High year round

Table 2. Regions and sample size of smallholder households interviewed in nine SSA countries and four water-control classes.

Category	BFS	NGR	MAL	GH1	GH2	TNZ	MLW	KEN	NGA	ETH	TOTAL
WCC1 Rainfed farming	27	27	29	30	80	18	30	62	66	30	**399**
WCC2 Gravity-flow irrigation	34	33	32	30	40	47	30	53	31	30	**360**
WCC3 Manual-pump irrigation	48	43	18	25	99	48	45	41	21	25	**413**
WCC4 Motorized-pump irrigation	28	30	50	50	08	22	40	24	80	50	**362**
TOTAL (1554)	**137**	**133**	**129**	**135**	**227**	**135**	**145**	**180**	**198**	**135**	**1554**

Note: BFS, Burkina Faso; NGR, Niger; MAL, Mali (Torou & Barbier, 2011); GH1, Ghana (Owusu et al. 2013); GH2, Ghana (Dittoh et al. 2013); TNZ, Tanzania (Kashaigili, 2011); MLW, Malawi (Kamwamba, 2011); KEN, Kenya (Kipkoech et al. 2011); NGA, Nigeria (Tijani et al. 2011); ETH, Ethiopia (Ayenew et al. 2011).

rainfed farmers have larger holdings compared to the irrigators. It is likely that in the former countries, rainfed farmers are too poor and disadvantaged to gain access to irrigation; but in the latter countries, it is possible that large landowners find extensive rainfed farming with family labour on their larger plots a better option than intensive irrigated farming. There might also be special situations, such as in Malawi, where extensive cultivation of rainfed tobacco for the market offers as good an option to smallholders as intensive irrigated farming of garden crops.

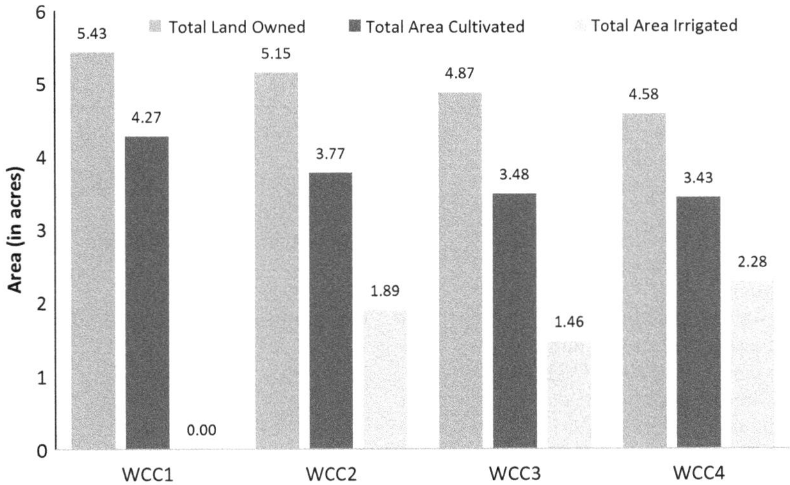

Figure 2. Total land owned, area cultivated and area irrigated per household for different water-control classes.

How do women-headed households respond to irrigation?

In all the study locales, bar the Upper East Region of Ghana and Nigeria, women-headed households actively pursue irrigated agriculture. The survey data suggest that while female-headed households have, on average, smaller land holdings, they cultivate and irrigate a slightly larger proportion of the same compared to male-headed households. However, these differences in average cultivated and irrigated area were not found to be statistically significant (Table 3). Also, most women-headed households are using either

Table 3. Comparison across female- and male-headed households.

Particulars	Female-headed households ($N = 183$)	Male-headed households ($N = 1028$)	$\|t\|$	Sig.
Average land holding (acres)	2.8	5.4	5.548	0.000*
Proportion of land holding which is cultivated (%)	83.8	73.6	0.211	0.934
Proportion of land holding which is irrigated (%)	35.3	26.7	0.714	0.647
Percentage of households owning motor pumps	8.7	25.9	5.102	0.000*
Percentage of households owning manual pumps	12.6	8.9	1.586	0.002*
Percentage of households owning buckets	56.8	32.1	6.535	0.000*
Percentage of households owning flexible pipes	16.4	26.4	2.882	0.000*
Net value added per acre (USD, rounded to nearest dollar)	360	456	0.744	0.201
Net value added per family worker (USD, rounded to nearest dollar)	351	422	0.580	0.243

Note: $|t|$ represents the modular value for Student's t-test; Sig. represents the p-value. *Values with an asterisk are significant at the 95% confidence level.

Table 4. Profiles of manual and motor-pump irrigators.

Particulars	Manual irrigators ($N = 347$)	Motor-pump irrigators ($N = 381$)
Avg. investment in irrigation equipment (USD, rounded to nearest dollar)	49 (range: 15–150)	1016 (range: 350–2650)
Most manual irrigators found manual lift irrigation profitable	Yes, especially in Nigeria and Ethiopia	
Most manual-pump or bucket irrigators got subsidies from the government or from donors	No, except in Malawi and Ethiopia	
Most motor-pump irrigators bought irrigation equipment (motor pumps and irrigation pipes) off the shelf, without any government support	Yes, except in Ghana (GH2) and Ethiopia	
Do motor pump irrigators have:		
a. Younger heads of households	a. Not necessarily	
b. More years of education	b. Not necessarily	
c. Males as heads of households	c. Yes, except in Nigeria	
d. More family workers in agriculture	d. Not necessarily	
e. Greater wealth and assets	e. Yes, with some exceptions (see Figure 3)	
Energy sources for irrigation pumps	Electricity	7%
	Diesel	40%
	Petrol	53%

manual (mostly bucket) irrigation or gravity-flow irrigation, suggesting socio-economic or other barriers that prevent them from adopting motor-pump irrigation.

Profiles of manual irrigators and motorized irrigators

Manual irrigation, using human labour, is characterized by its low capital investment, which ranges from USD 15 for a bucket (used by a large number of the sample farmers in the Upper-East Region of Ghana, Burkina Faso, Niger and Mali) up to USD 150 for high-end treadle pumps coupled with overland flexible pipes (found mostly in Tanzania). Contrary to the authors' expectation of aggressive promotion of manual irrigation devices in SSA by local and international NGOs, donors and government programmes with high levels of subsidy, in reality, most manual irrigators in the survey had bought whatever they needed off the shelf with their own funds at market prices, except in Ethiopia and Malawi, where subsidies played a significant role in uptake (Table 4). A majority of the manual irrigators in the sample used buckets, which need neither promotion nor much capital investment. Most manual-pump irrigation was found to be profitable, especially in Ethiopia (where roughly two-thirds of the manual-pump irrigators reported profits) and in Nigeria (where all manual-pump irrigators reported profits).

High initial capital investment is a defining aspect of motor-pump irrigation. As Table 4 shows, investing in a motor pump, pipe and the complete assembly entails sums ranging from USD 350 to USD 2650. We saw earlier (Figure 2) that motor-pump irrigators on average enjoy the highest area under irrigation, both in absolute terms as well as in terms of proportion of total land holdings. It was also mentioned that female-headed households are less likely to be motor-pump irrigators. Are these also the farming households headed by younger members exposed to modernity or educated individuals with an interest in

Table 5. Wealth status of rainfed and irrigator households in East and West African samples.

	Ghana, Kenya and Ethiopia				Burkina Faso, Niger and Mali			
	WCC1	WCC2,3,4	$\|t\|$	Sig.	WCC1	WCC2,3,4	$\|t\|$	Sig.
Sample size	202	475			83	316		
Average total land holding (acres)	3.44	3.70	0.791	0.097	9.92	6.32	4.065	0.002*
Average cultivated area (acres)	2.76	2.52	1.175	0.537	7.63	4.90	4.665	0.007*
Average number of work animals	1.40	1.28	0.691	0.113	1.17	0.76	2.536	0.437
Percentage of households owning a motor car	0.3	1.9	0.869	0.083	2.4	0.6	1.446	0.004*
Percentage of households owning a colour TV	9.9	18.3	2.758	0.000*	25.3	18.4	1.413	0.008*
Percentage of households owning a radio	57.9	82.9	6.996	0.000*	78.3	59.2	3.251	0.000*

technology? The survey suggests that the adoption of motor-pump irrigation has little to do with the age or the education of the head of the household (Table 4). If anything, however, motor-pump-irrigating households are in general wealthier and have larger endowments of family assets. More of them have motorcycles, colour TVs and radios (Table 5).

In Ethiopia, Eguavoen et al. (2011) found that well owners acquired more cattle than other farmers. However, this survey indicates that well owners, on average, have fewer livestock compared to rainfed farmers and gravity-flow irrigators (Figure 3). The data on household assets suggest distinct patterns in different groups of countries. In Ghana, Kenya and Ethiopia, rainfed farmers are generally worse off compared to irrigating households. In contrast, in Burkina Faso, Niger and Mali, rainfed farmers have larger land holdings, livestock herds and other assets. Here there is a distinct possibility that large landowners

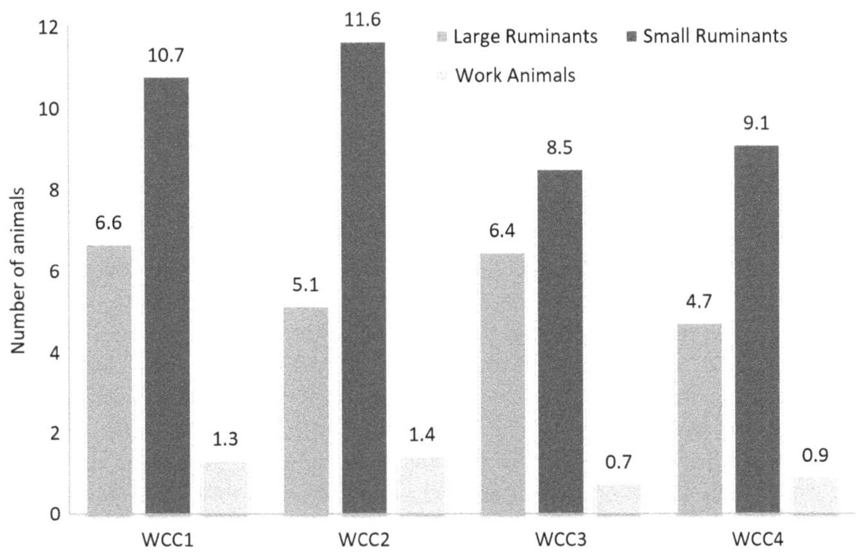

Figure 3. Average number of livestock owned by households in different water-control classes.

practise rainfed farming by choice rather than compulsion. Table 5, which compares rainfed farmers (WCC1) with irrigating farmers (WCC2, 3 and 4) in the two groups of countries, gives credence to this contention. In the former sample, rainfed farmers (WCC1) are poorer in terms of several assets, particularly in ownership of TVs and radios. In the latter, rainfed farmers are significantly wealthier compared to irrigators in terms of most assets. These rainfed farmers could easily have made the transition to motor-pump irrigation if they wanted to. This suggests that they pursue rainfed farming out of choice.

Expectedly, electric pumps were rare in all the countries sampled (7% of all pumps; see also Carter & Howsam, 1994); by far the majority of pumps ran on petrol (53%), fewer on diesel (40%) – except in Tanzania, where all the sampled motorized-pump irrigators had diesel pumps. As was the case with manual-pump irrigators, most motor-pump irrigators bought irrigation equipment off the shelf, without government subsidies or support. The only exceptions to this were found in Ethiopia (where as many as 84% of the sample received some support or subsidy) and in the Upper East Region of Ghana (where a little over half the respondents received support).

Irrigation impact on crop choices

The literature suggests that access to irrigation expands a smallholder's opportunity set in several different ways (Chambers et al. 1989). It mitigates production risk arising from drought or mid-season dry-spells: farmers with access to supplemental or crop-saving irrigation can manage rainfed cropping patterns more successfully than rainfed-only farmers without access to supplemental irrigation. It reduces production risk and thereby encourages farmers to make more intensive use of capital-intensive inputs, such as improved seeds, fertilizers, pesticides and irrigation. It encourages smallholders to make more intensive use of land by growing multiple crops during a year, within and outside of the rainy season, depending on water availability. Access to reliable "on-demand" irrigation in turn encourages smallholders to turn all or part of their farming over to high-value, market-oriented garden crops (Shah, 2009).

To understand the nature of the constraints facing rainfed farmers, farmers were asked the number of crops they can grow and the production risk facing them. Barring Nigeria, Kenya and Ethiopia, the sample rainfed farmers everywhere else can grow only a single rainfed crop in a year; and their subjective probability of a normal rainfall lies between 0.40 and 0.70 (Table 6).[2] In such a situation, it is natural that rainfed farmers would be apprehensive about input intensification in their farming. For gravity-flow irrigators, reservoirs offer irrigation for a few months of the year. Only the Kenya sample claimed, on average, 11 months of irrigation availability from gravity-flow irrigation systems; in Ghana, it was only 3–4 months. Irrigation service fees were negligible in East African sites but substantial in West African locales (Table 6).

Does access to irrigation change the crops smallholders grow in SSA? Irrigation-induced crop-choice differences are deeper and more sweeping with smallholders in Sub-Saharan Africa than is the case throughout Asia. In Asia, irrigation is commonly used by smallholders to increase land-use intensity and to curtail production risk in field crops such as rice and wheat. In the SSA sample, rainfed farming is largely associated with subsistence food crops. Gravity-flow irrigation is associated with field crops for subsistence as well as for market. In contrast, lift irrigation, whether manual or motorized, is strongly associated with garden crops for the market. Rarely did the SSA sample encounter farmers who grow rice with lift irrigation, either manual or motorized – something quite common in

Table 6. Rainfall regimes, irrigation access and crop choices for different water-control classes.

Particulars	Response
Average number of rainfed crops possible in a year	1.0, except in Nigeria (1.7), Kenya (2.0) and Ethiopia (2.0)
Sample farmers' subjective probability of good rainfall	Tanzania, 0.40; Kenya, 0.44; Burkina Faso, Niger and Mali, 0.45; GH2, 0.46; GH1, Malawi and Ethiopia, 0.60; Nigeria, 0.70
Months of the year when gravity irrigation is available	Ghana, 3–4; Ethiopia, 5; Nigeria and Malawi, 6–7; Burkina Faso, Niger, Mali and Tanzania, 7–8; Kenya, 11
Average irrigation service fee for gravity-flow irrigation	Ethiopia, free; Ghana, Tanzania and Malawi, <USD 5/acre; Kenya, USD 10/acre; Nigeria, USD 45/acre; Burkina Faso, Niger and Mali, >USD 100/acre
Do rainfed farmers (WCC1) tend to choose a subsistence field crop for food as their most important crop?	Yes: less than 15% of rainfed farmers in the pooled sample reported having a cash crop as their most important crop. (Exception: in Malawi, rainfed tobacco dominates.)
Do gravity-flow irrigators (WCC2) tend to choose subsistence field crops for food as their most important crop?	Not always: 44% of gravity-flow irrigators in the sample reported having a subsistence crop as their most important crop; but they also grow a significant number of market crops.
Do pump irrigators (WCC3 and WCC4) choose to grow market crops as their most important crops?	Yes: 81% of manual-pump irrigators and 75% of motor-pump irrigators reported having a market crop as their most important crop.

Asia. This pattern has also been noted by other researchers on African irrigation (Eguavoen et al. 2011).

Rainfed cultivation is dominated by subsistence food crops, except in Malawi, where rainfed tobacco and green maize dominate, and in Kenya where green maize dominates. Contrariwise, lift irrigation is used mostly for garden crops for the market. Gravity-flow irrigation shows a mixed picture: where it is more reliable and offers some water control, farmers use it for garden crops; but much of gravity-flow irrigation is used in subsistence food crops like rice and maize for food. Based on a survey of 400 households, in a study in Burkina Faso, Niger and Mali, Torou & Barbier (2011) concluded that the choice of crops largely depends on the level of water control. The authors argued that for Sahelian farmers it does not make economic sense to irrigate grain crops such as millet and sorghum with motor pumps. As the extent of water control improves, farmers prefer to cultivate higher-value crops – cabbage, carrot, green beans and other vegetables – despite the public-policy nudge towards rice cultivation.

Input intensification and productivity

Do differing levels of water control result in differentiated levels of input intensification (as given by cash expenditure per acre) in SSA, as they do with smallholder irrigation elsewhere in the developing world? If this is true, lift irrigators of all types, who enjoy a higher level of control compared to gravity-flow irrigators, should intensify more and generate more value added per acre as well as per farm worker. As Table 7 suggests, this indeed turns out to be the case. The difference in input expenditure per acre between manual-pump

Table 7. Average net value added per acre and per family worker. All values in USD, rounded to the nearest dollar.

Particulars	WCC1 Rainfed farmers ($N = 203$)	WCC2 Gravity-flow irrigators ($N = 239$)	WCC3 Manual irrigators ($N = 300$)	WCC4 Motor-pump irrigators ($N = 160$)
Gross value of output per acre	320	358	348	881
Cash expenditure per acre	59	72	84	178
Net value added per acre	261	286	265	703
Gross value of output per family worker	405	414	398	1413
Value added per family worker	319	325	307	1092

and gravity-flow irrigators was large everywhere in the sample, except in Malawi, where it was marginal. The input-intensification effect (expressed as higher value added per acre as well as per farm worker) was much more pronounced in the case of motor-pump irrigators. Torou and Barbier (2011) found that in their sample of smallholders in Mali, Burkina Faso and Niger, motor-pump irrigators incurred the highest expenses on inputs per acre but also enjoyed the highest degree of water control and tended to earn higher incomes per acre.

One might think that motor-pump owners enjoy the same level of water control as manual-lift irrigators but would be able to bring larger areas under garden crops per family farm worker, compared to manual irrigators, who have to divide their labour between tending crops and fetching water. The latter would be limited not only by the smaller volumes of water manually lifted per hour but also by having less labour time available for non-irrigation farming operations. This would mean that motor-pump irrigators would be able to generate more value added per family farm worker. This hypothesis is well supported by the survey data (Table 7). What is surprising is that there is no significant difference in gross or net value added per acre or per family worker between WCC1, 2 and 3.

Based on Asian experience, it might be surmised that gravity-flow irrigation would be highly unreliable and offer much less water control when compared to lift irrigation. However, the data suggest that, on average, gravity-flow irrigators in this sample enjoy lower costs while extracting marginally higher gross value of output per acre as well as per family worker compared to manual-pump irrigators (Table 7). But these averages across fairly large geographic regions do not show the complete picture. Plotting individual data points reveals several instances where the value added per acre as well as per family worker is higher in rainfed conditions. We surmise that these are cases where rainfed conditions permit hardy and profitable market crops like tobacco and green maize.

The distribution of samples across different earning brackets (Figures 4 and 5) reveals that: (1) each category has a 5–8% sample of households where the net value added per acre as well as per family worker is negative; (2) for WCC1, 2 and 3, the majority of sample households fall in the lowest-earning (USD 0–200) category, while the proportion for this category is much lower for pump irrigators (WCC4); and (3) pump irrigators also have the highest proportion of sample households in the highest-earning category (>USD 800 per acre and per family worker).

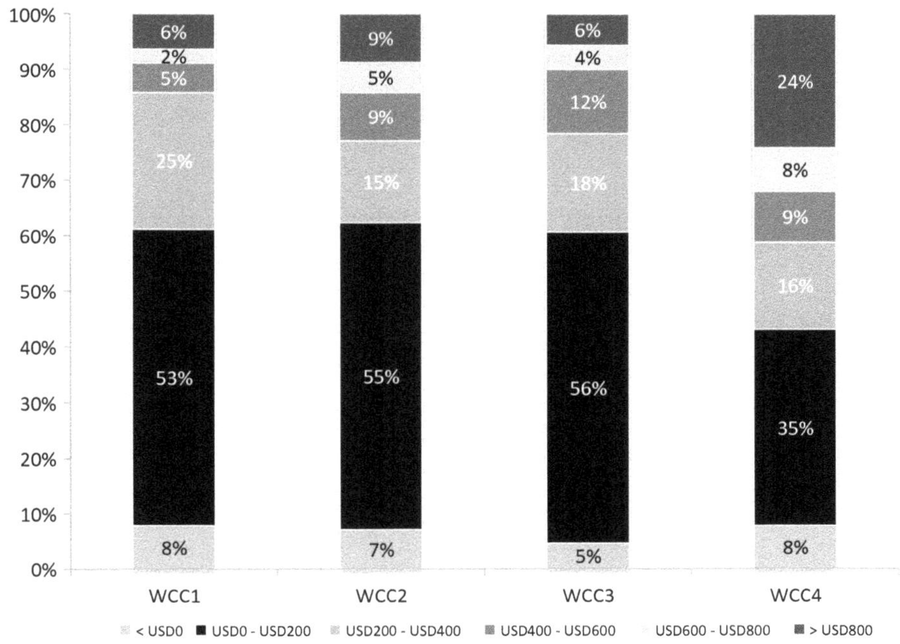

Figure 4. Net value added per acre across water control classes.

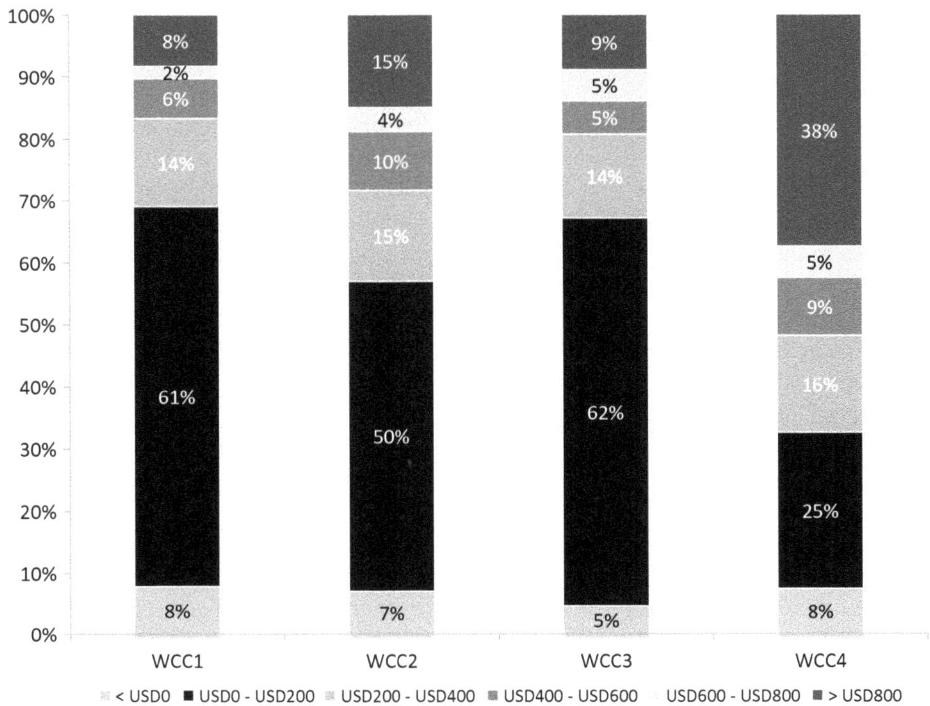

Figure 5. Net value added per family worker across water control classes.

Table 8. Average annual productivity of manual and motor-pump irrigators from surface-water and groundwater sources. All values in USD, rounded to nearest dollar.

	Manual irrigators from surface water (WCC3a) ($N = 95$)	Manual irrigators from groundwater (WCC3b) ($N = 204$)	Motor-pump irrigators from surface water (WCC4a) ($N = 97$)	Motor-pump irrigators from groundwater (WCC4b) ($N = 54$)
Gross value of output per acre	331	321	1039	598
Cash expenditure per acre	69	98	182	170
Value added per acre	262	233	857	427
Gross value of output per family worker	508	320	1697	903
Value added per family worker	419	233	1382	570

It might also be surmised that as compared to groundwater, surface-water sources (normally small reservoirs) would reduce the drudgery of lifting water for manual irrigators; as a result, manual irrigators dependent on surface sources would irrigate larger areas and add greater value per family worker. This was found to be true overall (Table 8), but not in Ghana, Kenya or Malawi. Likewise, motor-pump irrigators lifting water from surface reservoirs, overall, were able to generate higher value added per family worker, but not in Kenya or Malawi (Table 8).

Land tenure and irrigation investments

Perception of insecurity of land tenure is often cited as one of the reasons preventing increased investments in land improvement and irrigation assets (Hagos, 2012). Security of land tenure may stem from formal as well as informal rights, either individual or communal. Sample households were asked to specify whether they felt their land tenure was secure. In all, 800 households in four countries (Ethiopia, Ghana, Kenya and Tanzania) responded to this question, and 60% considered their tenure to be secure. No clear trends were seen in security of tenure across the various WCCs (Figure 6). However, the responses differed significantly across the survey locales. While more than 90% of the respondent households in Ethiopia reported secure land tenure, less than 30% responded similarly in Kenya. In Ghana and Tanzania, the figures were roughly 40% and 70% respectively.

The next question is whether there were significant differences in ownership of irrigation assets and in capital investments in pumps and pipes between the two groups of households with contrasting land tenure. Table 9 shows the key differences between the two groups.

Roughly one-fourth of the sample households in both groups are rainfed farmers. These households are unlikely to own any irrigation assets. On the whole, there are statistically significant differences in means between land-tenure classes with respect to ownership of boreholes, permanent wells, motor pumps, manual pumps, buckets and flexible pipes. It follows that respondents with secure land holdings make larger investments in fixed irrigation assets, particularly permanent wells. Specifically, among the manual and motor-pump irrigators, the average investment in pumps and pipes of households with secure land tenure is more than three times that of households with insecure land tenure.

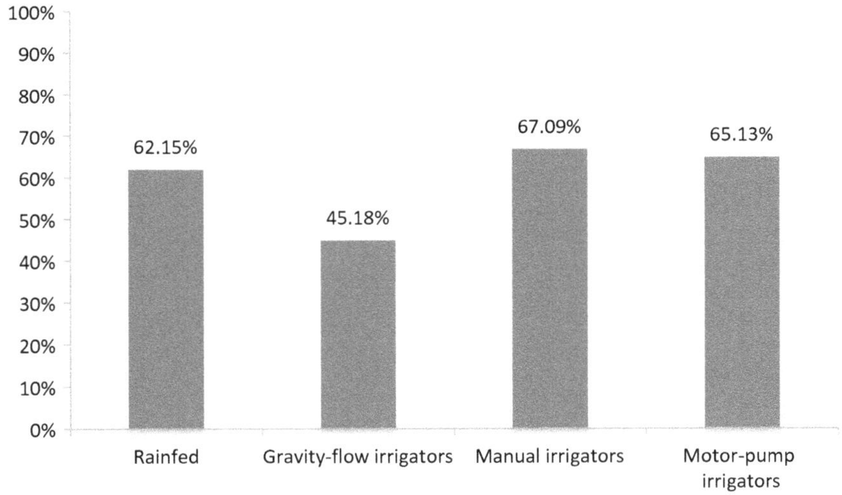

Figure 6. Percentage of respondent households that perceived land tenure to be secure.

Table 9. Security of land tenure and irrigation investments.

Particulars	Secure tenure (per cent) ($N = 480$)	Insecure tenure (per cent) ($N = 320$)	$\|t\|$	Sig.
Rainfed farmers (WCC1)	133 (27.7)	081 (25.3)		
Gravity-flow irrigators (WCC2)	089 (18.5)	108 (33.8)		
Manual-pump irrigators (WCC3)	159 (33.1)	078 (24.4)		
Motor-pump irrigators (WCC4)	099 (20.6)	053 (16.6)		
Percentage owning boreholes	4.4	2.5	1.390	0.005*
Percentage owning permanent wells	9.6	2.8	3.735	0.000*
Percentage owning temporary wells	31.7	30.9	0.217	0.662
Percentage owning motor pumps	21.5	17.8	1.263	0.011*
Percentage owning manual pumps	10.2	15.3	2.161	0.000*
Percentage owning buckets	35.4	49.4	3.966	0.000*
Percentage owning flexible pipes	30.4	15.0	5.057	0.000*
Average investment in irrigation equipment among pump (manual and motorized) irrigators (rounded to the nearest dollar)	USD 603	USD 174	1.588	0.011*

Constraints and aspirations facing SSA smallholders

While pump irrigation does not automatically and always translate into higher land or labour productivity compared to gravity-flow irrigation, it does inspire confidence in farmers to intensify more, take greater risks and open up to the market. What keeps gravity-flow irrigation from small and large reservoirs from inspiring similar confidence among farmers? What are the most important constraints faced by gravity, manual-pump and motor-pump irrigators? Table 10 summarizes the responses from the survey. As mentioned, gravity-flow irrigation systems offer irrigation for only a few months of the year. Among the constraints facing gravity-flow irrigators, therefore, "insufficient water availability" featured prominently, along with "working-capital shortage", "land scarcity" and

Table 10. Severity index (on a scale of 0 to 5) for constraints faced by households in different water-control classes.

	WCC2 Gravity-flow irrigators	WCC3 Manual irrigators	WCC4 Motor-pump irrigators
Insufficient water	2.14	1.73	0.00
Unreliable water supply	0.98	0.00	0.00
High water fee	0.00	0.00	0.00
Land scarcity	2.08	1.92	1.83
Shortage of family labour	1.01	0.97	1.02
Working-capital shortage	3.64	3.77	2.53
Lack of access to output markets	0.69	0.83	0.82
Crop predators	0.21	0.26	0.00
Repair and maintenance costs	0.00	0.11	1.56
High fuel costs	0.00	0.00	4.06

Note: During the field surveys, respondents in different water control classes were asked to rank the constraints they faced on a scale of 0 to 5. These results were aggregated by our co-researchers in different countries. The severity index used here was computed by weighing rank against proportion in the overall sample and taking the inverse. A higher value therefore depicts greater severity, and vice-versa.

"shortage of family labour". Somewhat surprisingly, "high water fees" did not seem to bother the sample gravity-flow farmers (Table 10), even in West African locales, where canal water rates are relatively high (Table 6).

The constraints perceived by manual irrigators in SSA are somewhat different. It might be expected that availability of family labour would be an important constraint; however, this was the case only in some of the locales, and felt with severity only by the sample farmers in Tanzania. In informal discussions, the drudgery involved also emerged to be an important constraint, given that "irrigating with a treadle pump requires the farmer to 'climb' the effort equivalent of 1000 meters each day to irrigate half a hectare" (Lankford, 2009). Most manual irrigators in the sample, however, did not have a half-hectare to irrigate. The constraints more widely faced were those of "working-capital shortage", "land scarcity" and "insufficient water".

The high cost of fuel to run motor pumps was expectedly mentioned as a severe constraint by farmers in most countries. Carter and Howsam (1994) suggested that cheap petrol powering irrigation pumps drove away the *shaduf*[3] in Nigeria during the 1970s and 1980s but that rising petrol prices may bring it back. High and rising fuel prices may constrain the pace with which SSA smallholders are able to move from manual to mechanical pumps. However, many observers, NGOs and donors have also bemoaned the absence of technical support for quick and cost-effective repair and maintenance of motor pumps as a powerful constraint (Allaire, 2009; Carter & Howsam, 1994). The motor-pump irrigators in the sample did not think so; some did mention it, but far more mentioned "working-capital shortage" and "high fuel costs" as more severe constraints limiting their irrigated agriculture.

Most rainfed farmers have some idea of the pros and cons of different modes of irrigation. Farmers in the sample were asked what mode of irrigation they would prefer to receive if given a choice to move to irrigated farming. "Own well, motor pump and 500 feet of flexible pipe" had the highest overall score (Table 11); it was ranked the most preferred option in four locales and the second-most preferred in two locales. "Government canal irrigation scheme" scored second-highest overall; it was the most preferred in three West African countries (Burkina Faso, Niger and Mali) and in Malawi, but very low elsewhere. "Open

Table 11. Irrigation wish-list of rainfed smallholders in SSA.

Modes of irrigation	Overall score
a. Government canal irrigation scheme	0.10
b. Canal drawn from a small reservoir	0.08
c. Own well, motor pump and 500 ft. of flexible pipe	0.15
d. Open well, treadle pump and 500 ft. of flexible pipe	0.06
e. Motor pump and 500 ft. of flexible pipe to be used on a small reservoir or stream, canal, or community pond	0.07

Note: During the field surveys, rainfed respondents were asked to rank the modes of irrigation on a scale of 1 to 5 (1 representing the most preferred mode and 5 the least). In the different locales, these ranks were cumulated and an average rank was determined by our co-researchers. An overall score was then computed by multiplying the inverse of the sum of average local ranks (weighted by their respective sample sizes) by 100 $\left(\text{score} = \frac{100}{\sum \text{weighted ranks}}\right)$. Thus, higher overall scores represent greater preference, and vice versa.

well, treadle pump and 500 feet of flexible pipe" scored the least overall and was ranked the lowest (least preferred) or second-lowest by rainfed farmers everywhere except in Kenya. "Canal drawn from a small reservoir" or a "motor pump and pipe on a small reservoir" both scored quite low, presumably suggesting apprehension among farmers regarding the year-round reliability of small-scale reservoirs as a source of irrigation water, besides the issue of coordination among the multiple users of these water bodies.

Summary and conclusion

The survey of 1554 smallholders in nine countries of SSA shows that irrigation, especially with motor pumps, improves agricultural productivity (measured as value added per acre as well as per worker) and transforms the farming systems of the adopters. Net value added per acre did not show large differences between WCC1, 2 and 3. However, net value added per acre as well as per family worker was nearly three times larger for motor-pump irrigators compared to WCC1, 2 and 3. High fuel cost was a major determinant of the economics of motor-pump irrigation. While manual-lift irrigators were partly able to substitute family labour for expensive petrol or diesel, motor-pump irrigators brought larger acreage under lift irrigation compared to manual-lift irrigators.

A gender divide was evident in the sample. Male-headed households had more land and dominated the rainfed sample. Because they operated significantly larger holdings, rainfed farmers generated valued added per family worker nearly similar to that of gravity-flow and manual-lift irrigators. Female-headed households had less land but cultivated and irrigated a larger proportion of their holdings. These relied more on manual and gravity-flow irrigation and made minimal use of motor-pump irrigation, possibly because of the high capital investment required and high fuel cost for operating it. Motor-pump-irrigator households tended to be male-headed and wealthier in terms of land and household assets.

When asked to choose a mode of irrigation, the sample of rainfed farmers gave maximum preference to "own well, motor pump and 500 feet of flexible pipe". The runner-up was a government canal. The least preferred was "own well, a treadle pump and 500 feet of flexible pipes". These preferences perhaps explain why it is sometimes argued that simply making motor pumps available at an affordable price would accelerate irrigation development in SSA. However, the survey also suggests that smallholder irrigation in SSA is influenced by a complex web of factors such as the feminization of farming, size of operational land holding, family labour endowment, security of land tenure, availability of non-farm sources of income, access to term loans and working capital, fuel cost and canal irrigation fees. The present analysis suggests that in addition to making affordable pumps more readily available, improving working-capital availability, enhancing security of tenure and providing fuel at affordable prices are all likely to accelerate the pace of smallholder irrigation development in SSA.

Acknowledgement

This paper is a product of the 3-year research project on "Groundwater in sub-Saharan Africa: Implications for food security and livelihoods" supported by the Rockefeller Foundation. The authors acknowledge the additional support provided by the Rockefeller Foundation to bring researchers from the study countries to the Bellagio Center, Italy for a 3-day writing workshop. We also gratefully acknowledge contributions by reviewers, particularly Dr Karen Villholth who made valuable comments on several drafts of the paper.

Notes

1. Any analysis of such a diverse data-set is bound to have some internal contradictions, given the range of physical, socio-economic and cultural landscapes covered. It may be expected that the country-level studies, which are duly referenced, will provide a richer description and analyses of the dynamics at that scale, whilst this paper focuses on a broader perspective. However, important country-level issues have been flagged at appropriate places.
2. "A subjective probability describes an individual's personal judgement about how likely a particular event is to occur. It is not based on any precise computation but is often a reasonable assessment by a knowledgeable person" (Easton and McColl, n.d.). In our survey, respondents were asked, "In a typical five-year period, how many years receive normal rainfall?"
3. Hand-operated water-lifting device originally developed in ancient Egypt and still used in parts of Africa and Asia.

References

Adams, W. M., & Carter, R. C. 1987. Small-scale irrigation in sub-Saharan Africa. *Progress in Physical Geography*, 11(1), 1–27.

Alam, M. 1991. Problems and potential of irrigated agriculture in Sub-Saharan Africa. *Journal of Irrigation and Drainage Engineering*, 117(2), 155–173.

Allaire, M. 2009. Drought mitigation in semi-arid Africa: The potential of small-scale groundwater irrigation. Paper presented at the SustainUS's program to promote Sustainable Development. Washington, USA. Available online (last accessed: March 29, 2013). http://sustainus.org/docs/citsci/winners/2009/Maura%20Allaire.pdf

Ayenew, T., Mamo, S., Ali, E., Abdu, A., GebreEgziabher, M., & Desalegn, A. (2011). Hydrogeological and groundwater potential evaluation, water quality assessment and socio-economic surveying for small-scale irrigation development in the Raya-Kobo valleys, Northern Ethiopia. Unpublished report to the International Water Management Institute (IWMI), September 2011.

Barnett, T. 1984. Small-scale irrigation in Sub-Saharan Africa: Sparse lessons, big problems, any solutions?. *Public Administration and Development*, 4(1), 21–47.

Brown, E. P., & Nooter, R. 1992. "Successful Small-Scale Irrigation in the Sahel". World Bank Technical Paper # 171.

Carter, R. C. 1989. The development of small-scale irrigation in Sub-Saharan Africa. *Public Administration and Development*, 9(5), 543–555.

Carter, R., & Howsam, P. 1994. Sustainable use of groundwater for small scale irrigation-with special reference to Sub-Saharan Africa. *Land Use Policy*, 11(4), 275–285.

Chambers, R., Saxena, N. C., & Shah, T. (1989). *To the hands of the poor: Water and trees*. Delhi: Oxford University Press and IBH Publishing.

Dittoh, S., Awuni, J. A., & Akuriba, M. A. (2013). Small pumps and the poor: A field survey in the Upper East Region of Ghana. *Water International*, 38(4), 449–464. DOI: 10.1080/02508060.2013.819454.

Easton, V. J., & McColl, J. H. (n.d.) Statistics glossary v1.1. STEPS glossary web version (revised and updated September 1997 by Stuart G. Young). University of Glasgow. Retrieved June 9, 2013, from http://www.stats.gla.ac.uk/steps/glossary/probability.html#subprob

Eguavoen, I., Derib, S. D., Deneke, T. T., McCartney, M., Otto, B. A., & Billa, S. S. 2011. Digging, damming or diverting? Small-scale irrigation in the Blue Nile basin, Ethiopia. ZEF Working Paper 84, Bonn.

Ertsen, M. W. 2006. Colonial Irrigation: Myths of Emptiness. *Landscape Research*, 31(2), 147–167.

FAO. 2012. The role of agricultural growth in economic growth, and poverty and hunger reduction". Rome: Food and Agriculture Organization of the United Nations (FAO). Available online (last accessed: June 5, 2013). http://www.fao.org/docrep/016/i3027e/i3027e04.pdf

FARA. (2003). *Securing the future for Africa's children: Building Sustainable Livelihoods through Integrated Agricultural Research for Development.* Sub-Saharan Africa, Challenge Program Proposal. Accra: Forum for Agricultural Research in Africa (FARA).

Foster, V., & Briceño-Garmendia, C. (Eds.). 2009. *Africa's Infrastructure: A Time for Transformation.* Washington D.C.: *Agence Française de Développement* and the World Bank. Available online (last accessed: December 17, 2012). http://www.infrastructureafrica.org/system/files/AIATT_Consolidated_smaller.pdf

Hagos, H. G. 2012. "*Tenure (In)security and Agricultural Investment of Smallholder Farmers in Mozambique*", Mozambique Strategy Support Program, Working Paper 5. Washington: International Food Policy Research Institute. Available online (last accessed: June 9, 2013). http://www.ifpri.org/sites/default/files/publications/mozsspwp5.pdf

ICID. 2010. *Irrigation Development Challenges for the Least Developed Countries in Africa.* Report of the ICID Taskforce on "Priority Issues of the Least Developed Countries in Africa", International Commission on Irrigation and Drainage (ICID).

Kamwamba, G. H. 2011. *Small pumps and agrarian livelihoods in Malawi.* Unpublished report to the International Water Management Institute (IWMI).

Kashaigili, J. J. 2011. *Smallholder irrigation in Tanzania: Current practices, constraints and opportunities.* Unpublished draft report. Colombo: International Water Management Institute (IWMI).

Kipkoech, A., Iyanje, L., & Mwakubo, S. 2011. *Economic Impacts of smallholder agricultural water management on livelihoods in Kenya.* Unpublished draft report. Colombo: International Water Management Institute (IWMI).

Lankford, B. 2009. The Right Irrigation? Policy Directions for Agricultural Water Management in Sub-Saharan Africa. *Water Alternatives*, 2(3), 476–480.

Merrey, D., & Sally, H. (2008). "Micro-agricultural water management technologies for food security in southern Africa: Part of the solution or a red herring?. *Water Policy*, 10 (2008) 515–530.

Owusu, V., Asante, A. V., & Pavelic, P. (2013). Assessing the factors influencing groundwater irrigation technology adoption in Ghana. In M. Hossain Ali (Ed.), *Irrigation Management, Technologies and Environmental Impact* (Chapter 7, pp. 181–192) Hauppauge, USA: Nova Science Publishers.

Perret, S. R. 2002. Water policies and smallholding irrigation schemes in South Africa: A history and new institutional challenges. *Water Policy*, 4(3), 283–300.

Shah, T. (2009). *Taming the Anarchy: Groundwater Governance in South Asia.* Washington D.C: Resources for the Future Press.

Shah, T., Koppen, B. V., Merrey, D., Lange, M. D., & Samad, M. 2002. *Institutional alternatives in African smallholder irrigation: Lessons from international experience with irrigation management transfer.* Colombo, Sri Lanka: IWMI. 24p (IWMI Research Report 60).

Tijani, M. N., Obayelu, A. E., Olatunji, A. S., Sobowale, A., Aladejana, J., & Oke, S. 2011. Livelihood impacts of improved on-farm water control in Sub-Saharan Africa: An empirical investigation of three modes of small-holder agricultural water management. Unpublished report to the International Water Management Institute (IWMI), December 2011.

Torou, B. M., & Barbier, B. 2011. *Small-holder irrigation in West Africa: Studies in Mali, Niger and Burkina Faso.* Unpublished draft report. Colombo: International Water Management Institute (IWMI).

Toulmin, C., & Gueye, B. 2005. Is there a future for family farming in West Africa?. *IDS Bulletin*, 36(2), 23–29.

Vaishnav, T. 1994. Increasing food production in Sub-Saharan Africa through farmer- managed small-scale irrigation development. *Ambio*, 23(8), 524–526.

Do policy and institutional factors explain the low levels of smallholder groundwater use in Sub-Saharan Africa?

Srinivas Chokkakula and Mark Giordano

This article examines the policy and institutional constraints on smallholder adoption of groundwater irrigation practices in Sub-Saharan Africa. The analysis departs from the unilateral focus on the promotion of technologies and probes not only the issues of groundwater governance but also those policies related to other enabling factors such as access to credit, energy and agricultural pricing policies and land-tenure security. The paper argues that the region may be missing an opportunity by not ensuring at least neutral policy towards agricultural groundwater development and addressing other constraints which hold back not only agricultural groundwater use but smallholder agriculture development in general.

Background and introduction

The agriculture sector accounts for 66% of the labour force in Sub-Saharan Africa (SSA) (Economic Commission on Africa [ECA], 2009), though contributing only 16% of the GDP in 2008 (Livingston, Schonberger, & Delaney, 2011). About 80% of farmers in SSA are smallholders (holding 0.25 to 3 ha of land) (Collier & Dercon, 2009; Nagayets, 2005), and more than 70% of the poor in SSA live in rural areas (ECA, 2009). Contrast this to the extent of irrigation in SSA: about 4% of the total cultivable area is irrigated, of which groundwater irrigation accounts for only about 10% (Frenken, 2005). High levels of rural poverty, the large number of smallholders, low productivity and low irrigation levels – this gives a broad sense of the challenges and opportunities for agricultural development in SSA.

The constraints on groundwater irrigation do not appear to be directly related to resource availability. Groundwater reserves in Africa appear substantial and have been estimated at 0.66 million km^3, more than 100 times the annual renewable freshwater resources (MacDonald, Bonsor, Dochartaigh, & Taylor, 2012), though their geographic distribution is uneven. Smallholder-driven groundwater-based agricultural growth in other parts of the world, especially in South Asia (Shah, 2009), could provide at least part of a model for SSA to achieve the multiple goals of economic growth and poverty alleviation.

Improving smallholder productivity is a widely shared growth strategy for SSA (see Dethier & Effenberger, 2012; Lipton, 2011; Wiggins, Kirsten, & Llambí, 2010), and groundwater could clearly play a role. But the low levels of groundwater irrigation in the presence of widely accessible groundwater reserves pose a paradox. Why do smallholders not practise groundwater irrigation, when it appears to offer greater income opportunities[1] and is reasonably accessible, at least in some locations? The paper examines at this paradox.

Adoption of groundwater irrigation is often treated as a problem of improving access to groundwater or the appropriateness of groundwater-irrigation technologies (e.g. Carter & Bevan, 2008; Sijali & Mwago, 2011). The present analysis departs from this conventional notion of the problem and instead probes not only the policy and institutional issues of providing or regulating groundwater access, but also other issues affecting a smallholder's transition to groundwater irrigation. These include indirect factors such as those associated with credit access, energy and agriculture commodity policies, training, and land-tenure policies. This article discusses how the corresponding policy and institutional environments around these factors in SSA countries are situated for greater smallholder adoption of groundwater-irrigation practices.

The paper builds directly on case studies of two countries, Ghana in West Africa and Ethiopia in East Africa, as well as on information derived from a larger set of country studies carried out as part of an International Water Management Institute research project (supported by the Rockefeller Foundation) to examine the possibility of increased groundwater-irrigation use in SSA. The two case studies draw on recently conducted field visits, which included site visits and key informant interviews with a range of stakeholder groups including policy makers, NGOs, government functionaries and smallholders.

This article presents findings on policy and institutional challenges associated with direct factors (e.g. groundwater and irrigation policy) related to agricultural groundwater use and then indirect factors (e.g. commodity price controls) which may impact groundwater adoption by smallholders. It then situates these challenges in the broader political economy of SSA, explaining how irrigation-development priorities are set. It concludes with some possible directions to allow greater smallholder groundwater irrigation.

Direct factors: groundwater irrigation development for smallholders

The policy environment: encouraging groundwater development or regulating use?

Many countries in Africa have recently formulated water policies. A common feature of the policy documents is their enthusiastic endorsement of IWRM (integrated water resource management) principles. Advocated largely by external funders, IWRM derives from principles in regions of advanced water resource development, where the focus is rightly on coordination in water resources planning as well as demand management and regulation (Petit & Baron, 2009; Shah & van Koppen, 2006).[2] It is a rather paradoxical approach for SSA, where water resources development in most countries is still low by global standards. The situation for groundwater specifically is not different; still, groundwater policies in these documents, when they do exist, reflect the focus on regulation rather than development.

For example, the Water Resource Commission (WRC) of Ghana's strategy is to adopt IWRM for sustainable utilization of nation's water resources; its mandate is to regulate and manage water resources while implementing relevant policies (Wiggins et al., 2010). In Ethiopia, the water policy does not explicitly mention IWRM, but the policies and principles of the National Water Policy (MoWR, 2001) embody all IWRM principles as normally applied,[3] including regulation of water use, river-basin-wide planning, water as an economic good, water rights and participatory water management. However, the proclamations (legislation) in place for implementing the policy are categorical in their endorsement of IWRM. The proclamation (No. 534/2007) on setting up river-basin councils begins, "Whereas, the Ethiopian Water Resources Management Policy, envisages the establishment of river basin councils and authorities as one of the main instruments to implement integrated water resources management which is actually the pillar of the policy . . . " (The Federal Democratic Republic of Ethiopia [FDRE], 2007: 1). There is no exclusive groundwater governance policy, but the intent is subsumed in the water resources management proclamation, No. 197 of 2000 (FDRE, 2000).[4] It enunciates basin-level management using permits: "Management of the water resources of Ethiopia shall be in accordance with a premit [*sic*] system" (p. 3). It makes a permit mandatory for any use of water (surface or groundwater). Similarly, Ghana's WRC has recently published guidelines for drilling licences and water-use permits. Licenses target drilling service providers as well as seekers of these services. It is mandatory for a drilling service provider to possess a licence, and the users of services are obliged to inform the WRC of their intent to drill a borehole (WRC, n.d.b). Water-use permits, on the other hand, are required for both groundwater and surface-water use, and for all kinds of end uses, including irrigation and drinking-water supply.

It is true that in both Ghana and Ethiopia, there are licensing exemptions for some smallholder groundwater use. In Ghana, licensing is exempted for preventing fire, any kind of manual abstraction, mechanical abstraction not exceeding 5 L/s and subsistence agriculture not exceeding one hectare (WRC, n.d.c). Proclamation No. 197 of 2000 of Ethiopia exempts use for traditional animal rearing, water mills, abstraction from hand-dug wells, and traditional irrigation (defined as subsistence farming not exceeding 1 L/s or for one hectare per peasant). However, the overall policy framework is still one of regulation rather than support for new use.

A review, done as part of this study, of formal groundwater-related policy throughout SSA showed a similar tendency towards regulation and against promotion, as shown in Table 1. The table was generated by reviewing policy and plan documents for water, sometimes including other sector policies such as irrigation and agriculture (whenever they engage directly with groundwater use). This table indicates the existence or absence of policies and institutions for the explicit purpose of regulating or promoting groundwater use. See Appendix 1 for further notes on the corresponding policies and institutions for each country.

Of course, caution in groundwater development is not unwarranted, given the problems of overdraft seen in other parts of the world (Giordano, 2009), and could be at least theoretically justified in terms of the precautionary principle. However, the widespread and high-level institutional focus on regulation in an environment of high poverty, high dependence on agriculture and low groundwater utilization seems asynchronous. Such a focus on regulation at such early stages of development is perhaps unprecedented globally.

Policies for groundwater-irrigation promotion

While most groundwater policies in SSA focus on regulation, there is at the same time a desire to further develop irrigation. However, groundwater seems not to be a priority in irrigation plans. In Ghana and Ethiopia, policy documents do recognize groundwater abstraction as a form of irrigation practice. However, groundwater does not figure prominently in strategies for agriculture development (MoFA & GIDA, 2011; MoWR, 2001; MoWRWHG, 2007).[5] Instead, government-supported irrigation development is heavily biased in favour of surface water, which receives government budget allocations regularly.

Table 1. Formal policy and legislation related to groundwater promotion and regulation in SSA.

		Regulate groundwater use		Support for groundwater irrigation	
Country	Year of latest policy/legislation	Policies (a)	Institutions (b)	Policies (c)	Institutions (d)
Burkina Faso	2003	Yes	Yes	No	No
Ethiopia	2001	Yes	Yes	Yes	Yes
Ghana	2007	Yes	Yes	No	No
Kenya	2002	Yes	Yes	No	No
Malawi	2006	Yes	Yes	No	No
Mali	2002	Yes	n/a	No	No
Mozambique	2007	n/a	n/a	No	No
Niger	1993	Yes	n/a	No	Yes
Nigeria	1993	n/a	n/a	No	Yes
Somalia	n/a	n/a	Yes	n/a	Yes
South Africa	1998	Yes	Yes	n/a	n/a
Tanzania	2002	Yes	Yes	No	Yes
Uganda	1998	Yes	n/a	No	No
Zambia	2006	Yes	Yes	No	No
Zimbabwe	2002	Yes	Yes	No	No

Notes: 1. The table is derived from Web search and country reports prepared for a Rockefeller Foundation–supported project at the International Water Management Institute (individual reports are listed in the Appendix). There may be recent initiatives to promote groundwater irrigation through special programmes or ordinances not captured by this review. 2. N/a: no information available. 3. Wherever available, the policy/law documents were downloaded from the respective government websites. Where the websites are not available, the review relies on legislation in the Food and Agriculture Organization of the United Nations database (FAO, 2012) of water laws and policies (http://faolex.fao.org/faolex/waterlex.htm) and other sources such as the International Environmental Law Research Centre. 4. Letters (a, b, c, d) serve to index the details provided in the Appendix.

Ghana's water policy recognizes water for food security as one of its focus areas. Promotion of small-scale irrigation is one of the action points in the policy, but it does not single out the source of irrigation (surface or groundwater). The irrigation policy, announced around the same time as the water policy, categorizes smallholder groundwater irrigation as "informal irrigation" (MoFA & GIDA, 2011). It acknowledges that in the Kumasi region alone (one of the 10 regions of Ghana), there are at least 12,700 smallholders irrigating more than 11,900 ha of land.[6] It further notes that this is more than the area irrigated by formal (surface-water) irrigation in the entire country. Yet, groundwater irrigation is not identified as a development strategy, though it does receive attention as a part of environmental management to "promote access to safer groundwater or safer irrigation practices where only marginal-quality water is available" (MoFA & GIDA, 2011, p. 16). This reflects the general emphasis on management rather than development. Interviews suggest that this language originates from the involvement of external participants in the writing process and that the move from language to action is unlikely.

The few groundwater-irrigation projects are typically smaller-scale pilots and externally funded.[7] The only government-supported programme in Ghana for promoting groundwater use was an experimental project supported by the World Bank on a short-term basis.[8] MoFA offered drilling services to communities on a cost-sharing basis. To the authors' knowledge, the only other similar effort to encourage groundwater irrigation was the supply of pumps under agricultural-mechanization programmes. MoFA's (2009) data show the provision of just 95 water pumps to farmers (79 in 2006 and 16 in 2008) – all provided under a Japanese-assisted project. Other than these limited examples, most external support for irrigation is also in surface systems. Since the 1990s, the Japan International Cooperation Agency's (JICA's) technical-cooperation programme has consistently focused on improving the management and performance of existing public (surface-water) irrigation systems (Miyoshi & Nagayo, 2006). The majority of bilateral assistance goes into improving public irrigation systems. But there are also many projects supported by external assistance and NGOs focusing on private smallholder irrigation. The success of these projects is prompting external-aid agencies to rethink their investments in agricultural water management (Abric et al., 2011).

The situation in Ethiopia has some encouraging features for groundwater irrigation. Ethiopian water policy inventories groundwater resources but does not articulate a development strategy. However, Ethiopia has set up a separate Directorate of Groundwater Management (DGM) to develop and manage groundwater resources in the country. In addition, under its accelerated economic-development programme, a Ministry of Agriculture agency, the Agriculture Transformation Agency (ATA), is implementing an irrigation-promotion programme targeting smallholders, the Household Irrigation Technology (HIT) project, which includes groundwater.

A closer look at both these initiatives does raise questions about their long-term sustainability. The DGM promotes groundwater irrigation through implementation of pilot projects based on community-oriented deep-borehole schemes similar to the existing Raya Valley and Kobo-Girana pressurized-irrigation schemes. All the current projects implemented by the DGM are externally funded and receive no government budget allocation. But the HIT project, even though not supported by an explicit policy of groundwater-irrigation promotion, is government funded. It aims at improving access to manual and mechanized pump technologies for about 650,000 smallholders in 83 *woreda*s (districts in Ethiopia). This is a Ministry of Agriculture programme and does not involve the DGM. Rather, it is implemented by the ATA in partnership with the Regional Bureaus of Agriculture (ATA, n.d.). The project is in its early stages, with groundwater mapping for five *woreda*s completed so far (ATA, 2012). The initiative is time-bound, and there is an element of uncertainty as to whether it will continue beyond the project period of 2011–2015. The absence of links between the DGM and the HIT project is another concern. Nevertheless, a policy-sanctioned project to increase the income opportunities of smallholders through improved access to pumps is indeed encouraging.

Institutional mandates for supporting groundwater irrigation

As one would expect given the limited policy attention to groundwater irrigation, institutions focused on promoting or supporting groundwater irrigation are either in the early stages of development or non-existent (Table 1). Typically, groundwater irrigation support is either presumed to occur as part of overall irrigation support or is allocated as additional responsibility given to existing irrigation-development agencies. These institutions often already suffer from such factors as minimal financing, limited qualified personnel, and ill-defined roles. In Ghana, though the WRC is the nodal agency for managing water resources in the country, its role is largely regulatory. The other irrigation-development agency, GIDA, has a primary focus on surface-water irrigation development. According to the National Irrigation Development Master Plan, GIDA's goal in the medium term is to bring

500,000 ha of land under irrigation. It does not specify whether it is to be surface-water or groundwater irrigation, but interactions with GIDA confirmed that groundwater irrigation development had not been considered as part of the strategy.[9]

Ethiopia's formal groundwater development is restricted to that discussed earlier. But it has plans for strong groundwater management, for example to "ensure that the exploitation of groundwater shall be based on abstraction of the maximum amount equal to the sustainable yield as determined by competent authorities and establish regulatory norms" (MoWR, 2001: 15). Such plans require highly skilled personnel and strong institutions, which are not now present. Even institutional connections are problematic. The DGM has to work through the regional arms of the Ministry of Water and Energy and the Regional Bureaus of Water Resources for groundwater-irrigation development and promotion. But these links are not actually established, and the DGM is so far involved only in the implementation of pilot projects, which it implements in collaboration with local partners such as the river-valley development agencies. The other agency with an important role in regulating groundwater abstraction is the Environmental Protection Agency (EPA). The EPA is a recently created institution with a rather large mandate – from formulating policies for genetically modified organisms to considering climate change, including regulating groundwater pollution.[10] The agency has yet to fully establish itself, even at the federal level. It is poorly equipped and short of quality personnel.[11] In some regions, the regional arms of the EPA have not been created yet.

Extension systems

There is accumulated evidence on the crucial role of extension systems in the adoption of treadle or motorized pumps among smallholders (Adeoti, 2009; Adeoti, Barry, Namara, Kamara, & Titiati, 2007). Effective extension systems are needed not just for promoting groundwater irrigation but also for improving access to other inputs like seeds and fertilizers. But research and extension systems are not a strong priority in SSA countries (Lipton, 2011). SSA governments' public expenditure in general is higher (29% of GDP) compared to that in Asia (22%) and Latin America (26%), but public expenditure on agriculture is low – even less than the 10% most SSA countries have set themselves as a target (Fan, Omilola, & Lambert, 2009). Expenditures on extension are obviously even smaller.

Interviews in the Keta region of Ghana provided no evidence of any extension services. Bob Nanes of iDE attributes the high costs of agricultural equipment in part to poor extension systems and the dominant role of middlemen in the market.[12] In contrast, Ethiopia has a considerable history of emphasis on improved extension systems. The Growth and Transformation Plan (GTP) of Ethiopia identifies the strengthening of extension systems for smallholders as one of the three important strategic directions for increasing agricultural productivity (Ministry of Finance and Economic Development [MoFED], 2010). The extension systems may not be active with respect to groundwater-irrigation technologies, the HIT project being an exception. But since the 1970s, a series of development programmes has focused on extension activities for other agriculture inputs. During the 1980s and 1990s, the Peasant Agriculture Development Project led to about 3000 service cooperatives for providing agriculture inputs, credit and extension services to smallholders. Later, following the "agricultural development–led industrialization" strategy, the Government of Ethiopia took up various macroeconomic reforms for the intensification of agricultural-production systems. In spite of the positive impacts of these reforms in overall grain production, Ethiopia's extension systems still fall short of improving smallholders' access to inputs, technology and information for achieving higher and stable agricultural production (Spielman, Kelemwork, & Alemu, 2011).

Indirect factors: why isn't groundwater irrigation spreading more quickly?

Clearly there is limited policy or institutional support for the development of groundwater irrigation in SSA. However, one would still think that since groundwater resources were available, the atomistic nature of groundwater extraction would allow at least some use by smallholder farmers based on their own initiative. In fact, that is the case to some extent. In a recent study of micro-watersheds in the White Volta Basin of Ghana's Upper East Region, for example, it was found that more than 61% of households depend on irrigated agriculture; of these, 90% depend on shallow groundwater for their irrigation needs (Namara, Horowitz, Nyamadi, & Barry, 2011). In Ethiopia, too, there are reports of increasing groundwater irrigation by smallholders (Gebregziabher, 2012). The Rift Valley, one of the areas visited for this paper, is one such region, with extensive tracts under shallow-groundwater irrigation. Much of this spread is through the individual initiative and entrepreneurship of farmers, who buy pumps on their own from private suppliers. A study of pump use in the Oromia, Tigray and Amhara regions of Ethiopia found that farmers bought some 75,000 pumps (motor pumps, treadle pumps, electric pumps, etc.) during 2004–2011 (AGWATER Solutions, 2011).

But, as mentioned at the outset and despite some recent growth, overall use in Sub-Saharan Africa is still extremely low. Are there other factors that hold back private initiative? The next section explores some possible indirect factors.

Input costs

In advanced groundwater economies, the development and adoption of groundwater irrigation by smallholders happened largely through individual initiative and entrepreneurship. Governments stepped in much later, improving access to affordable pumping technologies and offering credit and fuel subsidies (Shah, 2009, Ch. 2). The high cost of pumping and drilling technologies and services may be one of the reasons this take-off has not occurred in SSA. Boreholes 40 m deep cost from USD 4000 to as much as USD 10,000 in SSA; this is unaffordable for smallholders.[13] Comparable costs in India might be USD 1000. Similarly, a 5 hp pump might cost USD 100–150 in India, but twice that in much of SSA.

Some factors contributing to these high costs are poor marketing systems and trade barriers on equipment related to groundwater abstraction. The overall quality of pumps available within Ghana is low, and an inefficient marketing chain means that prices for the same pump vary widely between locations (Namara, Hope, Sarpong, de Fraiture, & Owusu, forthcoming). Though both Ghana and Ethiopia have 100% exemption of import duty on agricultural and industrial machinery,[14] it has been reported in the case of Ghana that the process of actually obtaining the exemption is both lengthy and costly, with the result that the tariff exemption seems not to apply in practice (Namara et al., forthcoming).[15] Even when efforts have been made to reduce tariffs, in practice, the impacts have not been as anticipated because of administrative issues, or sometimes policy sequencing problems. In Ghana, though the private sector was allowed to provide mechanization services, the importing and supply of machinery remained under government control. Red tape and inefficiency, combined with restrictions on the private sector, led to the failure of subsidy stimulus in generating demand and improving supply chains (Diao et al., 2012). As an example that some liberalization coupled with the right policy sequencing could make a difference, take the case of seed policy in other parts of Africa. By modifying the seed-policy environment to include harmonization of import-export policies and encouraging participation of seed trade associations, East African and Central African countries have been able to reduce the seed-variety release period from three or more years to just two seasons.[16] This has led to increased private-sector participation and improved access of farmers to seed varieties (Waithaka, Nzuma, Kyotalimye, & Nyachae, 2011).

While the role of subsidies in such processes can be debated (Lee, 2005), it is clear that reducing prices encourages adoption. In Niger, treadle-pump adoption rates among smallholders rose when subsidies were offered and dropped when they were withdrawn (ANPIP, 2005, cited in Adeoti et al., 2007). Similarly, the widespread growth of small-scale pump irrigation in the 1980s in Northern Nigeria was attributed to provision of well and pump subsidies and very low fuel costs (Carter & Howsam, 1994). Power subsidies in India seem to be what launched its groundwater revolution.

The price of energy to fuel pumps is also high in most of SSA. Diesel prices are usually higher than electricity, but there is no other option, given the lack of electrification in rural areas and often unreliable supplies when it is available. According to International Energy Agency (IEA) estimates, the rural electrification rate (the percentage of the population with access to electricity) is just 14% in SSA.[17] Ghana is comparatively better off, with 61% electrification (IEA, 2011). In Keta (southern Ghana), one of the field sites visited, the availability of shallow groundwater and better access to electricity led to extensive use of groundwater by smallholders and others. Electrification in Ethiopia is only 17%, meaning that those who do use pumps mostly use diesel, as is the case in the Rift Valley.

A final point is that even when there is a system to reduce input prices for farmers, it is sometimes not designed with the prospects for groundwater irrigation in mind. For instance, fertilizer subsidies in Ghana are restricted to only one season in a year. Smallholders in Keta – where extensive shallow-groundwater irrigation is practised – repeatedly complained about the unavailability of these incentives for the multiple cropping seasons made possible by groundwater to produce high-valued crops.

Input and output markets

The costs of inputs remain high, and returns low, also because of immature and poorly functioning markets. Overall transaction costs impact farmers' investment decisions in irrigation pumps (see Blein & Longo, 2009). In Nigeria's Fadama II project, implemented during 2005–2009, provision of subsidies on the cost of irrigation alone was not adequate. Unobserved transaction costs in acquiring pumps such as travel time, transportation, and the costs of obtaining information and services were also found to be significant deterrent factors in farmers' investing in pumps. Further, poor transportation infrastructure and access to markets added uncertainty about their returns from selling produce. These factors affected farmers' decisions with respect to investing in groundwater-irrigation equipment (Takeshima, Adeoti, & Salau, 2010). In Zambia, it was found that underdeveloped supply chains influenced smallholders' decisions about investing in motorized pumps. Unreliable availability of pumps in proximate locations and poor access to repair facilities added to smallholders' costs (Colenbrander & van Koppen, 2012).

When the extension systems are poor and market links are weak, as they are in the areas studied in Ghana and Ethiopia, traditionally strong local middlemen capture the markets and keep the costs of pumps and other technologies (e.g. seeds) high.[18]

Also, government efforts to control prices for consumers have sometimes had negative impacts on producers. For instance, in January 2011, the Ethiopian government decided to control prices for 18 items, including sugar and oil, to check spiralling prices. In the subsequent months, the control led to shortages of items; the government was forced to relax the controls in April 2011. The prices soon shot up in July.[19] Such interventions affect producers, especially smallholder producers, badly (Hassan, 2011). Price caps force producers to sell their produce at low prices, reducing returns. They also create incentives to produce commodities not likely to be subject to price caps, further distorting those markets by driving down prices.[20]

Credit access

Access to credit is perhaps the most crucial factor in farmers' transition to groundwater irrigation, especially when initial costs are high (sometimes artificially, as just described). Interest rates available to farmers are also often high. Banking systems in SSA generally are not strong, and agricultural finance institutional infrastructure is poor (Sacerdoti, 2005). Complex land-tenure systems (see below), absence of collateral, high risks in rainfed agriculture, and price fluctuations make access to credit difficult for smallholders (Sacerdoti, 2005). In Ghana, even though inflation has been in the single digits, lending rates for agriculture ranged between 27% and 45% in 2009 and have been in that range for at least 10 years (MoFA, 2009). Smallholders may also have trouble accessing capital, even at these high rates, due to lack of collateral. Ethiopia has made significant progress in improving credit access for the poor in the last decade or so. There are many well-functioning rural credit institutions (Brislin & Dlamini, 2006); however, the extent and coverage of these institutions are not clear, and it is estimated that 70% of rural credit demand is still unmet (Sacerdoti, 2005).

Land-tenure security

The other crucial factor in smallholders' investment decisions and transition to groundwater irrigation is secured land tenure. Land-tenure relations across SSA are highly complex and deeply entrenched in local cultural contexts. These vary not just across countries but also within countries and are highly subject to indigenous power structures and social relations. The need for land-tenure reforms to improve land productivity and economic growth has been stressed by several studies (e.g. Devereux, 2000; Firmin-Sellers & Sellers, 1999).

In Ghana, land-tenure relations are family- and clan-centred and largely administered by local and indigenous institutions. The government's attempt to take control of land administration was resisted by the indigenous institutions of earth priests and chiefs (Ubink & Amanor, 2008). The terms and arrangements of tenant relations vary across the 10 regions of the country. More than 80% of the land in Ghana is under the administration of customary laws (Sarpong, 2006). This complexity poses at least two kinds of challenges: (1) any disputes and their arbitration are subject to heterogeneous tribal regulations, but not to courts of law; and (2) tenure relationships are prone to disputation because ownership rights are defined by familial relations. The rights are not always clearly defined; any sale or lease can be challenged and subject to dispute. Vulnerability to disputes and insecure rental arrangements discourage smallholders from making long-term investments.

The Ethiopian case is a contrast to Ghana. Land ownership by private individuals is not allowed under the constitution. Land is owned by the president on behalf of the state. Land use rights are given in the form of certificates of licence to cultivate. Regional Bureaus of Agriculture issue licence certificates for farming. Short-term and uncertain licensing procedures typically discourage long-term investments. For example, the Tigray region's land-tenure policy allows renting land. If it is for farming, the licence is issued for only two years. The rent agreement may be renewed after two years, but is not assured in any way. However, if it is rented for a commercial purpose other than farming, a licence can be issued for 10 years (Regional State of Tigray, 1989). The intent of such variations in renting land is apparently to ensure equity in accessing land. But the insecurity in renewal of land certificates discourages farmers from making long-term investments, such as in groundwater-access structures.

The constitutional deterrence of private ownership of land appears to have helped attract investments in commercial farming. Ironically, while private land tenure for smallholders is problematic, land acquisition and leasing for large-scale private investors is easy. Land leases for commercial farming are given for periods of 25 to 100 years, encouraging large capital investments in infrastructure such as that required for groundwater irrigation. In Ethiopia, the government allocated 1.17 million ha of land to commercial farming from 2007 to 2010 and has a target of 7.4 million ha by 2013 (Abbink, 2011). The commercial investments on these lands are in high-value crops and typically require good water control, which has come in the form of groundwater – showing that the potential is there.

Political economy

Perceptions and policy priorities

One explanation for the under-utilization of groundwater for irrigation is the lack of capacities and knowledge in SSA about practising groundwater irrigation (Carter & Howsam, 1994). However, a knowledge gap also appears to exist in the form of misperceptions amongst decision makers. In Ethiopia and Ghana, despite policies focused on regulation, popular perceptions and policy makers' worldview include belief in abundant water resources. As Ben Ampomah, chief of the WRC in Ghana, observes: "In Ghana, water is assumed [to be abundant]."[21] Such notions hold back the idea that this abundant water, and in particular groundwater, needs to be developed, as was done in earlier groundwater economies, where (1) groundwater irrigation increased smallholder productivity and (2) smallholder agriculture development significantly contributed to overall economic growth (Shah, 2009). Such perceptions also reinforce conventional strategies of focusing on surface-irrigation development. Though development based on farmer initiative has begun in some places, another related constraint is sometimes farmers' own apprehensions, often well justified, about the profitability of investing in the transition to groundwater irrigation (Gebregziabher, 2012). For example, in the Kobo-Girana Valley Development Project in Ethiopia, assuaging smallholders' apprehensions about groundwater irrigation as a viable option was a major transition block (Abate, 2006; KGVDP, 2008).

However, these notions and perceptions are changing. Both policy makers and farmers have begun to appreciate the potential benefits of groundwater irrigation. There are pockets of use by farmers, and interest in groundwater irrigation is growing among researchers and donor agencies. For example, there has been a recent increase in academic publications on groundwater for irrigation in Africa (e.g. Braune & Xu, 2010; de Graaff, Kessler, & Nibbering, 2011; MacDonald et al., 2012). Some donors have specifically focused on agricultural groundwater use. The funding behind this project (which supported this study) is one example, but there have also been recent calls from the Department for International Development (DFID), the Natural Environment Research Council (NERC) and the Economic and Research Council (ESRC) for specific research on unlocking groundwater's potential for the poor in SSA.

Politics of power and populism

Many countries in SSA are recently transitioned democracies which have yet to consolidate institutions allowing space for inclusive participation of all interest groups. Political terrains tend to be uneven and favour powerful and organized political interest groups (ECA, 2009). Perceptions of abundant water resources appear to be framed in these kinds of power structures and policy thinking. For example, in Ghana, the southern parts are prosperous and endowed with better rainfall and water resources. Rich and powerful political leaders from the south dominate policy making and have a tendency to focus on apparently easy access through surface-water irrigation structures. The other factor is related to the primacy of cocoa in Ghana. Historically, cocoa was the primary driver of the economy and the main earner of foreign exchange. Cocoa farming is concentrated in the south and adds to the political clout of leaders from the south. Reliance on cocoa exports helped reduce poverty, but this benefited only cocoa farmers, accounting for 19% of rural households in Ghana (Chamberlin, Diao, Kolavalli, & Breisinger, 2007). This is much less than the large proportion of smallholders in Ghana. More than 70% of farms in Ghana are less than 3 ha in size (Chamberlin et al., 2007).[22]

Ghana has strong and organized cocoa lobbies, active in negotiating concessions with the state (Kolavalli & Vigneri, 2011). Fishing communities along the coast are another well-organized political force, negotiating for their interests with the government.[23] This domination of cocoa farmers and fishing communities may be a reason for the lack of attention to groundwater irrigation; neither of these powerful groups has an interest in groundwater irrigation. Smallholder farmers, on the other hand, do not have an organized political presence to negotiate for their particular interests, e.g. access to fertilizer subsidies in multiple seasons, rural electrification, fuel subsidies, or improved access to groundwater irrigation.

Structural challenges

The discourse about institutional capacities often begins with the lack of hydrogeological information for promoting groundwater irrigation. But groundwater-irrigation development may not wait for comprehensive information, especially at the present low levels of groundwater development. The challenge of institutional capacities needs to be understood at a more fundamental level: the capacity of personnel at lower levels of governance (e.g. the district) to promote groundwater irrigation among smallholders. This is a major challenge, as it appears from Ghana and Ethiopia. In Ethiopia, the GTP identifies capacity building as a critical challenge at the federal, regional and local levels (MoFED, 2010). Similarly, in Ghana, a comprehensive review by Miyoshi and Nagayo (2006) of the JICA's long-term irrigation-development partnership with Ghana with respect to capacity building lists a series of endogenous and exogenous maladies with respect to irrigation development in the country. These include prominently: reduction in the number of personnel; aging and inertia at local-level irrigation institutions; and excessive external dependency. GIDA's personnel strength fell from about 1500 in the 1980s to about 300 in the 2000s, supposedly to allow farmers' organizations to take over irrigation management. But Miyoshi and Nagayo also found that 6 out of 22 irrigation districts became dysfunctional over this period. This suggests larger structural challenges. In this state of affairs, groundwater-irrigation promotion cannot occur in the absence of committed policy direction, which does not exist at the moment. However, the problem of personnel is deeper. Availability of skilled human resources is itself a major challenge across the countries in SSA.

Further, most countries in SSA witnessed active water policy changes in the last decade that go far beyond groundwater. This has involved the restructuring of institutions and reallocation of their responsibilities. The frequent changes and interim arrangements create uncertainty in institutional coordination and functioning, a problem identified as an important challenge in Ethiopia (Ayenew, Masresha, & Awulachew, 2011).

Conclusions

While there seems to be some growth in groundwater irrigation in SSA by smallholders, as documented elsewhere in this issue, use is still low and appears to be less than what it could or "should" be. This article began with the question of whether policy and institutional factors explain these low levels of smallholder groundwater irrigation in SSA. Research suggests that the answer to this question is at least a partial yes. Water and irrigation policies either tend to ignore groundwater or are somewhat biased against its development, focusing instead on regulation. While the precautionary principle is not without merit, its emphasis, rather than policies to promote more use, in a region with low groundwater utilization and high poverty, is questionable. Compounding this effect for groundwater, those policies which do exist for the promotion of water resources development (i.e. irrigation) tend to overlook groundwater as an option, and the institutions to support groundwater irrigation within the formal institutional structure are understaffed and underfunded. One can argue of course that water policies in Sub-Saharan Africa are not necessarily implemented, and so this negative bias does not have direct impacts. Still, a negatively biased discourse can have some effect on decisions and does not make it more likely that funds and staff will be put in place.

In addition to the direct policies, a second set of factors related to the broader economy and policy environment probably reduce agricultural groundwater use. The costs of groundwater access are high due to trade barriers; the costs of operation are high due to poor rural electrification; and the value of groundwater-produced output is reduced by poor supply chains and weak market linkages. The high cost of credit makes access to technology difficult, and periodic price controls reduce incentives for investment. Finally, land tenure makes investment in fixed assets such as groundwater wells risky. While these factors have a particular impact on the expansion of groundwater irrigation, they also have negative implications for the overall agricultural sector.

A first key recommendation for facilitating greater use of groundwater in irrigation is to ensure that groundwater irrigation receives reasonable consideration vis-à-vis surface irrigation within water resources development policies and plans by both national governments and donors. While the conditions under which groundwater may or may not be a choice preferable to surface irrigation will vary by location, groundwater should at least be part of the basic discussion of how to increase irrigation and food security in SSA. To support this, water, irrigation and agricultural policy and institutions should also be designed to include consideration of groundwater alongside surface water. Groundwater is clearly prone to overexploitation, for a myriad of reasons highlighted elsewhere, but this is not in itself a reason to ignore it in water resources development.

A second key recommendation is to review the indirect policies which may be holding back groundwater development, especially by smallholders. Many of the groundwater schemes mentioned in this paper are deep boreholes dependent on community management and requiring more formal involvement of the state or donors. Shallow-groundwater irrigation practice is largely atomistic and driven by smallholders' own enterprise. However, many indirect policies, including those related to pump imports, rural electrification, and other factors discussed in this paper make it difficult for smallholders to invest. The more formalized deep-bore well schemes have their place, but creating an environment in which farmers can profitability invest in their own groundwater control offers many advantages. This could include the relatively easy task of lowering taxes on equipment and reducing red tape in its import or the more difficult tasks of improving rural electrification and market efficiency. These tasks probably cannot be justified simply from the benefits of groundwater use, but fortunately they would be equally beneficial to the overall rural economy.

Acknowledgements

The authors want to thank Paul Pavelic, Fitusm Hagos and Gerard Forkuor for their kind and generous support during this research; and Karen Villholth, Shilp Verma and the anonymous external reviewer for their comments on earlier drafts.

Notes

1. For example, KGVDP (2008), Schraven (2010) and Villholth (2013) show how groundwater irrigation generates greater income opportunities for smallholders.
2. For more critical reviews of the IWRM paradigm and practices, see the special issue of *Natural Resources Forum* (2009).
3. See Shah and van Koppen (2006) for a discussion of how these features constitute IWRM in the context of India.
4. Other sources refer to regulatory instruments in the 1960 Civil Code and the federal Wells Drilling, Construction and Rehabilitation Directive; but most of these regulatory provisions were incorporated in Proclamation 197 of 2000.
5. Ministry of Food and Agriculture (MoFA), Ghana; Ghana Irrigation Development Authority (GIDA); Ministry of Water Resources, Works and Housing, Ghana (MoWRWHG); Ministry of Water Resources (MoWR), Ethiopia.
6. These figures are not entirely reliable. The document defines a smallholder as one holding less 0.5 ha of land. By this definition, the given number of smallholders (12,700) cannot irrigate more than 6500 ha. The policy document also acknowledges about 2000 ha of land under smallholder groundwater irrigation in south-eastern Ghana, where Keta is located – one of the sites visited under this study.
7. Personal interviews, Tesfaye Tadesse, 31 May 2011; Ben Ampomah, 12 July 2011; GIDA deputy chief, 13 July 2011.
8. There are conflicting interpretations of this programme. In this World Bank–designed scheme, MoFA procured 20 drilling rigs and offered drilling services to communities for irrigation and livestock on a cost-sharing basis. MoFA's website claims it to be a successful programme, resulting in 1000 boreholes across the country (MoFA, n.d.). But local interactions suggest that the scheme failed to attract communities, and that the rigs are now lying defunct.
9. Personal interview, chief of GIDA, 13 July 2011.
10. Personal interview, Berhanu Ayalew and Dereje Agonafir (EPA), 31 May 2011.
11. The recent draft strategic framework for managed groundwater development by MoWR and GW-MATE (2011) says that the EPA is a *one-person* agency.
12. Personal interview, Bob Nanes, 17 July 2011.
13. Compiled from country reports.
14. Ghana Investment Promotion Centre (http://www.gipcghana.com); Ethiopian Investment Agency (http://www.ethioinvest.org).
15. Though according to the Ethiopian Investment Agency there is 100% exemption of import duty on agricultural equipment, Gebregziabher's (2012) study found 37% of the cost of pumps going to government taxes.

16. The harmonization and rationalization agreements between countries reduced the number of varied protocols and procedures required to release a seed variety, such as evaluations, trials, registration, etc.
17. The overall (urban and rural) electrification rates for the countries of focus here, Ethiopia and Ghana, are 17% and 60.5%, respectively.
18. Personal interview, Bob Nanes, 17 July 2011.
19. See the following news items to trace this cycle of price control–shortage–high prices: http://ethiopiaforums.com/ethiopia-food-prices-spike-after-govt-intervention-food-becomes-unaffordable-for-many; http://ethiopiaforums.com/ethiopia-price-shock-continues-despite-ceiling; http://ethiopiaforums.com/ethiopia-govt-start-selling-cooking-oil-and-sugar.
20. http://ethiopiaforums.com/the-futility-and-damaging-effects-of-ethiopian-government-price-control
21. Personal interviews, Tesfaye Tadesse, 31 May 2011; Ben Ampomah, 12 July 2011.
22. MoFA's (2009) estimate of small farms is much higher: more than 90% are less than 2 ha in size.
23. See e.g. http://www.ghanaweb.com/GhanaHomePage/regional/artikel.php?ID=200776, one of several reports on online forums about fuel subsidies to fishing communities.

References

Abate, A. (2006). *Smallholder farmers' experience on pressurized irrigation systems in Kobo valley*. Paper presented at the Conference Papers MoWR/MoARD/USAID/IWMI Workshop. Retrieved from http://publications.iwmi.org/pdf/H039816.pdf

Abbink, J. (2011). 'Land to the foreigners': Economic, legal, and socio-cultural aspects of new land acquisition schemes in Ethiopia. *Journal of Contemporary African Studies, 29*(4), 513–535.

Abric, S., Sonou, M., Augegard, B., Onimus, F., Durlin, D., Soumaila, A., . . . Gadelle, F. (2011). *Lessons learned in the development of smallholder private irrigation for high-value crops in West Africa* (Joint Organizational Discussion Paper – Issue 4). Washington, DC: The World Bank.

Adelana, S. M. A. (2011) *The hydrogeology and groundwater situation in Nigeria*. Country report to the International Water Management Institute.

Adeoti, A. I. (2009). Factors influencing irrigation technology adoption and its impact on household poverty in Ghana. *Journal of Agriculture and Rural Development in the Tropics and Subtropics, 109*(1), 51–63.

Adeoti, A., Barry, B., Namara, R., Kamara, A., & Titiati, A. (2007). *Treadle pump irrigation and poverty in Ghana*. IWMI.

AGWATER Solutions. (2011, July). *Motorised water lifting in Ethiopia*. AGWATER Solutions. Retrieved from http://awm-solutions.iwmi.org/Data/Sites/3/Documents/PDF/waterliftinginethiopia.pdf

ANPIP (Agence Nigerienne de Promotion de l'Irrigation Privee). (2005). *An assessment of treadle pump use invegetable production in Niger*. A report by ANPIP, Niger, Draft.

ATA (Agriculture Transformation Agency). (2012). Household irrigation. *Ethiopian ATA Program Update, 1*(2), 9.

ATA (n.d.). Household irrigation. Retrieved August 31, 2012, from http://www.ata.gov.et/programs/system-programs/household-irrigation-hhi-project/

Ayenew, T., Masresha, P., & Awulachew, S. B. (2011). *Status report on groundwater in Ethiopia*. Country report to the International Water Management Institute, November, 2011.

Blein, R., & Longo, R. (2009). *Food price volatility–how to help smallholder farmers manage risk and uncertainty*. Paper presented at the Discussion Paper for Round Table 1, Round Table Organized during the Thirty-Second Session of IFAD's Governing Council, Palazzo dei Congressi, Piazzale J.F. Kennedy 5 (EUR), Rome.

Braune, E., & Xu, Y. (2010). The role of ground water in Sub-Saharan Africa. *Ground Water, 48*(2), 229–238.

Brislin, N., & Dlamini, P. (2006). *Amhara Credit and Savings Institute: Ethiopia* (microREPORT #58). USAID. Retrieved from http://www.ruralfinance.org/fileadmin/templates/rflc/documents/1196417194973_Amhara_credit_and_savings_institute___Ethiopia.pdf

Carter, R. C., & Bevan, J. E. (2008). Groundwater development for poverty alleviation in sub-saharan africa. In S. M. A. Adelana & A. M. MacDonald (Eds.), *Applied groundwater studies in Africa* (pp. 25–42). Leiden: CRC.

Carter, R. C., & Howsam, P. (1994). Sustainable use of groundwater for smallscale irrigation with special reference to sub-Saharan Africa. *Land Use Policy, 11*(4), 275–285.
Chamberlin, J., Diao, X., Kolavalli, S., & Breisinger, C. (2007). *Smallholder agriculture in Ghana* (GSSP Discussion Brief 3). Accra: International Food Policy Research Institute.
Chavula, G. M. S. (2011). *Status report on groundwater in Malawi*. Country report to the International Water Management Institute, November, 2011.
Colenbrander, W., & van Koppen, B. (2012). *Improving the Supply Chain of Motor Pumps to Expand Small-scale Private Irrigation in Zambia*. IWMI Working Paper 154.
Collier, P., & Dercon, S. (2009). *African agriculture in 50 years: Smallholders in a rapidly changing world?* Paper presented at the How to Feed the World in 2050. Proceedings of a Technical Meeting of Experts, Rome, Italy, 24–26 June 2009, pp. 1–13.
Cotula, L. (Ed.). (2006). *Land and water rights in the Sahel: Tenure challenges of improving access to water for agriculture*. Issue Paper No. 139. London: IIED.
de Graaff, J., Kessler, A., & Nibbering, J. W. (2011). Agriculture and food security in selected countries in sub-Saharan Africa: Diversity in trends and opportunities. *Food Security, 3*(2), 195–213.
Dethier, J. J., & Effenberger, A. (2012). Agriculture and development: A brief review of the literature. *Economic Systems, 36*, 175–205.
Devereux, S. (2000). *Food insecurity in Ethiopia*. Paper presented at the DFID Ethiopia Seminar, London. Retrieved from http://www.addisvoice.com/wp-content/uploads/2010/03/FoodSecEthiopia4.pdf
Diao, X., Cossar, F., Houssou, N., Kolavalli, S., Jimah, K., & Ohene, P. (2012). *Mechanization in Ghana: Searching for sustainable service supply models*. IFPRI Discussion Paper 01237.
ECA (Economic Commission for Africa). (2009). *African governance report* (No. II). Oxford: Oxford University.
FADAMA III. (2013). *Third national Fadama development project*. Retrieved from http://www.fadama.net/HTML/index.php
Fan, S., Omilola, B., & Lambert, M. (2009). *Public spending for agriculture in Africa: Trends and composition*. Regional Strategic Analysis and Knowledge Support System (ReSAKSS)', Working Paper (28).
FAO. (2012). FAO Legal Office's database of treaties, laws and regulations on food, agriculture and renewable natural resources [Online database]. Retrieved from http://faolex.fao.org/
FDRE (The Federal Democratic Republic of Ethiopia). (2000). Proclamation no 197/2000: Ethiopian Water Resources Management Proclamation.
FDRE (The Federal Democratic Republic of Ethiopia). (2007). Proclamation no. 534/2007: River Basin Councils and Authorities Proclamation.
Firmin-Sellers, K., & Sellers, P. (1999). Expected failures and unexpected successes of land titling in Africa. *World Development, 27*(7), 1115–1128.
Frenken, K. (2005). *Irrigation in Africa in figures: AQUASTAT survey, 2005*. Rome: FAO.
Gebregziabher, G. (2012). *Water lifting irrigation technology adoption in Ethiopia: Challenges and opportunities*. Ethiopia: IWMI.
Giordano, M. (2006). Agricultural groundwater use and rural livelihoods in Sub-Saharan Africa: A first-cut assessment. *Hydrogeology Journal, 14*(3), 310–318.
Giordano, M. (2009). Global Groundwater? Issues and Solutions. *Annual Review of Environment and Resources, 34*, 153–178. doi:10.1146/annurev.environ.030308.100251.
Hassan, S. (2011). The futility and damaging effects of Ethiopian government price control. Retrieved from http://nazret.com/blog/index.php/2011/01/09/ethiopia-the-futility-and-damaging-effects-of-price-control?blog=15
IEA. (2011). *Energy for all: Financing access for the poor*. Paris: IEA.
Kashaigili, J. J. (2011). *Status report on groundwater in Tanzania*. Country report to the International Water Management Institute, November, 2011.
KGVDP (Kobo Girana Valley Development Program). (2008). Impact assessment of pressurized irrigation projects. *Tebita Magazine*.
Kolavalli, S., & Vigneri, M. (2011). Cocoa in Ghana: Shaping the success of an economy. In P. Chuhan-Pole & M. Angwafo (Eds.), *Yes Africa can: Success stories from a dynamic continent* (pp.201–218). Washington, DC: The World Bank.
Lee, D. R. (2005). Agricultural sustainability and technology adoption: Issues and policies for developing countries. *American Journal of Agricultural Economics, 87*(5), 1325–1334.

Lipton, M. (2011). *Learning from others: Increasing agricultural productivity for human development in SSA*. Working Paper: WP 2012-007: January 2012. Regional Bureau of Africa, UNDP. Retrieved from http://web.undp.org/africa/knowledge/WP-2012-007-Lipton-Agriculture-Productivity.pdf

Livingston, G., Schonberger, S., & Delaney, S. (2011). Sub-Saharan Africa: The state of smallholders in agriculture. In *IFAD Conference on New Directions for Smallholder Agriculture*, Rome.

MacDonald, A., Bonsor, H., Dochartaigh, B. É. Ó., & Taylor, R. (2012). Quantitative maps of groundwater resources in Africa. *Environmental Research Letters, 7*(2), 024009.

Masiyandima, M. (2011). *Status report on groundwater in Zimbabwe*. Country report to the International Water Management Institute, November, 2011.

Miyoshi, T., & Nagayo, N. (2006). *A study of the effectiveness and problems of JICA's technical cooperation from a capacity development perspective: Case study of support for the advancement of Ghana's irrigated agriculture*. Tokyo: JICA.

MoFA. (2009). *Agriculture in Ghana: Facts and figures 2009*. Accra: MoFA.

MoFA. (n.d.). Latest achievements. In *About MoFA*. Retrieved January 30, 2012, from http://mofa.gov.gh/site/?page_id=335.

MoFA & GIDA. (2011). *National irrigation policy, strategies and Regulatory measures*. Accra: GIDA.

MoFED. (2010). *Growth and transformation plan (GTP) 2010/11-2014/15 (Draft)*. Addis Ababa: MoFED.

MoWR. (2001). Ethiopian Water Sector Policy.

MoWR & GW-MATE. (2011). *Ethiopia: Strategic framework for managed groundwater development (DRAFT)*.

MoWRWHG. (2007). National Water Policy.

Nagayets, O. (2005). Small farms: Current status and key trends. In *Proceedings of the research workshop on the future of small farms* (pp. 355–366). Information brief presented at Wye, UK, June 26–29. IFPRI.

Namara, R. E., Hope, L., Sarpong, E. O., de Fraiture, C., & Owusu, D. (forthcoming). Adoption and poverty reduction potential of water lifting technologies for agricultural production in Ghana. *Agricultural Water Management. Special Issue on Smallholder Private Irrigation*.

Namara, R. E., Horowitz, L., Nyamadi, B., & Barry, B. (2011). *Irrigation development in Ghana: Past experiences, emerging opportunities, and future directions*. Accra: IFPRI, GSSP Working Paper no. 0027.

Natural Resources Forum. (2009). Special issue: Integrated water resources management in water-stressed countries. *Natural Resources Forum, 33*(1), 1–97.

Ndiritu, P. G. (2011). *Status report on groundwater in Somalia*. Country report to the International Water Management Institute, November, 2011.

Ndiritu, P. G., & Githae, I. T. (2011). *Status report on groundwater in Kenya*. Country report to the International Water Management Institute, November, 2011.

Nonde, A. K. (2011). *Status report on groundwater in Zambia*. Country report to the International Water Management Institute, November, 2011.

Nwachukwu, I. N., Agwu, N. M., Ezeh, C. I., Mbanasor, J. A., Onyenweaku, C. O., & Kamalu, C. E. (2008). *Evaluation of second national Fadama development project in Nigeria: A rapid policy appraisal*. MPRA. Retrieved from http://mpra.ub.uni-muenchen.de/12914/1/MPRA_paper_12914.pdf

Obuobie, E., & Barry, B. (2011a). *Status report on groundwater in Burkina Faso*. Country report to the International Water Management Institute, November, 2011.

Obuobie, E., & Barry, B. (2011b). *Status report on groundwater in Ghana*. Country report to the International Water Management Institute, November, 2011.

Obuobie, E., & Barry, B. (2011c). *Status report on groundwater in Mali*. Country report to the International Water Management Institute, November, 2011.

Obuobie, E., & Barry, B. (2011d). *Status report on groundwater in Niger*. Country report to the International Water Management Institute, November, 2011.

Pavelic, P., Giordano, M., Keraita, B., Rao, T., & Ramesh, V. (2012a). *Groundwater availability and use In Sub-Saharan Africa. A review of 15 countries*, 255pp. Colombo, Sri Lanka: International Water Management Institute (IWMI).

Petit, O., & Baron, C. (2009). Integrated water resources management: From general principles to its implementation by the state. The case of Burkina Faso. *Natural Resources Forum, 33*(1), 49–59. doi:10.1111/j.1477–8947.2009.01208.x.

Regional State of Tigray. (1989). Proclamation to Decide on Utilization of Rural Lands.
Sacerdoti, E. (2005). *Access to bank credit in sub-Saharan Africa: Key issues and reform strategies*. IMF (International Monetary Fund) Working Paper WP/05/166.
Sarpong, G. A. (2006). *Improving tenure security for the rural poor*. Ghana Country Case Study.
Schraven, B. (2010). *Irrigate or migrate? Local livelihood adaptation in northern Ghana in response to ecological changes and economic challenges* (PhD dissertation). Centre for Development Research, University of Bonn.
Shah, T. (2009). *Taming the anarchy: Groundwater governance in South Asia*. New Delhi: Routledge.
Shah, T., & van Koppen, B. (2006). Is India ripe for integrated water resources management? Fitting water policy to national development context. *Economic and Political Weekly, 41*(31), 3413.
Sijali, I. V., & Mwago, M. G. (2011). MoneyMaker pumps: Creating wealth in sub-saharan africa. In P. Chuhan-Pole & M. Angwafo (Eds.), *Yes, Africa can: Success stories from a dynamic continent* (pp. 319–336). Washington, DC: World Bank.
Spielman, D. J., Kelemwork, D., & Alemu, D. (2011). *Seed, fertilizer, and agricultural extension in Ethiopia*. Ethiopia Strategy Support Program II (ESSP II) ESSP II Working Paper.
Takeshima, H., Adeoti, A. I., & Salau, S. (2010). *Measuring the effect of transaction costs for investment in irrigation pumps: Application of unobserved stochastic threshold model to the case of Nigeria*. International Food Policy Research Institute (IFPRI).
Tindimugaya, C. (2011). *Status report on groundwater in Uganda*. Country report to the International Water Management Institute, November, 2011.
Tiware, D. D. (2011). *Status report on groundwater in South Africa*. Country report to the International Water Management Institute, November, 2011.
Ubink, J. M., & Amanor, K. (2008). *Contesting land and custom in Ghana: State, chief and the citizen*. Amsterdam: Amsterdam University Press.
Villholth, K. G. (2013). Groundwater irrigation for smallholders in Sub-Saharan Africa – A synthesis of current knowledge to guide sustainable outcomes. *Water International, 38*(4), 369–391.
Waithaka, M., Nzuma, J., Kyotalimye, M., & Nyachae, O. (2011). *Impacts of an improved seed policy environment in Eastern and Central Africa*. Policy Analysis and Advocacy Programme (PAAP) of the Association for Strengthening Agricultural Research in Eastern and Central Africa (ASARECA) Entebbe, Uganda.
WE Consult. (2011). *Assessment of groundwater for food security and livelihoods in Mozambique*. Country report to the International Water Management Institute, March 2010.
Wiggins, S., Kirsten, J., & Llambí, L. (2010). The future of small farms. *World Development, 38*(10), 1341–1348.
WRC. (2013). Water Resource Commission of Ghana. Retrieved from http://wrc-gh.org/en/
WRC. (n.d.a). *Managing the water resources of Ghana for socioeconomic development [Brochure]*. Accra: WRC.
WRC. (n.d.b). *Water use permits [Brochure]*. Accra: WRC.
WRC. (n.d.c). *Drilling license and groundwater development [Brochure]*. Accra: WRC.

Appendix 1

Table A1. Details supporting the content of Table 1. Letters (a, b, c, d) refer to columns in Table 1.

Country	Notes	Sources (besides FAOLEX)
Burkina Faso	(a) The government is politically committed to implementing IWRM in the country. The ministry's IWRM action plan (2003) focuses largely on monitoring and regulation. No specifics on authorization to exploit groundwater are given. (b) The Burkina National Council for Water, under the Ministry of Environment, is responsible for implementation.	Obuobie and Barry (2011a); IWRM Action Plan 2003[1]
Ethiopia	(a) Permits needed by any user of groundwater, with exemptions to smallholders. (b) The EPA is responsible for developing guidelines to monitor groundwater use. (c) This is a partial yes. Water and irrigation policies do not include groundwater development as a strategy. But the Growth and Transformation Plan does include it as a strategy. This apparently led to the Ministry of Agriculture's Household Irrigation Technology project aiming at promoting smallholder groundwater irrigation. This is a project in its early stages and can potentially become a policy in the future. However, the plan does not have any influence on the mandate of the Directorate of Groundwater Management. (d) A separate Directorate of Groundwater Management exists for implementing groundwater irrigation projects.	Ayenew et al. (2011); MoWR (2001); FDRE (2000); MoFED (2010)
Ghana	(a) Users need permits, and drilling service providers must possess licences. (b) The Water Resources Commission is responsible for regulation of use.	Obuobie and Barry (2011b); WRC (2013, n.d.a, n.d.b, n.d.c)
Kenya	(a) Groundwater abstraction through wells and boreholes requires a permit. (b) The Water Resource Management Authority is responsible for regulating use.	Ndiritu and Githae (2011); Kenya National Water Development Report 2006[2]
Malawi	(a) Regulation is through the issuing of rights. (b) The Groundwater Division of the Ministry of Irrigation and Water Development is responsible for issuing abstraction rights.	Chavula (2011); Malawi National Water Policy[3]
Mali	(a) Permits are required for any water use involving groundwater abstraction of more than 1 m^3 per hour.	Obuobie and Barry (2011c)
Mozambique	No policy or institutions for promoting groundwater irrigation.	WE Consult (2011)

(*Continued*)

Table A1. (Continued).

Country	Notes	Sources (besides FAOLEX)
Niger	(a) A licence or an authorization is required for any non-domestic use of groundwater, including agriculture. (d) A central-government parastatal organization was responsible for borehole construction till the 1980s. After this, borehole construction was privatized.	Obuobie and Barry (2011d); Cotula, 2006
Nigeria	(d) This is a partial yes. There is no groundwater irrigation policy, but the World Bank–funded Fadama project, with its major focus on shallow-groundwater irrigation for improving farmer incomes, is in its third phase. A national organization, the National Fadama Coordination Office, implements the project. In the absence of policy, there is uncertainty about the continuation and success of the project when the donor discontinues support.	Adelana (2011); FADAMA III (2013)
Somalia	(b), (d) A central organization, the Water Development Agency, under the Ministry of Minerals and Water Resources, is responsible for groundwater development. But its current status is not known.	Ndiritu (2011)
South Africa	(a) With its advanced groundwater resources development, South Africa's National Water Act of 1998 allows for the formal allocation of tradable water rights. Allocations are in the form of periodically renewable licences. (b) The Department of Water Affairs and Forestry is responsible for the development and management of groundwater.	Tiware (2011)
Tanzania	(a) Permits are needed for abstracting groundwater for irrigation purposes. (b) Basin Water Boards are responsible for granting permits. (d) A central government parastatal, the Drilling and Dam Construction Agency, used to be solely responsible for the construction of boreholes. Per the National Water Policy of 2002, private service providers are allowed.	Kashaigili (2011)
Uganda	(a) A landowner has to obtain a permit for either abstraction or recharging of groundwater.	Tindimugaya (2011)
Zambia	(a) Irrigation, categorized as secondary use under the law of 1949 (revised 2006), requires permission. (b) Permissions are granted by the Water Development Boards under the Ministry of Energy and Water Development.	Nonde (2011)
Zimbabwe	(a) Groundwater users for irrigation have to apply for permits. (b) Permits are granted by the local catchment council.	Masiyandima (2011)

Notes: [1]http://cap-net.org/sites/cap-net.org/files/wtr_mngmnt_tls/40_action_plan_IWRM_BF.pdf
[2]http://www.source.irc.nl/url/19650
[3]http://www.moafsmw.org/ocean/docs/Policy%20Documents/National%20Water%20Policy%20FINAL.pdf

Small pumps and poor farmers in Sub-Saharan Africa: an assessment of current extent of use and poverty outreach

Regassa E. Namara, Gebrehawaria Gebregziabher, Meredith Giordano and Charlotte De Fraiture

The expansion of irrigation in Sub-Saharan Africa has been slow. In Asia, the rapid expansion of smallholder irrigation systems was attributed in part to the availability and affordability of motorized pumps. This paper appraises the current extent of pump-based irrigation in Sub-Saharan Africa; profiles the socio-economic and demographic attributes of current pump adopters; and assesses the poverty outreach of small-pump technology. It shows that private smallholder irrigation is practised mainly by the wealthier farmers. The development of groundwater irrigation requires targeted and deliberate public-policy interventions and institutional support focusing on the more marginal farmers.

Introduction

Africa is characterized by uneven distribution and availability of water resources and fragile soils. In spite of this, there are substantial untapped water resources. Africa accounts for only 2% of the total irrigated area worldwide (Thenkabail et al., 2006). Further, only about 4.9% of the total cultivated area of 183 million ha is irrigated. The expansion of irrigation has been slow as compared to the rest of the world; only 4 million ha of new irrigation has been developed over the last 40 years (FAO, 2003). This situation is often claimed to be one of the reasons for the absence of a green revolution in most farming systems of Sub-Saharan Africa (SSA). In two-thirds of the part of Sub-Saharan Africa that lies outside of the equatorial humid zone, water is the key constraint to farming (Woodhouse & Ganho, 2011). In this situation, returns on investment in high-yield varieties and fertilizer technologies, which are often considered the crux of the green revolution, are meagre or even negative. For example, in Malawi, a farmer growing hybrid maize could expect to produce 10,000 kg/ha, provided reliable water supply is ensured and other conditions of production are met; but an average Malawian farmer would only get 2195 kg/ha, a 1:4.6 ratio of actual to potential yield (MIWD, 2010).

To reverse this scenario, there are ongoing national, regional, continental and international initiatives. The New Partnership for Africa's Development, in its Comprehensive Africa Agriculture Development Programme, aims to develop 16.1 million ha of irrigation (14.2 million ha in small-scale and 1.9 million ha in large-scale schemes[1]) and rehabilitate 3.6 million ha of large-scale irrigation by 2015. The total resources required to realize this development target are USD 36.9 billion (USD 14.4 billion for small-scale, USD 13.6 billion for large-scale, and USD 8.9 billion for rehabilitation) (NEPAD/CAADP, 2009). Note that the resources required for small-scale and large-scale systems are about the same size, but with resources committed to small-scale systems about 7.5 times more irrigated area can be developed, meaning that the development cost per hectare is lower for small-scale irrigation systems.

There is a genuine endeavour to increase large-scale public and private investment in the sector. But the ongoing smallholder-driven private irrigation development that is largely based on water-lifting technologies is not yet fully appreciated in the plans. The major premise of this paper is that smallholders themselves can play a significant role in realizing the irrigation-development plans of Sub-Saharan African countries at higher resource efficiency, provided they have access to water-lifting technologies, especially small motorized pumps.

The specific objectives of this paper are to:

- highlight the potential for small-scale pump-based irrigation in SSA
- provide an initial appraisal of the current extent of small-scale pump-based irrigation
- profile the socio-economic and demographic attributes of the early adopters
- assess the poverty outreach of water-lifting technologies
- suggest strategic and policy recommendations with the intent of fostering wider uptake of water-lifting technologies in SSA

The paper draws upon reviews of literature relevant to Sub-Saharan Africa and data generated by the Agricultural Water Management Solutions project (Evans, Giordano, & Clayton, 2012a, 2012b, 2012c, 2012d, 2012e) led by the International Water Management Institute.

Key concepts and definitions

A multi-dimensional poverty index. The measurement of poverty often relies on income, which is just one in a suite of indicators (such as health, education, living standard, social status, peace, etc.) that may be used to assess the poverty status of beneficiaries of technologies or development interventions. This paper describes a poverty index that takes into consideration the multi-dimensional nature of poverty to classify sampled farm households into five relative classes: very poor, poor, middle, rich and very rich.

Poverty outreach. This is simply a measure of whether a technology or a project reaches the poor. In the present context, poverty outreach refers to the proportion of small-pump adopters in the sample who are poor. If the proportion of adopters who are relatively poor is high, we may conclude that small pumps have good poverty outreach or are appealing to the poor.

Potential of small-scale pump-based irrigation in SSA

Irrigation potential in Africa

Irrigation potential depends on the availability and suitability of the physical resources such as soils and water, and irrigation water requirements, which are determined by cropping patterns and climate, existing environmental and socio-economic constraints, and the type of irrigation (i.e. sprinkler, drip or surface irrigation systems). When the criteria used to determine the potential are limited to physical resources and irrigation water requirements, the resulting estimate is called the physical irrigation potential (FAO, 1997). If systems other than surface irrigation, such as sprinkler systems, are considered, the physical irrigation potential is expected to be much higher. If, however, socio-economic criteria are imposed, the possibilities for irrigation development could be less than the physical irrigation potential. Considering the physical resources and assuming surface irrigation systems, the irrigation potential of Africa has been estimated at 46.7 million ha (FAO, 1997).

The result of an assessment by the International Food Policy Research Institute (IFPRI) that included economic and physical criteria reduced the physical irrigation potential estimate of the FAO by about half. The IFPRI estimate was about 23.6 million ha, 16.3 million ha of which were potential large-scale irrigation with an internal rate of return (IRR) of 6.61% and 7.3 million ha of small-scale irrigation with an IRR of 28% (You et al., 2010). The total investment requirement was estimated at USD 70 billion (USD 38 billion for small-scale irrigation, USD 32 billion for large-scale irrigation). The profitability and potential for irrigation expansion of both large- and small-scale irrigation are quite sensitive to underlying assumptions, in particular investment cost and IRRs. Considering returns on investment (i.e. 12% as a cut-off point for IRR), the irrigation potential for Africa would be only 6.1 million ha, with a total investment cost of USD 26 billion, over an investment horizon of 50 years (You et al., 2010).

Pump-irrigation potential

Disaggregated information on irrigation development potential by irrigation typology beyond the usual large–small categorization is scant in SSA. However, the IWMI, IFPRI and FAO have jointly tried to assess the potential of four smallholder agricultural water management (AWM) systems (motor pumps, treadle pumps, communal river diversions and small reservoirs) using an integrated modelling framework that combines GIS data analysis, biophysical and economic predictive modelling (using the SWAT and DREAM models) and crop mix optimization tools (Xie, You, Wielgosz, & Ringler, 2013). The study showed that Sub-Saharan Africa possesses a large potential for smallholder AWM, with profitable area of up to 21 million ha for motor pumps, 17 million ha for treadle pumps, 1.1 million ha for communal river diversion and 0.4 million ha for small reservoirs. However, application areas were calculated separately for each AWM intervention and are not mutually exclusive, meaning that some locations are suitable for one or more of the AWM interventions. Up to 160 million people could benefit from these technologies, reaping profits of up to US 7.5 billion per year. It has been reported that about 122 million people could adopt motor pumps (Xie et al., 2013). However, rapid expansion of smallholder irrigation will impose stress on aquatic environments, which calls for regulation of smallholder AWM activities.

A more pragmatic and in-depth analysis of potential application area and beneficiaries for small motorized pumps have been conducted in selected SSA countries and one state

Table 1. Potential application area and beneficiaries of small motorized pumps.

Country	Potential beneficiaries (min–max)		Potential application area (min–max)	
	Rural people (thousands)	% of total rural population	Area (1000 ha)	% of total agricultural land
Burkina Faso	2,800–3,350	24–29	445–540	6.3–7.6
Ghana	4,400–5,500	30–38	880–1,100	11.4–14
Ethiopia	9,000–18,500	13–27	1,400–2,850	4.5–9
Tanzania	4,000–6,000	12.5–18.5	625–925	2–2.9
Zambia	710–2,900	8.9–36	120–490	1.6–6.8
West Bengal	15,200–23,700	26.6–41.5	2,700–4,200	41–64.5

Source: Evans et al. (2012a, 2012b, 2012c, 2012d, 2012e).

in India (Table 1). The results show that there is significant potential for expanding small-motor-based irrigation in SSA.

Estimates of pump-irrigated area in SSA

A robust estimate of Africa- or SSA-level irrigated area by small and large motorized pumps is currently non-existent, and even if it were available, it would not be reliable. However, a rough and conservatively low estimate can be made of the significance of pump irrigation based on the level of groundwater irrigation because groundwater irrigation always involves pumping or lifting (Table 2). Africa's share in the global groundwater-irrigated area is very low, and a large proportion of that (about 83.5%) is in North Africa. There is about 2.1 million ha of pump-irrigated area in Africa, and only 0.34 million ha in SSA (Siebert et al., 2010).

The estimated 340,134 ha of pump-irrigated area depicted in Table 2 represents only the least possible extent, because groundwater irrigation as a proxy for pump irrigation underestimates the significance of pump-based irrigation. Other water sources can also be pumped for irrigation but this information is not available. Some regional, national and sub-national studies confirm that the extent of pump-based irrigation in SSA is already considerable. Fifteen experts drawn from nine countries of Southern Africa (Zambia, South Africa, Malawi, Tanzania, Lesotho, Botswana, Madagascar, Namibia and Swaziland) were interviewed in 2006 regarding the current most widely used smallholder water-lifting technologies in their respective countries. They reported that small motorized pumps are the most widely used, followed by varieties of manual pumps including treadle pump, rope-and-washer pump, elephant pump and hand pump. The experts also indicated that small motorized pumps were the most promising technologies for market production and food security (IWMI, 2006).

On its website (http://www.kickstart.org/products) as of October, 2011, KickStart claims to have sold 186,804 treadle pumps in Kenya, Tanzania and Mali, creating 120,200 new jobs, generating USD 102 million in new profits and wages and moving 600,900 people out of poverty. Assuming that the sold pumps are in use and that each pump can irrigate 0.81 ha (about 2 acres) (Kickstart, 2012), the total area irrigated in SSA by KickStart-distributed treadle pumps alone is potentially about 151,311 ha.

In West Africa, lift irrigation has indigenous roots with traditional technologies such as the *shaduf*, an ancient irrigation or water lifting tool still in use in many areas of Africa and

Asia. Traditional irrigation systems, many of which are lift irrigation, cover an estimated 1.1 million ha in West Africa, supporting approximately 6.5 million people (Purkey & Vermillion, 1995). In six countries (Burkina Faso, Mali, Mauritania, Niger, Nigeria and Senegal), about 871,662 ha of lift irrigation systems have been recorded, 92.3% of which are found in Nigeria due to the low cost of fuel for pumping (FAO, 1986).

For the past 15 years, smallholder private irrigation has been heavily promoted in West Africa. More sophisticated drilling, pumping, and water-distribution techniques have been introduced and are now practised in 55% of the irrigated area in Niger and 75% of the irrigated area in Nigeria. In 1992, a review of irrigation use showed that there were successful privately operated small-scale irrigated areas in Niger, northern Nigeria, Mali, Mauritania, Chad, Senegal and Burkina Faso (Abric et al., 2011).

The significance of lift irrigation in SSA can be demonstrated by considering the situation of a selection of countries representing the main regions of SSA. Three countries, namely Malawi (representing Southern Africa), Ethiopia (representing East Africa) and Ghana (representing West Africa), were selected for this purpose.

The major irrigation systems in Malawi are gravity-fed systems, motorized pump, treadle pump and watering can. The total irrigated area in Malawi as of 2010 was about 39,618 ha, about 19,862 ha (50.1%) of which was lift-irrigated. The area under motorized pumps increased from about 935.8 ha in 2005 to 2819 ha in 2010, an increase of 201.2%. The area under treadle pumps increased from 6731.9 ha in 2005 to 12,890 ha in 2010, an increase of 91.5%. The area under buckets increased from 2660.8 ha to 4153 ha in the five-year period. The number of treadle pumps distributed also increased. In 2010, according to Malawi's Department of Irrigation, the total number of treadle pumps distributed in the country was over 60,000; based on estimated treadle-pump irrigation potential (0.3 ha/pump) this would irrigate around 20,000 ha of land. However, reports indicate

Table 2. Global and regional groundwater-irrigated area as a proxy for pump-irrigated area.

Category	Region	Area (ha)	%
Total area equipped for irrigation	Africa	13,576,142	4.5
	SSA	7,198,316	2.4
	North Africa	6,377,826	2.1
	Global	300,895,091	
Total area equipped for groundwater irrigation	Africa	2,505,954	2.2
	SSA	413,758	0.4
	North Africa	2,092,196	1.9
	Global	112,936,434	
Total area actually irrigated	Africa	11,527,882	4.6
	SSA	5,510,258	2.2
	North Africa	6,017,624	2.4
	Global	252,579,046	
Total area actually irrigated by groundwater	Africa	2,157,978	2.2
	SSA	340,134	0.4
	North Africa	1,817,844	1.9
	Global	97,821,180	

Note: The percentage figures are calculated with respect to global values corresponding to each of the items in column 1. The percentage actually irrigated can be and is higher that the equipped area because of irrigation initiatives underway outside of the purview of public domain.
Source: Siebert et al. (2010).

that only about 13,000 ha of land was under treadle-pump irrigation in the year 2009–2010 (MIWD, 2010).

In Ethiopia, a census involving 25,696 farm households in 8 districts (2 regions) has shed light on the extent of pump-based irrigation.[2] Some 7845 households (30.5%) used water-lifting technologies. Of these, 7200 households revealed the kind of water-lifting equipment they use: 6068 (84.3%) used different kinds of buckets and 1116 (15.5%) used motor pumps, often in conjunction with buckets and other types of water-lifting technologies. Motor-pump users constituted only 4.34% of the total 25,696 farm households included in the census. Diverse ownership patterns were observed. Of the 1116 households with access to motorized pumps, 370 farmers owned privately, 335 owned communally, and 312 rented from others; 21 had access to communally owned pumps and at the same time had their own pumps. Some 75 farmers did not clarify the ownership of pumps, though they claimed to have access to motorized pumps. Farmers used different energy sources, including diesel (56%), petrol (27%), electricity (7%) and a combination of these (9.5%). The choice of energy depends on the type of pump available and the availability and cost of energy. Diesel fuel is cheaper than petrol. Only 1 out of the 100 farmers had private connections with an electric pump connected to the grid with its own meter. The other 99% of farmers jointly or communally owned a pump and shared the electricity cost. The net area irrigated by small motorized pumps in this census alone was about 4072 ha. According to Ethiopian government statistics, about 800,000 motor pumps were imported into the country between August 2004 and December 2010. More disaggregated data from the regions show that at end of 2009 alone, the regional bureaus of water resources distributed 19,338 pumps in Oromia, 20,916 in Amhara and 18,348 in Tigray. This translates to 117,204 ha of pump-irrigated area in these three regions alone, assuming that one pump can irrigate about 2 ha (Gebrehaweria, 2011).

In Ghana, the number of small motorized pumps was estimated at 169,624 in 2008–2009. Official import data from Ghana's Customs Excise and Preventive Service indicate that 65,275 pumps and accessories worth USD 8 million were imported into Ghana between 2003 and May 2010 alone. It is estimated that 186,000 ha of land is irrigated by water-lifting technologies. For comparison, the total area irrigated by public irrigation systems (including small reservoirs and dugouts) in 2009 was about 13,301 ha (Namara, Hope, Sarpong, De Fraiture, & Owusu, 2011).

Socio-economic profiles of adopters: insights from Ethiopia and Ghana

In both Ghana and Ethiopia, small-motorized-pump adopters tend to be younger and predominantly male, with a high child-dependency ratio and lower aged-dependency ratio (Table 3). The dependency ratio is an age-population ratio of those not economically active and those in the labour force. Taking the proportion of women in the overall sample as well as in the rainfed-dependent sample as a bench-mark, we may conclude that women are under-represented in the small-pump-adopter sample (Table 3). Similar inferences can be made for the case of Ghana, except that women are over-represented in the gravity or canal-dependent farmer sample. The higher proportion of female-headed households in the canal-irrigation sample in Ghana may be due to deliberate targeting of women farmers in irrigated-land allocation in the public schemes.

In Ghana, water-lifting-technology adopters tend to have higher levels of education. This is particularly so for electric pump users, about 30.8% of whom had received college or university-level training. The proportion of illiterate farmers in the rainfed-dependent category is higher. Farmers practise diverse livelihood strategies. Arable crop farming and

Table 3. Comparison of socio-economic and demographic characteristics of adopters and non-adopters in Ethiopia.

Category	Female-led households (%)	Age of the household head (years)	Household size	Education (years)
Rainfed farmers	15.7	40.4	5.0	2.8
Bucket users	16.6	41.1	5.4	4.3
Treadle-pump users	12.5	41.0	4.5	8.7
Motorized-pump users	6.0	38.4	5.6	3.8
Gravity irrigators	9.4	38.1	5.1	1.6
Overall	15.2	40.4	5.2	3.1
Statistics and significance	$\chi^2 = 108.8$, $p < .001$	F = 19.9, $p < .001$	F = 33.7, $p < .001$	F = 127.5, $p < .001$
N	25,711	25,696	25,439	25,696

Source: Namara et al. (2011).

non-farm activities are the major occupations of the sample farmers. The importance of non-farm activities in the overall livelihood strategies of the sample farmers is remarkably high (Namara et al., 2011). In Ethiopia, too, water-lifting-technology adopters had higher levels of education compared to non-adopters (Table 3).

Poverty outreach of water-lifting technologies

To determine the welfare status of the current crop of pump adopters, a composite multi-dimensional poverty index (MPI) was developed using principal component analysis of primary data collected for adopters and non-adopters from Ethiopia and Ghana.[3] This index was used to categorize the sample farmers into quintiles (very poor, poor, middle, rich and very rich). First, the non-adopter samples were ranked based on their relative MPI to create the five economic groupings described above. This approach was used because non-adopters are considered to represent an unbiased sample of the general farming population. The cut-off values for the quintiles of non-adopters were determined and used to group water-lifting-technology adopters. The socio-economic status of current adopters (implicitly, the poverty outreach of pumps) was assessed by constructing bar graphs of the distributions of the water-lifting-technology adopter and non-adopter samples by MPI quintiles and visually inspecting the distribution patterns. By default, the bars for non-adopters are expected to be equal in size across the MPI groups; if water-lifting technologies are poverty neutral, the distribution of adopters is expected to follow a similar pattern to that of non-adopters.

The MPI was developed for every sample farm household using various farm-household welfare or poverty indicators collected from surveys in Ghana and Ethiopia. The welfare indicators encompass all dimensions of household welfare such as household human resources (e.g. household demographic structure and level of education), food security and vulnerability, dwelling characteristics, and assets.

The cumulative distributions of MPI by farm-household categories are presented in Figure 1 for Ghana and Figure 2 for Ethiopia.[4] Similar distributional patterns were observed for the two countries, with slight differences in the distributions of bucket and treadle-pump users. In Ghana, the water-lifting-technology adopters are better off than the

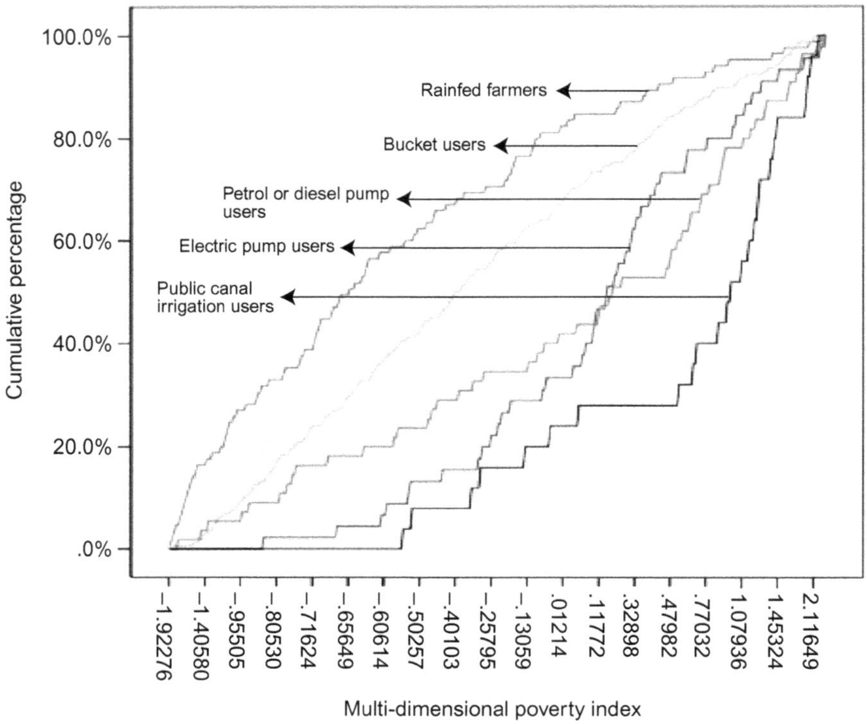

Figure 1. Cumulative distributions of MPI by farm-household category in Ghana.

purely rainfed farmers in terms of poverty status. Also in Ghana it is interesting to note that farmers with access to public-canal irrigation are at a higher socio-economic rank as compared to those with access to water-lifting technologies.

In Ethiopia, the conclusion regarding the wealth status of manual-irrigation technology adopters vs. small motorized pump users was not clear as the line for treadle pump and bucket (manual irrigation technology) users stradles the line for the diesel and petrol pump users (see Figure 2). It can be observed that electric-pump users have a completely different and significantly higher wealth status.

The plot of the cumulative distribution of MPI by water-lifting technology adoption status category indicated above is not conclusive for electric, diesel, and petrol pump users in Ghana and treadle, diesel, and petrol pump users in Ethiopia. Thus, the poverty outreach of water-lifting technologies was further assessed by drawing bar graphs of the distributions of the water-lifting technology adopter and rainfed-dependent farmer samples by MPI quintiles and visually inspecting the distribution patterns (Figures 3 and 4). By default, the bars for rainfed-dependent farmers are expected to be equal in size across the MPI groups; if water-lifting technologies are poverty neutral or equally accessible to all categories of farmers, the distribution of adopters is expected to follow a similar pattern to that of entirely rainfed-dependent farmers. However, this was not true in the present case, as can be observed in Figures 3 and 4. In both Ghana and Ethiopia, at the moment, the richest 20% of the farming population are the most prominent adopters of water-lifting technologies.

Finally, the MPI was used to check for differences in the relative poverty of the surveyed regions within Ghana and Ethiopia. The results, displayed in Figure 5, suggest that

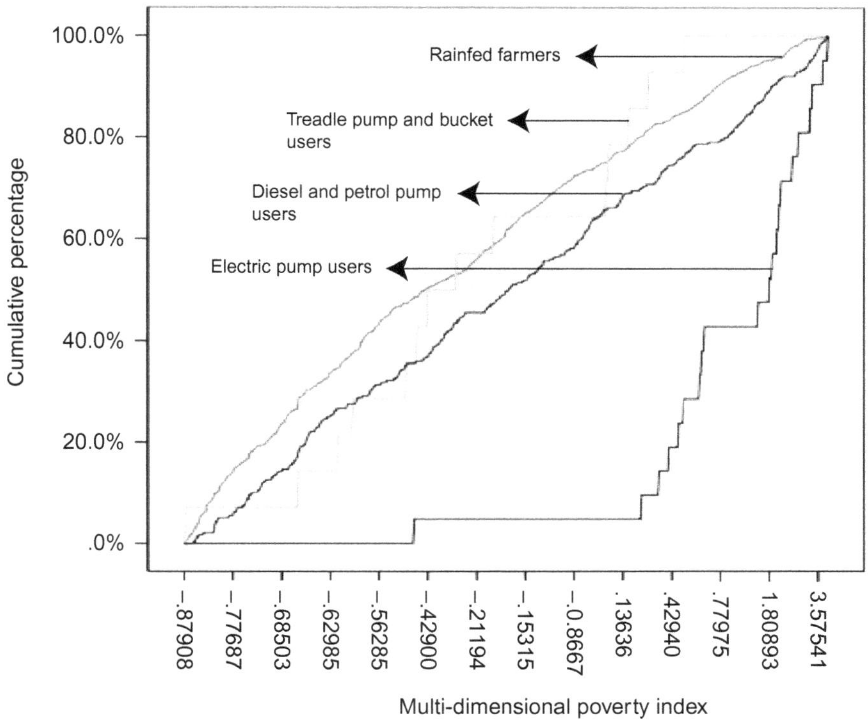

Figure 2. Cumulative distributions of MPI by farm-household category in Ethiopia.

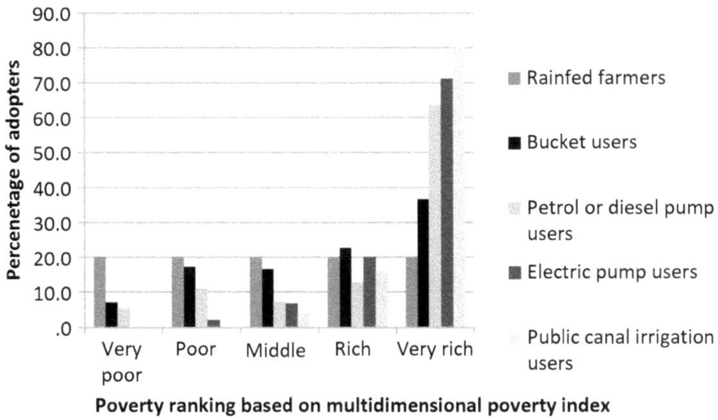

Figure 3. Depth of poverty outreach of water-lifting technologies in Ghana.

overall wealth levels are lower among farm households in Volta and Ashanti and higher in the Greater Accra region. In addition, water-lifting technologies seem to attract richer households in all three regions in Ghana (except for buckets in Volta); or alternatively, the adoption of these technologies may have lifted the farmers out of poverty.

Results from Ethiopia indicate that overall wealth levels among farm households are lower in the Amara, SNNP and Tigray Regions and higher in the Oromia Region. Sample farmers from Amara are the worst in terms of relative poverty. In addition, the results

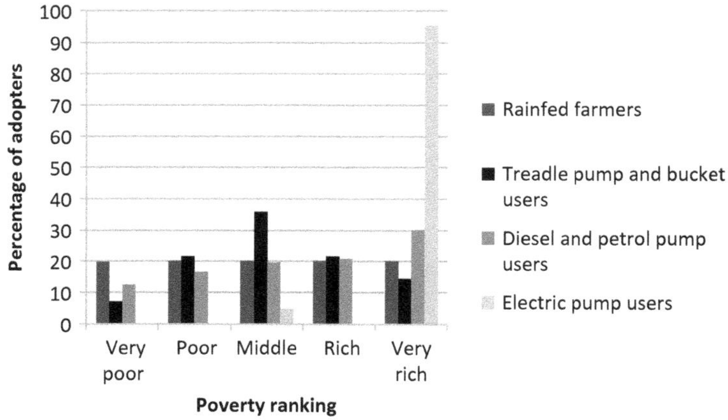

Figure 4. Depth of poverty outreach of water-lifting technologies in Ethiopia.

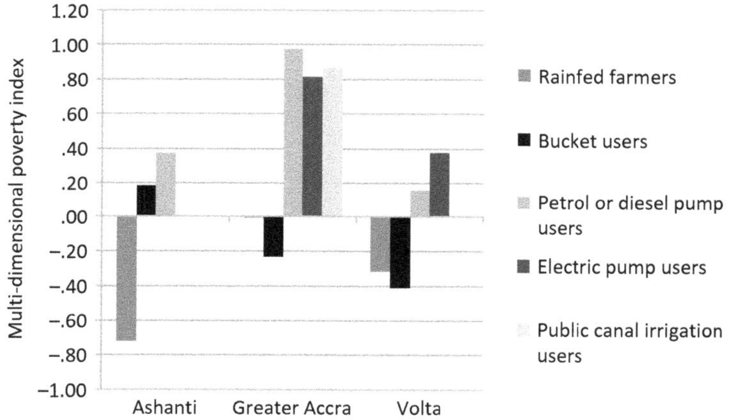

Figure 5. Average MPI disaggregated by survey region and water-lifting technology adoption status in Ghana.

depicted in Figure 6 indicate that, overall, the poverty outreach of water-lifting technologies is higher in the Amara, Tigay and SNNP Regions. In Oromia, buckets seem to attract poorer farmers, while small motorized pumps attract richer farm households.

Conclusions and recommendations

This paper reviewed the irrigation-development potential and the current level of realization of that potential in Sub-Saharan Africa. It is evident that SSA has great potential but is lagging behind in terms of irrigated-agricultural development, which is paradoxical given that water is a key constraint to farming in most parts of SSA. The effect of this scenario on the agricultural sector is far-reaching. For example, it decreases the adoption rate of other yield-boosting technologies such as high-yield varieties and fertilizers.

While the low level of irrigation development in SSA is well known, the strategy for addressing this challenge is not so obvious. This paper makes the case that smallholders themselves can significantly contribute to the development of the irrigation sector, provided that supportive policies are put in place. It shows that smallholder private irrigation

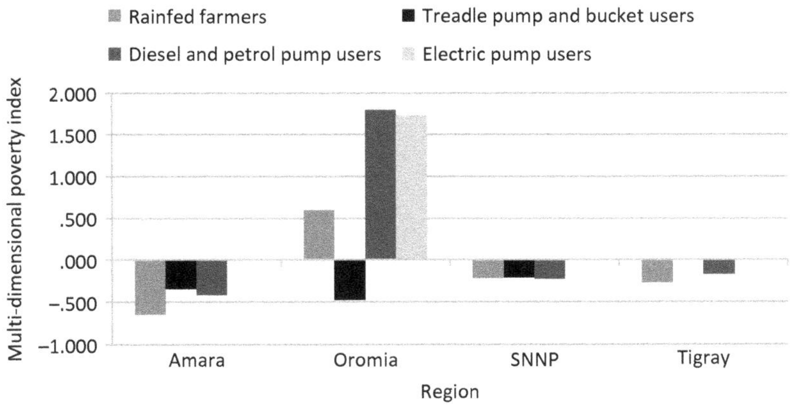

Figure 6. Average MPI disaggregated by survey region and water-lifting technology adoption status in Ethiopia.

based on pump technologies is already considerable. The current adopters of water-lifting technologies tend to be richer, younger, predominantly male, and with higher levels of education. Broader adoption of water-lifting technologies is constrained by: (1) lack of access or availability; (2) high cost; (3) weak equipment supply chain; (4) high energy cost; and (5) output marketing challenges.

Improving the uptake and poverty outreach of lift-irrigation technologies will require concerted efforts across the entire value chain. The specific possible policy interventions are:

- allowing more competition in the provision of services and products by encouraging the participation of public agencies, private firms and NGOs
- instituting subsidies, particularly for the poor
- encouraging imports with minimal or no duties or controls
- introducing innovative financing mechanisms to enhance accessibility
- sensitizing relevant ministries to the potential food-security and poverty-reduction benefits (e.g. through well-crafted policy briefs)
- creating awareness by engaging mass-media outlets
- improving the supply chain of pumps, for example through instituting a quality-assurance system
- developing information systems, including registries of importers, dealers and retailers
- incorporating relevant material into the agricultural curricula of schools and technical colleges
- preparing training manuals for use by extension professionals, dealers, retailers and maintenance service providers
- improving the capacity of farmers to optimally utilize pumps through targeted training.

Notes

1. Irrigated area can be classified by the size of the scheme – small, medium, or large. Different criteria are used for this classification in different parts of Africa. In many countries, irrigation schemes having a command area of 100 ha or less are considered small-scale.

2. This was a summary from primary data collected by the first author through surveys in Ethiopia in 2011.
3. For technical details of how the MPI is constructed and applied, refer to Zeller, Sharma, Henry, & Lapenu (2003).
4. The poverty status comparison of the different categories of farmers is indicated by the relative position of the curves. Curves closer to the upper left-hand corner indicate a high incidence of poverty among the farmers.

References

Abric, S., Sonou, M., Augeard, B., Onimus, F., Durlin, D., Soumaila, A., & Gadelle, F. 2011. Lessons learned in the development of smallholder private irrigation for high-value crops in West Africa. Joint Organizational Discussion Paper – Issue 4. The World Bank.

Evans, A. E. V., Giordano, M., & Clayton, T. (Eds.). 2012a. *Investing in agricultural water management to benefit smallholder farmers in Burkina Faso. AgWater Solutions Project country synthesis report*. Colombo, Sri Lanka: International Water Management Institute (IWMI). 30p. (IWMI Working Paper 149).

Evans, A. E. V., Giordano, M., & Clayton, T. (Eds.). 2012b. *Investing in agricultural water management to benefit smallholder farmers in Ghana. AgWater Solutions Project country synthesis report*. Colombo, Sri Lanka: International Water Management Institute (IWMI). 37p. (IWMI Working Paper 147).

Evans, A. E. V., Giordano, M., & Clayton, T. (Eds.). 2012c. *Investing in agricultural water management to benefit smallholder farmers in Ethiopia. AgWater Solutions Project country synthesis report*. Colombo, Sri Lanka: International Water Management Institute (IWMI). 35p. (IWMI Working Paper 152).

Evans, A. E. V., Giordano, M., & Clayton, T. (Eds.). 2012d. *Investing in agricultural water management to benefit smallholder farmers in Zambia. AgWater Solutions Project country synthesis report*. Colombo, Sri Lanka: International Water Management Institute (IWMI). 37p. (IWMI Working Paper 150).

Evans, A. E. V., Giordano, M. & Clayton, T. (Eds.). 2012e. *Investing in agricultural water management to benefit smallholder farmers in Tanzania. AgWater Solutions Project country synthesis report*. Colombo, Sri Lanka: International Water Management Institute (IWMI). 34p. (IWMI Working Paper 146).

FAO. 1986. Irrigation in Africa south of the Sahara. Investment center technical paper No.5. Rome.

FAO. 1997. Irrigation potential in Africa: A basin approach. FAO Land and Water Bulletin 4.

FAO, 2003. Food supply situation and crop prospects in Sub-Saharan Africa. Africa Report No. 23. December 2003. Retrieved from http://www.fao.org/docrep/006/j1119e/J1119e00.htm

Gebrehaweria, G. (2011). *Motorized water-lifting in Ethiopia*. AgWater Solutions Project Case Study Report. Addis Ababa, Ethiopia: International Water Management Institute (IWMI).

IWMI. 2006. Agricultural water management technologies for small-scale farmers in southern Africa: An inventory and assessment of experiences, good practices and costs. Final report produced by the international Water Management Institute (IWMI) for office of foreign disaster assistance, Southern Africa regional office, United States Agency for International Development and Investment Centre of the Food and Agriculture Organization of the United Nations. International Water Management Institute (IWMI) Southern Africa Sub-Regional Office.

Kickstart, 2012. Moneymaker irrigation pumps. Retrieved from http://www.kickstart.org/news/KickStart_B2B_BrochureJan09.pdf

Ministry of Irrigation and Water Development (MIWD). 2010. Annual Progress Report 2009–2010. Kasungu Irrigation Services Division, Kasungu. Malawi.

Namara, R. E., Hope, L., Sarpong, E. O., De Fraiture, C., & Owusu, D. 2011. *Adoption of water lifting technologies for agricultural production in Ghana: implications for investment and public policy*. Unpublished Report.

New Partnership for Africa Development (NEPAD)/Comprehensive Africa Agriculture Development Programme (CAADP). 2009. Sustainable land and water management: The CAADP Pillar I Programme.

Purkey, D. R., & Vermillion, D. 1995. Lift irrigation in West Africa: Challenges for sustainable local management. International Irrigation Management Institute (IIMI). Working Paper No. 33.

Siebert, S., Burke, J., Faures, J. M., Frenken, K., Hoogeveen, J., Döll, P., & Portmann, F. T. (2010). Groundwater use for irrigation – a global inventory. *Hydrology and Earth Systems Sciences,* 14, 1863–1880.

Thenkabail, P. S., Biradar, C. M., Turral, H., Noojipady, P., Li, Y. J., Vithanage, J., . . . Dutta, R. (2006). *An irrigated area map of the world (1999) derived from remote sensing*. Research Report 105. Colombo, Sri Lanka: International Water Management Institute.

Woodhouse, P., & Ganho, A. S. 2011. *Is water the hidden agenda of agricultural land acquisition in sub Saharan Africa?* Paper presented at the international conference on Global Land Grabbing, 6–8 April, Institute of Development Studies, University of Sussex.

Xie, H., You, L., Wielgosz, B., & Ringler, C. (2013). What is the potential for smallholder agricultural water management in Sub-Saharan Africa? An integrated hydrologic-economic assessment. *Agricultural Water Management*, Forthcoming.

You, L., Ringler, C., Nelson, G., Wood-Sichra, U., Robertson, R., Wood, S., . . . Sun, Y. 2010. What Is the Irrigation Potential for Africa? A Combined Biophysical and Socioeconomic Approach. IFPRI Discussion Paper 00993.

Zeller, M., Sharma, M., Henry, C., & Lapenu, C. (2001). *An operational tool for evaluating poverty outreach of development policies and projects*. FCND Discussion Paper No.111. Food Consumption and Nutrition Division, International Food Policy Research Institute.

Small pumps and the poor: a field survey in the Upper East Region of Ghana

Saa Dittoh, Joseph A. Awuni and Margaret A. Akuriba

Despite public-sector attempts at developing irrigation, most smallholder farmers in Northern Ghana still rely on manual methods to irrigate dry-season vegetables on small acreages. A field survey was carried out to ascertain the potential of motorized pumps in irrigated agriculture in the region. The findings indicate that use of pumps is far more beneficial than other irrigation methods, and rainfed farming in general, and irrigators generally regard their pumps as "saviours". The major problem for the majority of farmers is the lack of financial means to purchase the pumps.

Introduction

Ghana is one of the West African countries that have made some strides in reducing poverty in recent years (ISSER, 2007; World Bank, 2007). Yet poverty reduction in Ghana has been skewed towards export-crop farmers and farmers in the southern regions of the country. Farmers in the northern regions have experienced not only continuing poverty but also increased inequality (Al-Hassan & Diao, 2007; Appiah, Demery, & Laryea-Adjei, 2000). One of the main reasons for this persistence is the short cropping season in a year. The need to overcome this limitation has been known for a long time by the farmers themselves. Successive governments, NGOs and international organizations have at various times promoted irrigated agriculture in Northern Ghana with the belief that it is critical in improving the livelihoods of the people. There are several large, medium and small-scale public irrigation schemes in the region. A number of NGOs have also supported the building of dams and other irrigation infrastructure; the International Fund for Agricultural Development (IFAD), in particular, has supported the rehabilitation of several small dams. These efforts do not, however, seem to have had much impact on the poverty of the majority of the people. Irrigated agriculture in Northern Ghana can best be described as very modest in extent, and it is still dominated by traditional methods. For many decades, farmers have cultivated dry-season vegetables such as onions, tomatoes, peppers, okra, leafy vegetables and other "high-value crops" on very small acreages using dug-outs, shallow wells and small dams, as well as the larger multi-purpose dams, and by pumping or diverting water onto small vegetable gardens close to rivers and streams.

There continues to be outmigration of youth, permanent as well as temporary, from the area because of very limited gainful activities during the long dry season. There is, however, evidence of great potential for irrigated agriculture to improve livelihood significantly in the area (Nyamadi, 2011a). Some farming households depend on the dry-season production for their livelihood rather than rainfed crops. There is potential for many households to be able to earn substantial income from irrigated production in the area if conditions for profitable irrigated production are created.

Over the years, the "bucket and calabash" system of irrigation has been dominant, and still is today. This system is sometimes adapted through the use of watering cans. The use of "bucket and calabash" technology by irrigators has continued to increase because many farmers, especially those in the more arid northern parts of the country, appear to have no choice but to practice the method of irrigation they know and can afford. A key question that needs to be asked is: *Why is irrigated agriculture still practiced using rudimentary technologies, in an area where the people seem to know the benefits of modern irrigation technologies?* A follow-up question is: *Why have investments in public and other relatively large irrigation (mainly gravity-flow) systems not had the desired effect?*

Shah (2009) has outlined several positive microeconomic benefits of pump irrigation over gravity-flow irrigation in India. Most are related to better control over the water, but there are also lower losses of water in conveyance and application (Aremu & Ogunwale, 1994). In arid and semi-arid areas with very erratic rainfall patterns, such as Northern Ghana, the ability to effectively control the water being used is critical. Could the lack of progress in the heavily invested gravity-flow irrigation systems in Ghana be due to some or even all the reasons advanced by Shah with regard to India? Should Ghana be looking towards more aggressive use of pumps by poor smallholder irrigators for improvements in their incomes and livelihoods? The appropriateness for women of pumps, or for that matter any irrigation technology, is also important since they, especially rural women, continue to be disadvantaged and yet carry great economic and social responsibility everywhere in Africa (Bortei-Doku Aryeetey, 2000). Women in the Sahel areas of West Africa have asserted that existing irrigation technologies are too expensive and too sophisticated and are asking for more women-friendly irrigation technologies (Akuriba, Dittoh, Issaka, & Bhattarai, 2010). Can the pump be such a technology? This paper critically analyzes some of these issues based on extensive literature review of irrigation in Ghana and the West African sub-region and a field survey of irrigated agriculture in two districts of the Upper East Region of Ghana.

The Upper East Region has the largest and the most advanced irrigation infrastructure in Ghana. The region is characterized by shallow and accessible groundwater resources (Anayah, 2012). It is nonetheless one of the two poorest regions in the country. The two chosen districts, Bawku-West and Talensi/Nabdam, are well known for irrigation activities, especially with regard to groundwater use. Dittoh (1997) estimated that irrigated agriculture in West Africa constitutes only 3% of the value of all crop production, with no evidence that this has improved over time. Indeed, it is more plausible that it has worsened (Dittoh, Bhattarai, & Akuriba, 2013).

Limited productive land, high population density and worsening climatic conditions pose a problem of perennial food deficit and perpetuation of poverty in the Upper East Region in general and the selected districts in particular. The two study districts are probably the most rural districts in the Upper East Region. Small-scale rainfed agriculture has been the mainstay of the people, but its productivity in Ghana and elsewhere in West Africa has been declining due to persistently erratic rainfall and other factors (Breisinger, Xinshen, Kolavalli, Al-Hassan, & Thurlow, 2011; Oriola, 2009). Livestock rearing is an important component of the farming system; every household has some livestock.

Both surface water and groundwater are utilized for irrigation in the Upper East Region. However, not much attention has been paid to groundwater use for irrigation by the government, NGOs and other development agencies as compared to the attention given to dams, dugouts, rivers and streams as sources of water for irrigation (Nyamadi, 2011b). Indeed, even though groundwater has long been used for agricultural purposes in Sub-Saharan Africa, total usage remains low (Giordano, 2005). This is so despite indications that public surface-gravity irrigation systems do not work in favour of smallholder farmers (Pant, 1992). Nanes (2011) has argued that there are far more groundwater irrigators than surface-water irrigators in the Upper East Region and also that the combined area under groundwater irrigation is much greater. It must also be pointed out that groundwater has been extensively exploited for domestic use in Ghana, especially in Northern Ghana, and to a much smaller extent, industrial use in the southern part of the country. Groundwater is utilized in the Upper East Region more than in any other region of the country, for both domestic use and irrigation. According to Mainoo and Dapaah-Siakwan (2007), the Upper East Region has the highest areal density of boreholes in the country.

Data-collection methods

The underdevelopment of irrigated agriculture in the study area and its practice within a very traditional agricultural setting makes socio-economic research quite challenging. There are very many complexities, such as those of land and water use arrangements between land owners and tenants; continuous interference by traditional authorities; positive and negative government and NGO interventions; and private-sector interests. These complexities make data collection challenging, and large sample sizes are difficult or even impossible. A largely non-parametric method was used to collect information from relatively small numbers of irrigators (see Table 1). It is now generally accepted that participatory impact assessment (PIA) methodologies are best in assessing the impact of

Table 1. Numbers of farmers interviewed, by irrigation typology.

Irrigation typology	District Bawku West	Talensi/Nabdam	Total	Women respondents (women respondents as % of total respondents in parentheses)
GW bucket-fetch irrigators	40	49	89	15 (16.9%)
SW bucket-fetch irrigators	10	0	10	1 (10.0%)
SW motor-pump irrigators[a]	5	3	8	0 (0.0%)
SW gravity-flow irrigators	40	0	40	5 (12.5%)
Rainfed farmers	40	40	80	26 (32.5%)
Total	135	92	227	47 (20.7%)

[a]These are relatively few in the study area so "case studies" were done rather than a usual sample survey. Also, no manual pumps were encountered in the study area.
Source: field survey (2010/2011).

technologies. Dietz et al. (2009, p. 19) stated that impact assessments of development and technological interventions have "shifted away from purely technocratic and expert-oriented towards stakeholder-inclusiveness and participatory assessment". This is also true for comparison of the impact of categories of existing technologies. Thus, the methodology adopted was participatory assessment of different irrigation technologies. Focus-group discussions (FGDs), key informant interviews, and interviews of various categories of irrigators were used to gather the data used in the analysis.

Data on production systems, costs, returns, constraints and challenges as well as opportunities were collected from a sample of 147 irrigators, divided into four categories, as indicated in Table 1. A sample of 80 non-irrigating (rainfed only) farmers was also taken for a control-group analysis that included obtaining reasons why they are not undertaking any of the different types of irrigation technologies even though they are situated within the same general environment. It must be emphasized that information was obtained through FGDs and from the key informants to assist in obtaining the best estimates of costs and returns as opposed to depending on only average values. In all cases, sampling for individuals to interview was done randomly from among those who took part in FGDs and at irrigation sites, without regard to gender, status or age. The very few women in the sample (Table 1) reflect the small proportion of women who own irrigation plots. Many female household members provide labour for irrigation but do not have their own irrigated plots. No woman irrigated with a motor pump, indicating probable gender bias.

Results and discussion

Relative importance of different irrigation typologies in the Upper East Region

There are numerous communities in the Upper East Region where the different irrigation typologies are practiced. The region has the two largest gravity-flow irrigation systems in the country, the Tono and Vea. It also has the largest number of small reservoirs used for irrigation. In contrast, there is not a single publicly supported groundwater-irrigation scheme, despite claims by experts that groundwater development has relatively low cost and does not require expensive infrastructure and that groundwater in the Upper East Region is "plenty, accessible and a safe water source" (Anayah, 2012). The Upper East Region is characterized by shallow aquifers with low storativity (Bannerman & Ayibotele, 1984) and moderate transmissivity (Carrier, 2008; Obuobi, 2008).[1] Comparison of groundwater recharge rates across Ghana by the Water Resources Commission (WRC/CIDA, 2010) indicated that the Upper East Region, and more specifically the Atankwidi River basin, has a reasonably high recharge rate of 4–13% of rainfall (Martin, 2006). There has however been public support for pumps for lifting water from surface-water sources. The government of Ghana has purchased and sold pumps to irrigators and farmer groups on a hire-purchase basis (Ayariga, 2011a). In the two selected districts, there are many small irrigators cultivating different types of crops using a mix of irrigation technologies.

Demographic and socio-economic profiles of irrigators and rainfed farmers in the study area

Household sizes, cultivated areas and major occupations of household members

Table 2 indicates that surface-water motor-pump households have the highest average household size of 9.9 persons, whilst groundwater bucket-fetch households have the lowest, an average of 8.8 persons. However, statistical analysis (primarily analysis of variance)

Table 2. Average household sizes, irrigated areas and rainfed cultivated areas (standard deviation in parentheses).[a]

Irrigation typology	Average household size	Average area under irrigation (acres)	Average rainfed cultivated area (acres)
GW bucket fetch	8.8 (2.44)	0.69 (0.16)	3.12 (0.98)
SW bucket fetch	9.1 (1.89)	1.16 (0.23)	3.65 (1.31)
SW motor pump	9.9 (2.46)	3.31 (1.05)	3.94 (1.52)
SW gravity flow	9.6 (2.11)	0.52 (0.15)	3.95 (1.50)
Rainfed	9.4 (3.45)	0.0	3.56 (1.51)

[a]F-tests (through analysis of variance) indicate that the averages of household sizes and rainfed areas for the various technologies are *not* significantly different, but the average irrigated areas are significantly different at the 5% level.
Source: field survey (2010/2011).

indicates that the average household sizes do not vary significantly between the different types of farmers in the study area.

Motor-pump irrigators, though far fewer in the two districts, irrigate far greater areas than all the other types of irrigators. While the average area irrigated by surface-water motor-pump irrigators is 3.31 acres, the next-highest average irrigated area, by surface-water bucket fetch, is only 1.16 acres (Table 2). It is quite surprising that surface-water gravity-flow irrigators cultivate the smallest average acreage, since gravity-flow irrigation is generally associated with large irrigated areas. This is due to high farmer demand for irrigating in the relatively very limited developed areas in the surface-water gravity-flow systems. Thus, there is considerable competition and therefore rationing of the available developed area.

All households in the study area cultivate rainfed crops, and, as shown in Table 2, there do seem to be significant differences in the rainfed farm sizes of irrigators and non-irrigators. Indeed, there is a tendency for irrigators to have higher areas under rainfed cultivation than non-irrigators. That is understandable since irrigated farming does not compete with rainfed production in the study area. To some degree, it enhances rainfed production in several households. Irrigators tend to have higher annual incomes and greater assets (see Table 4) and thus have greater potential to invest in rainfed production. Irrigators, for example, own more work animals, and are thus able to plough their rainfed farms on time.

Table 3 gives the major occupations of household members above 18 years of age. The table indicates that in all types of households in the study area, a large percentage of household members above 18 years old (between 56.4% and 90.6%) are arable-crop farmers and a sizeable percentage are also still in school. However, it is noteworthy that the proportion of surface-water motor-pump household members reporting arable farming as their major occupation is the lowest (56.4%). It is also interesting that about 10.3% of the surface-water pump household members claim to be unemployed. It is likely that some of them do not regard their activities in either rainfed or dry-season agricultural production as employment. Usually, if they do not have their own farms and irrigable areas but partake in the agricultural activities as household members, they do not regard themselves as being employed.

Ownership of household assets

The ownership of certain household assets tends to be indication of relative wealth. Table 4 shows that more homes of irrigators, as compared to rainfed farmers, are made of bricks

Table 3. Major occupations of household members above age 18.

	% by irrigation typology				
Major occupation	GW bucket fetch	SW bucket fetch	SW motor pump	SW gravity flow	Rainfed farmers
Arable crop farming	69.5	90.6	56.4	73.8	73.9
Petty trading	2.9	0.0	10.3	1.3	2.6
Salaried worker	4.7	0.0	2.6	1.3	2.0
Tradesman (carpenter, tailor)	2.3	9.4	2.6	0.6	5.6
Schooling	18.2	0.0	15.4	20.0	12.9
Artisan (basket weaver, porter)	0.0	0.0	2.6	3.1	0.0
Unemployed	0.6	0.0	10.3	0.0	0.3
Others	1.8	0.0.	0.0	0.0	0.7
Total	100.0	100.0	100.0	100.0	100.0

Source: field survey (2010/2011).

Table 4. Ownership of household assets.

	GW bucket-fetch irrigators	SW bucket-fetch irrigators	SW motor-pump irrigators	SW gravity-flow irrigators	Rainfed farmers
Sample size	89	10	8	40	80
Percentage who have homes with a thatched roof	100.0	100.0	100.0	100.0	100.0
Percentage who have homes made with cement and bricks	92.1	100.0	100.0	100.0	80.0
Percentage with formal (traditional) title to land	43.8	50.0	100.0	20.0	100.0
Average number of small ruminants (sheep and goats) owned	18	15	15	12	16
Average number of large ruminants (cattle) owned	10	10	12	9	5
Average number of work animals (bullocks and donkeys)	1.2	2.2	3.6	2	1.2
Percentage with dug wells for irrigation	100.0	N/A	N/A	N/A	N/A
Percentage owning motor pumps	N/A	N/A	100.0	N/A	N/A
Percentage owning bicycles	100.0	100.0	100.0	100.0	55.5
Percentage owning pesticide spray pumps	11.2	30.0	75.0	20.0	0.0
Percentage owning mobile phones	44.9	100.0	100.0	80.0	52.5
Percentage owning motorcycles	12.4	40.0	100.0	17.5	5.0
Percentage owning colour TV	16.9	20.0	100.0	27.5	12.5
Percentage owning transistor radios	89.9	100.0	100.0	95.0	37.5

Source: field survey (2010/2011).

and cement. Also, irrigators in general tend to own more household assets such as bicycles, motorcycles, television sets and mobile phones than non-irrigators. Among the irrigators, the table indicates that surface-water pump irrigators tend to own more of these assets. This is a clear indication that they are wealthier than other irrigators. However, Table 4 does not indicate whether the relative wealth of the motor-pump irrigators is a result of irrigation. Further probing does indicate that the pumps were bought from loans and most of the assets were obtained as a result of profits made from the use of the pumps in dry-season irrigated agriculture.

Following the World Bank–sponsored Fadama Development Projects in Nigeria, the Ghanaian government, through the Ministry of Food and Agriculture, implemented a Pilot Fadama Irrigation Project (Gbireh, 2011). Motor pumps and accessories were procured by the Ministry of Food and Agriculture and some district assemblies and given as loans to farmers for "river pumping irrigation", especially in the Upper East Region (Ayariga, 2011b). Four of the eight motor-pump irrigators in this study indicated that they obtained their first pumps through this government intervention. Two other irrigators obtained theirs through an NGO, while the other two bought theirs in the open market. None of the eight had most of the assets listed in Table 4 before they acquired their pumps. In particular, they had neither motor bikes nor colour TV sets. Now five of the eight have built houses in the towns of Bolgatanga and Zebilla, and all have "block rooms with zinc roofs" in their family houses in their villages. They attribute the increase in their assets and the ability to secure more areas for irrigation, pay children's school fees and assist family members to the use of motor pumps for irrigation.

Land use and land tenure

Land use and land tenure in the study area are serious issues of contention mainly because of population pressure and the increasing realization of the importance of irrigation. The areas suitable for irrigated agriculture are limited to lowlands and around dams, and those areas are owned by particular families or clans within the communities. There have to be some agreements between these "landowners" and tenants (land users). The land tenure is also such that a "landowner" does not have permanent ownership of the land. He holds it in trust for a large family or clan; thus, it cannot be easily passed on to his offspring. That creates problems with respect to agreements that might have been reached with land users. There are also issues of improvements on the land by land users as well as access to the land for use by women. These complexities in land use and land tenure need to be explicitly considered in attempts to develop irrigated agriculture in the study area. Some strategies employed in the area have included extensive consultations at the family, clan and community levels; involvement of traditional authorities such as chiefs and *tendanas*;[2] agreements on the area of the irrigated land that should be reserved for the use of the landowning family or clan members; and others.

As indicated in Table 4, it is rainfed farmers and surface-water pump irrigators who have 100% traditional titles to their lands. The rainfed farmers and some of the motor-pump irrigators obtained their lands by inheritance, but the other motor-pump irrigators had to negotiate with landowning families for "traditional" title to the land. Currently, agricultural land, including irrigated land, cannot be bought or sold in the study area.

Table 5 presents land ownership by the various categories of farmers. All of the sampled households (irrigated and rainfed) own and cultivate rainfed farm lands. Table 5 indicates that surface-water motor-pump irrigators own and cultivate more land under rainfed conditions than even the sampled rainfed farm households. This is again an indication of the

Table 5. Average land ownership by irrigation category, 2009/2010 season, in acres (SD in parentheses).[a]

Agricultural land description	GW bucket fetch	SW bucket fetch	SW motor pump	SW gravity flow	Rainfed
Own rainfed land	4.2 (1.06)	4.9 (1.22)	7.2 (1.86)	5.7 (1.31)	4.0 (1.32)
Own SW irrigable land	0.0	1.2 (0.48)	6.4 (2.01)	0.5 (0.17)	0.0
Own GW irrigable land	1.0 (0.25)	0.0	0.0	0.0	0.5 (0.19)
Cultivated rainfed land	3.7 (1.19)	4.7 (1.45)	6.4 (1.99)	4.9 (1.48)	3.5 (1.02)
Cultivated SW irrigable land	0.0	1.2 (0.62)	3.2 (0.88)	0.5 (0.22)	0.0
Cultivated GW irrigable land	0.7 (0.17)	0.0	0.0	0.0	0.0

[a]F-tests (through analysis of variance) indicate that the averages of the different land types for the various technologies are significantly different at the 5% level.
Source: field survey (2010/2011).

relative wealth and resourcefulness of surface-water motor-pump irrigators. These categories of irrigators did indicate that they use cash from irrigated farming to finance the cultivation of more land under rainfed conditions. As indicated earlier, rainfed farming and irrigation do not compete for farm labour because they are undertaken in different seasons. It seems that factors other than labour explain the non-involvement of some farmers in irrigation.

Livestock ownership

Livestock reared in the area include cattle, sheep, goats, chicken, guinea fowl and ducks. Other animals include donkeys, which are a very popular means for transporting goods to and from the urban centres. They are also used for ploughing. There is a marked difference between the irrigating and non-irrigating households in the area with respect to ownership of livestock. Relatively small quantities of livestock are owned by rainfed farmers (Table 6). Two main reasons can be advanced for the situation. First, livestock (particularly ruminants) owned by irrigators tend to have greater access to both water and feed in the

Table 6. Average numbers of livestock per household, by irrigation types (SD in parentheses).[a]

Type of livestock	GW bucket fetch	SW bucket fetch	SW motor pump	SW gravity flow	Rainfed
Cattle	4.4 (2.01)	5.9 (1.95)	6.6 (2.33)	4.9 (1.78)	3.8 (0.83)
Goats	7.4 (2.22)	7.8 (1.88)	10.6 (3.67)	6.1 (3.03)	5.9 (2.11)
Sheep	6.4 (2.93)	5.9 (1.69)	9.5 (2.30)	3.7 (1.09)	5.4 (1.45)
Donkeys	1.3	0.3	1.5	1.5	0.8
Horses	0.03	0.0	0.0	0.0	0.01
Pigs	1.3	0.0	0.0	2.3	1.7
Ducks	1.2	0.7	1.4	1.8	0.76
Guinea fowl	21.3 (5.56)	12.0 (3.12)	19.0 (6.55)	9.8 (3.26)	12.8 (4.22)
Chicken	17.7 (4.68)	12.3 (2.89)	21.3 (8.25)	16.7 (5.52)	14.4 (3.05)

[a]F-tests (through analysis of variance) indicate that the averages of the different types of main livestock for the various technologies are significantly different at the 5% level.
Source: field survey (2010/2011).

dry season. All types of irrigators consciously provide water and feed to their livestock in the dry season. Secondly, rainfed farmers (non-irrigators) are more likely to sell their livestock to meet household needs than irrigators. When respondents were asked how they obtain money to purchase food in times of shortages (during the "lean season"), all the non-irrigators mentioned the sale of livestock, as did about 25% of the bucket-fetch irrigators; but none of the motor-pump or gravity-flow irrigators indicated the sale of livestock as a source of money for food items or to meet other monetary obligations. Table 6 also indicates that, except for guinea fowl, motor-pump irrigators own larger numbers of the different types of livestock than other types of irrigators. This again goes to confirm the relatively wealthier status of motor-pump irrigators in the Bawku-West and Talensi/Nabdam Districts of the Upper East Region of Ghana.

Ayariga (2011b) has stressed the importance of developing multiple uses for irrigation water, including crop–livestock integration systems, to increase the profitability of irrigated agriculture. Smallholder irrigators can actively incorporate livestock production (albeit on a relatively small scale) as part of their irrigation activities to ensure more efficient use of irrigated water, labour and other resources as well as better use of irrigated-crop residues as feed for livestock.

Comparative economic analyses of different irrigation typologies in the study area

Comparative costs-and-returns analyses were undertaken for the different irrigation technologies as well as rainfed production by non-irrigators. Households sampled were of very similar socio-cultural and demographic characteristics. It must be stated that all irrigators are also rainfed farmers in the study area; thus, the cost comparison could have been undertaken using information from only irrigators. However, that was not done to avoid possible biases in the information given by the irrigators.

Table 7 gives some information on the relative cash expenditure and gross revenues per irrigator (irrigator farm family) and per acre for the different irrigation typologies. The superiority of surface-water (SW) pump and groundwater (GW) bucket-fetch irrigation over the other types of irrigation is quite evident. In terms of gross revenue, about USD2675 was realized from 3.31 acres by SW pump irrigators, or about USD807 per acre. For the GW bucket-fetch irrigators, the gross revenue was about USD610 from 0.69 acres or about USD884 per acre. It is noteworthy that the highest gross revenue per acre is obtained from "traditional" GW bucket-fetch irrigation. It is no wonder then that farmers in the Upper East Region have stuck to that system. Nonetheless, the average area under that system is very small, only about 0.7 acres, and that is why the income per farm family is less than that of SW motor-pump irrigators. The relatively higher profitability of the bucket-fetch technology is mainly due to low investment costs and the absence of fuel costs. The above analysis, however, obscures the fact that relative profitability per unit area does not necessary imply viability of the technology. Irrigators need to operate economically viable irrigated areas. Pumps can make that possible for the small farmer.

The superiority of irrigated agriculture over rainfed agriculture in income generation is obvious from Table 7. Rainfed production results in a gross revenue of only USD206.92 per unit area as compared to USD372.87 for SW gravity flow, the irrigation with the smallest average gross revenue. The rainfed production revenue is also only about 23.4% of the revenue of the system with the highest average gross revenue. Nonetheless, the gross revenue per rainfed farm family (USD724.22) is greater than GW bucket-fetch (USD610.56), SW bucket-fetch (USD717.14) and SW gravity-flow (USD192.03) irrigation systems, mainly

Table 7. Relative cash expenditure and gross revenues by irrigation type in the study area.[a]

Item	Type of irrigator				
	GW bucket fetch	SW bucket fetch	SW motor pump	SW gravity flow	Rainfed (non-irrigator)[b]
A. Average irrigated area per irrigator or farm family (acres)	0.69 (0.16)	1.16 (0.23)	3.31 (1.05)	0.52 (0.15)	3.56 (1.51)
B. Average cash expenditure on farm inputs (USD per irrigator)[c]	85.38 (16.23)	98.66 (25.05)	711.30 (67.70)	44.67 (23.45)	86.35 (45.97)
C. Average cash expenditure on farm inputs per acre (B/A, USD)	123.74	85.05	214.89	85.90	24.26
D. Average gross revenue per irrigator or farm family (USD)[d]	610.56 (90.25)	717.14 (100.06)	2674.73 (222.20)	192.03 (40.88)	724.22 (109.55)
E. Average gross revenue per acre (D/A, USD)	884.87	618.22	808.08	369.29	203.43

[a] F-tests (through analysis of variance) indicate that the various averages of the variables (areas, costs and revenues) for the various technologies are significantly different at the 5% level.
[b] All values under this column are with respect to rainfed production of non-irrigators. The column is for the purpose of indicating the relative efficiency of irrigated production vis-à-vis rainfed production.
[c] Includes mainly expenditure on planting and fencing materials, inorganic and organic fertilizers, herbicides, pesticides, fuel for pumps, and hired labour, but may include rentals of tractors, bullock and ploughs.
[d] Obtained by estimating total crop harvests and average market prices. Prices of vegetable crops vary very widely within and between seasons.
Source: field survey (2010/2011).

because of the small irrigated areas cultivated under those systems. It is not surprising that a major constraint identified by almost all irrigators in the area is the limited availability of land for irrigation.

Total costs and net revenue computations for the various irrigation types and rainfed production are presented in Table 8. Depreciated costs of fixed-cost items as well as the imputed value of family labour have been added to the cash expenditure on farm inputs to obtain the total farm cost values. Even though the average total cost per acre for the irrigated systems is many times that of the rainfed production system (as expected), the net revenue per acre for rainfed production is only USD129.46, less than all the irrigated production systems and only 14.6% of the system with the highest average net revenue. The highest average net revenue per farm family (USD1742.93) is obtained by SW motor-pump irrigators, but the highest per acre (USD631.51) by GW bucket-fetch irrigators. Again, if the latter could increase the area under cultivation it would be the best in terms of returns per family and per acre.

It must be emphasized that irrigated and rainfed production in the Upper East Region of Ghana are not competing in any way, so for almost all irrigating farm families, the annual income earned from irrigation in the dry season is simply added to that obtained from rainfed production in the rainy season. Thus, the annual revenue for the irrigating farm family is actually from irrigation plus rainfed production.

In the financial analysis above, as well as the earlier discussion on assets of farm families, it is obvious that motor pumps are playing a special role in income generation in the Upper East Region even though their use is not yet widespread. Many bucket-fetch irrigators do not use the pumps mainly because they cannot afford to purchase them, or possibly even to operate them, as will be discussed in the next section.

Constraints and opportunities in the use of the different technologies

Constraints and opportunities analyses were undertaken for each of the irrigation typologies. Table 9 gives a summary of the most serious constraints identified and solutions suggested by irrigators. The main constraints identified include inadequate working capital, insufficient irrigated land and high costs of production inputs. All irrigators clearly desire to own and use pumps to undertake irrigation using either groundwater or surface water from rivers and dams. They perceive motor pumps as saviours because of their convenience and the higher returns that are associated with their use. However, if some of the main constraints identified in Table 9 are not removed, most of the smallholder irrigators, and particularly women, will never be able to own and use motor pumps for irrigation. A major suggestion by farmers is the introduction of a subsidy on motor pumps (Table 9). That may be the preferred solution, but how optimal is it? There are now pumps on the market, especially from China, that are relatively cheap and perform relatively well (see Table 10). Access to credit for such pumps could prove to be a better solution, especially as the analysis in this paper has shown that the use of motor pumps is far more profitable than other irrigation technologies. A subsidy on pumps could prove to be wasteful and a disincentive to efficient performance.

Several opportunities for developing smallholder irrigation in the Upper East Region and Ghana were also identified by the farmers (Table 10). Many of them observed that in the past couple of years there have been considerable discussions, mainly through workshops, on smallholder irrigation and groundwater use, and that relevant government institutions and departments seem to be giving suggestions that emanate from them.

Table 8. Relative total costs and net revenues by irrigation types in the study area.[a]

Item	GW bucket fetch	SW bucket fetch	SW motor pump	SW gravity flow	Rainfed (non-irrigator)
A. Average depreciated cost of equipment per irrigator (USD)[b]	3.76 (1.01)	2.32 (0.78)	85.01 (4.88)	2.33 (0.63)	11.10 (2.89)
B. Average imputed family labour cost per farm family (USD)[c]	85.68 (17.23)	40.89 (13.64)	135.49 (30.66)	45.75 (12.33)	155.91 (45.44)
C. Average total cost per average area under irrigation (A + B + relevant costs in Table 7, USD)	174.82	141.87	931.80	92.75	263.36
D. Average total cost per acre (USD)	253.36	122.30	281.51	178.37	73.98
E. Average gross revenue per farm family (USD)	610.56	717.14	2674.73	192.03	724.22
F. Average net revenue per average area under irrigation (E – C, USD)	435.74	575.27	1,742.93	99.28	460.86
G. Average net revenue per acre (USD)	631.51	495.92	526.56	190.92	129.46

[a]F-tests (through analysis of variance) indicate that the average depreciated costs and imputed family labour costs for the various technologies are significantly different at the 5% level.
[b]Buckets, watering cans, hoes, cutlasses and sickles, as well as pumps in the case of pump irrigators.
[c]Obtained by estimating total family labour days used and average market wage rates.
Source: field survey (2010/2011).

Table 9. Main constraints in the use of the different technologies, and farmers' suggested solutions.

Irrigation typology	Main constraints in use of irrigation technology	% of farmers citing constraint as most serious	Farmers' suggested solutions
GW bucket fetch ($n = 89$)	Drudgery in digging and covering of wells	48.3	Assist farmers to construct permanent wells and buy pumps
	Land use problems associated with land tenure	15.7	More structured written agreements between landowners and users
	Seasonal drying up of wells (low water table) in some areas	11.2	Deeper wells; assist farmers to construct permanent wells and buy pumps
	Motor pumps are not affordable	24.7	Government should subsidize motor pumps
SW bucket fetch ($n = 10$)	Tedium in fetching water	50.0	Government should subsidize motor pumps
	Land use problems associated with land tenure	10.0	District assemblies should enact by-laws to protect irrigators
	Motor pumps are not affordable	40.0	Government should subsidize motor pumps
SW motor pump ($n = 8$)	Motor pumps are too expensive and too big	37.5	Government should subsidize motor pumps and procure more suitable ones for farmers to buy
	Cost of running pumps (fuel and maintenance) is too high	62.5	Electricity should be extended to irrigation areas so that electric pumps can be used; government should subsidize electric power for irrigation
SW gravity flow ($n = 40$)	Very badly constructed canals	42.5	Contractors should be prosecuted; canals should be rehabilitated
	Motor pumps are expensive (would prefer to use pumps to lift water from reservoirs)	27.5	Government should subsidize motor pumps
	Low prices of produce during harvests	17.5	Government and NGOs should help farmers to organize marketing and processing of produce
	Lack of maintenance and encroachment of catchment area	12.5	Chiefs and opinion leaders (water user association executives) should be more serious
All techno-logies (common problems identified)	Land use problems as a result of prevailing land tenure systems	N/A	District assemblies should enact by-laws to protect irrigators
	High cost of production inputs	N/A	Fertilizer subsidy should be extended to dry-season production
	Access to credit very difficult and expensive	N/A	Sustainable irrigation credit scheme
	Inadequate access to agricultural extension services	N/A	The Ministry of Food and Agriculture (MOFA) should train its personnel in irrigation technology

Source: field survey (2010/2011).

Table 10. Opportunities for improving irrigated agriculture in the Upper East Region and Ghana.

Irrigation typology	Opportunities identified by farmers in study area
GW bucket fetch	• There has recently been some attention, in terms of discussions, paid by the Ghana Irrigation Development Authority to smallholder groundwater irrigation. • There seems to be some interest in research into groundwater resources in recent years. • The government has removed taxes on pumps, and Chinese pumps are quite cheap. Low pump prices will help to replace bucket fetch with use of motor pumps.
SW bucket fetch	• The government has removed taxes on pumps, and Chinese pumps are quite cheap. Low pump prices will help to replace bucket fetch with use of motor pumps.
SW motor pump	• The government has removed taxes on pumps, and Chinese pumps are quite cheap. Low pump prices will lead more people to use motor pumps for irrigation. • There are recent discussions of provision of electricity for irrigated agriculture. A new hydroelectric dam being constructed in the northern part of the country will make use of electricity feasible. • There are promises of improved micro-credit schemes for agriculture by the government.
SW gravity flow	• There are promises of rehabilitation of canals. • Cheaper electric pumps will make gravity flow unnecessary.

Source: Field survey (2010/2011).

The way forward in smallholder irrigation in Upper East Region of Ghana

Rural poverty has been chronic in the Upper East Region for many reasons. One main one is the very limited opportunity for income-generating activities in the long dry season. The situation seems to be getting worse with climate change and the global food, energy and financial uncertainties (Akudugu, Dittoh, & Mahama, 2012). However, smallholder farmers in the Upper East Region and elsewhere in Ghana have undertaken irrigation on a very small scale for decades but with very many constraints and lack of scientific input. GW bucket-fetch irrigators, for example, have little or no idea of how much irrigated area their shallow wells can support or how long they may be able to continue to use shallow wells to irrigate.

The analyses in this paper clearly indicate the importance of dry-season irrigation in the Upper East Region in augmenting the incomes of farming families. Incomes from irrigation are additional to what is obtained in the rainy season; there are virtually no trade-offs between rainy and dry-season agricultural production in the area. Also, irrigated production, by whichever technology, has been shown to be more efficient with respect to the use of land and other inputs than rainfed production. However, most farm families cannot depend on only irrigated production because of very limited irrigated areas. Indeed, the total production, even in value terms, of irrigated crops can hardly be compared with rainfed crops in most cases because of the differences in areas under cultivation. This calls for improvements in irrigated production by removing constraints and taking advantage of opportunities. Fortunately, the farmers have clear understanding of their constraints and many of the opportunities that can be taken advantage of.

As shown in this paper, smallholder irrigators in the two districts of the Upper East Region believe that the increased use of pumps for lifting water from rivers, dams and wells for irrigation is the surest way to improve irrigated agriculture in the area. With increasing climatic uncertainty and the need to conserve available surface and groundwater resources, bucket-fetch and canal (gravity-flow) irrigation must give way to pump irrigation. Electric pumps could provide an attractive alternative to gas-fuelled pumps, but it has not yet been demonstrated that they provide a significant advantage.

Notes

1. Storativity and transmissivity are physical properties of aquifers that relate to the capacity to store and transmit groundwater.
2. *Tendanas* literally means "landowners", but they mainly play the role of spiritual custodians of the land. In the study area, land belongs to families and clans. The involvement of chiefs and *tendanas* helps to give consultations and decisions some traditional legitimacy.

References

Akudugu, M., Dittoh, S., & Mahama, E. (2012). The implications of climate change on food security and rural livelihoods: Experiences from Northern Ghana. *Journal of Environment and Earth Science*, *2*(3), 21–29.

Akuriba, M. A., Dittoh, S., Issaka, B. Y., & Bhattarai, M. (2010). *Women farmers' perspectives on micro-irrigation technologies in the West African Sahel.* Paper presented as a Poster at the 3rd Conference of the African Association of Agricultural Economists 2010, Cape Town, South Africa.

Al-Hassan, R., & Diao, X. (2007). *Regional Disparities in Ghana. Policy Options and Public Investment Implications* IFPRI Discussion Paper No. 00693. Development Strategy and Governance Division, Washington DC, 2.

Anayah, F. M. A. (2012). *Improving complementary methods to predict evapotranspiration for data deficit conditions and global applications under climate change* (Ph.D Thesis). Utah State University. http://digitalcommons.usu.edu/etd/1306

Appiah, K., Demery, L., & Laryea-Adjei, S. G. (2000). Poverty in a changing environment. In E. Aryeetey, J. Harrington and M. Nissanke (Eds.), *Economic reforms in Ghana: The miracle and the mirage.* Oxford, UK: Africa World Press, Asmara, Eritrea and James Currey Ltd.

Aremu, J. A., & Ogunwale, S. A. (1994). Comparative analysis of small and large-scale irrigation schemes in Northern Nigeria. In A. O. Sanda and S. B. Ayo (Eds.), *Impact of irrigation on Nigeria's environment* (pp. 165–185). Ibadan, Nigeria: Fact Finders International.

Ayariga, R. (2011a). Personal Communication. Northern Rural Growth Project, Tamale, Ghana.

Ayariga, R. (2011b). *Experiences of small scale irrigation financing and the way forward.* Paper presented at Agricultural Water Management Solutions Workshop on Financing and Marketing & Trade Workshop. Kumasi, Ghana. July 2011.

Bannerman, R. R., & Ayibotele, N. B. (1984). *Some critical issues with monitoring crystalline rock aquifers for groundwater management in rural areas.* IAHS Publication No. 144.

Bortei-Doku Aryeetey, E. (2000). The participation of women in the Ghanaian economy. In E. Aryeetey, J. Harrington & M. Nissanke (Eds.), *Economic reforms in Ghana: The miracle and the mirage.* Oxford, UK: Africa World Press, Asmara, Eritrea and James Currey Ltd.

Breisinger, C., Xinshen, D., Kolavalli, S., Al-Hassan, R., & Thurlow, J. (2011). A new era of transformation in Ghana: Lessons from the past and scenarios for the future. *IFPRI Research Monograph.* IFPRI, Washington, DC.

Carrier, M. -A. (2008). *Hydrogeological synthesis of Northern Ghana* (M.Sc. Thesis). Avant-Gard University, Canada.

Dietz, T., Obeng, F., Obure, J., & Zaal, F. (2009). Participatory development assessment: Subjective truths. *The Broker*, (15), 19.

Dittoh, S. (1997). Irrigation and increased food production in 21st century West Africa: Policies, research and development directions. In A. E. Ikpi and J. K. Olayemi (Eds.), *Governance and development in West Africa: Perspectives for the 21st century* (pp. 113–128). Arlington, USA: Winrock International/African Rural Social Sciences Research Networks.

Dittoh, S., Bhattarai, M., & Akuriba, M. A. (2013). Micro irrigation-based vegetable farming for income, employment and food security in West Africa. In M. A. Hanjra (Ed.), *Global food security* (pp. 178–199). New York, NY: NOVA Science Publishers Inc.

Gbireh, A. Z. (2011). *Shallow groundwater development for smallholder irrigation*. Paper presented at Agricultural Water Management Solutions Workshop on Technical Aspects of Small Dams, Wells, Pumps and Pipes. Tamale, Ghana.

Giordano, M. (2005). Agricultural groundwater use in sub-Saharan Africa: What do we know and where should we go?. *Water Policy*, *7*, 613–626.

ISSER (Institute of Statistical Social and Economic Research). (2007). *The state of the Ghanaian economy in 2006*. Legon: University of Ghana.

Mainoo, P. A., & Dapaah-Siakwan, S. (2007). *Groundwater inputs to grid cells covering the three Northern Regions of Ghana (Upper East, Upper West and Northern Regions of Ghana)*. Hydrogeological Assessment Project (HAP) of the Northern Regions of Ghana Project. CIDA Project 7038883. CSIR Water Resources Research Institute, Ghana.

Martin, N. (2006). *Development of a water balance for the Atankwidi catchment, West Africa: A case study of groundwater recharge in a semi-arid climate*. Ecology and Development Series, No. 41, Cuvillier Verlag Göttingen, 169 p.

Nanes, R. (2011). Personal Communication. International Development Enterprises (IDE), Accra, Ghana.

Nyamadi, B. (2011a). *Irrigation Development and Future Plans*. Paper presented at Agricultural Water Management Solutions Workshop on Financing and Marketing & Trade Workshop. Kumasi, Ghana. July.

Nyamadi, B. (2011b). Personal Communication. Ghana Irrigation Development Authority (Chief Executive), Accra, Ghana.

Obuobi, E. (2008). *Estimation of groundwater recharge in the context of future climate change in the Volta River Basin* (Doctoral Thesis). University of Gottingen, Germany.

Oriola, E. O. (2009). Irrigation agriculture: An option for achieving the millennium development goals in Nigeria. *Journal of Geography and Regional Planning*, *2*(7), 176–181.

Pant, N. (1992). *New trends in Indian irrigation: commercialization of groundwater*. Ashish Publishing House: New Delhi.

Shah, T. (2009). *Taming the anarchy: Groundwater governance in South Asia*. Resources for the Future, IWMI. Chapter 4.

World Bank. (2007). *World development report 2008: Agriculture for development* (p. 45). Washington, DC.

WRC/CIDA. (2010). *Hydrogeological assessment of the Northern Regions of Ghana (HAP)*. SNC Lavalin and University of Quebec (INRS-ETE).

Improving the supply chain of motor pumps to accelerate mechanized small-scale private irrigation in Zambia

Willem Colenbrander and Barbara van Koppen

During the past decade, smallholder groundwater irrigation with motor pumps has increased considerably in Zambia. This study analyzes an important but hitherto ignored factor for adoption: the supply chain of imported motor pumps. The main obstacles for farmers are identified as: the highly centralized supply chain and financing facilities in urban hubs; lack of information about prices, which vary significantly for the same make and model of pump; lack of information and training on proper use and maintenance; and lack of financing facilities. The Zambia National Farmers Union seems best placed to remove these obstacles.

Introduction

Rationale

Zambia has a substantial yet untapped potential for individual small-scale groundwater irrigation. Out of Zambia's 6 million ha under cultivation, only 200,000 ha (3%) is currently irrigated. At least half of this area (100,000 ha), which consists largely of lowlands with shallow groundwater tables, is estimated to be irrigated informally by individual smallholders for winter vegetables (GoZ, 2010). Most of this area (75,000 ha) is irrigated with buckets. On an estimated 10,000 ha, smallholders use river diversions (in particular in the northern, high-rainfall area of Zambia) or treadle pumps (in various other areas). Remarkably, an estimated 15,000 ha are irrigated with motor pumps, which represents an increase from virtually nil only 10 years ago (Colenbrander, Kabwe, & Van Koppen, 2011). Profit margins for smallholders using portable pumps are considerably higher than for smallholders using any other technology (Colenbrander et al., 2011). Moreover, the prices of the pumps have declined. Up to the end of the 1990s, only the more expensive pumps were available, which were only used by a few, better-off farmers. Since then, the retail prices of especially the Chinese pumps of 3–5 horsepower (hp) have come down to USD200 for the cheapest pump. This is about the same as the current retail price for treadle pumps in Zambia.

The potential for further adoption of motor pumps is immense. Zambia has an estimated 400,000–600,000 ha of land for further irrigation development (GoZ, 2010). Bucket irrigators may well step up from buckets to motor pumps. Indeed, a growth path similar to that of South Asia since the 1980s may be possible (Mukherji et al., 2009). There, large community and government-managed schemes are giving way to mechanized groundwater irrigation by individual smallholders (Shah, 2009). In India, more than 60% of the nation's irrigation now comes from atomistic motorized-pump irrigation, despite ongoing investments in government irrigation schemes.

Expansion of individual motor-pump irrigation would fully fit the policies of the government of Zambia. The Zambia Irrigation Policy and Strategy (GoZ, 2004) states as general objective for smallholders: "to empower smallholders to develop and use irrigation and other water management technologies so as to alleviate poverty and achieve food security". Its specific policy objectives are "to rehabilitate and upgrade public sector irrigation schemes, to initiate new public sector, farmer managed, irrigation schemes, to increase the productivity of rainwater through water harvesting and to promote sustainable dambo (lowlands, wetlands) utilization". The government has taken fiscal measures to support technology adoption in large- and small-scale farming. Furthermore, ongoing and planned investment in small-scale irrigation programs amounts to USD842 million over a period of about 10 years. However, out of this, only USD16 million (2%) is to be spent on traditional and low-cost technologies of individual smallholders (GoZ, 2010). Although policies are supportive, private smallholder irrigation has received limited attention as yet.

Past policy research on private smallholder irrigation has rightfully focused on the demand side and the marketing of irrigated produce (Hichaambwa, 2010). Produce markets are the "pull" factor for smallholder irrigation. Spontaneous smallholder irrigation is most frequent around the capital Lusaka and other urbanizing hubs, where national and international markets are flourishing. Yet, the "push" factors of improved supply of produce are also important. However, research and policies to support the supply of produce tend to focus on agronomic practices, in particular cultivation practices, fertilizers, choice of seed, water conservation and gravity or manual irrigation. Generally, limited research has been done on individual small-scale irrigation adoption patterns. For the supply chain of motor pumps in particular, the authors are unaware of any earlier research. Yet, a better understanding of the factors in the supply chain that contributed to this rapid increase, but also the obstacles that farmers still face, could further unlock the potential to mechanize smallholder groundwater irrigation. This study aims to fill this gap.

Research question and method

The research questions addressed are:

- What is the nature of the supply chain of portable motor pumps in Zambia, in terms of factors that enable adoption and obstacles that farmers face?
- How can farmer organizations, government, donors and NGOs address those obstacles to accelerate pump adoption by smallholders?

These questions were explored qualitatively and quantitatively through literature review and in-depth interviews with key actors in the supply chain. These included about two dozen small-scale irrigation farmers, as well as staff of the Zambia National Farmers Union, which has associations for smallholders in all districts of the country. In Lusaka and elsewhere, 21 retail shops were visited. Interviews were held with the Customs Department of the Zambia Revenue Authority (ZRA) under the Ministry of Finance and National Planning and with a customs clearing agent. Interviews were also held with the Ministry of

Agriculture and Cooperatives, which is responsible for formulation and implementation of irrigation policy. NGOs and micro-finance institutions (MFIs) that provide loans for agricultural equipment to smallholders, and one MFI membership organization, were visited as well. The study took place between November 2010 and November 2011 and covered the 2010–2011 rainy season and the 2011 dry season.

Conceptualization and research sub-questions

In order to define the research sub-questions, we conceptualized the supply chain as consisting of the entire network of interdependent actors that deliver products and services to the same customer, in this case the Zambian smallholder farmer. Adapted from the report by Research Into Use (RIU, 2007), Figure 1 shows how an agricultural input supply chain is an integral part of the agricultural production value chain.

A fully developed value chain and supply chain entail not only the flow of goods and services but also the flow of information about various aspects of the chain, such as importation procedures, prices, quality and customer preferences. The flows are bi-directional, upstream and downstream along the chain.

Focusing on the supply chain of one type of input, namely portable motor pumps for irrigation, Figure 2 details the manufacturers, importers, wholesalers, retailers (including after-sales services) and buyers (the farmers) of the motor pumps, and their transactions. The research sub-questions address each of these items.

The following section discusses the respective findings on: manufacturers and types of motor pumps; import, wholesale, retail and after-sales services; fiscal measures that are meant to improve this flow; retail prices; and financing facilities that enable the flow of payments. The last section discusses conclusions and provides recommendations.

Findings on the motor-pump supply chain in Zambia

Manufacturers and types of motor pumps

As the national electricity grid in Zambia only reaches 4% of the rural population (as opposed to 47% in cities and towns), this study focuses on pumps powered by petrol and diesel engines. Fuel stations are found in most districts in Zambia.

All motor pumps used in Zambia are imported. Most common are the Chinese makes King Max, Eagle, Asia, RATO, Kipor, Jiangdong, Changfa and NTC. Fewer pumps are

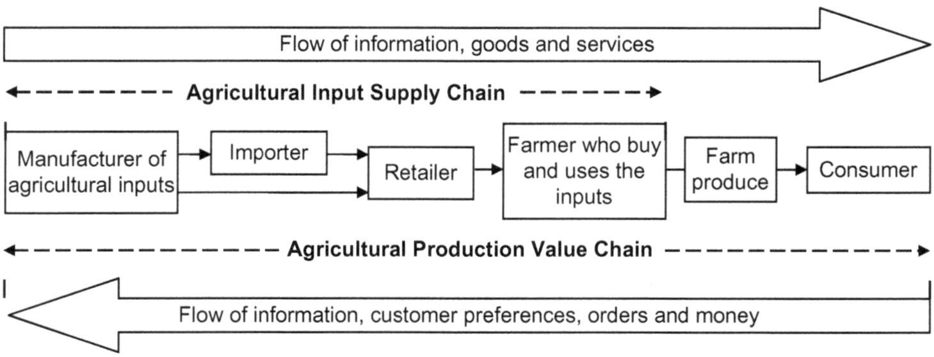

Figure 1. The input supply chain embedded in the production value chain (adapted from RIU (2007)).

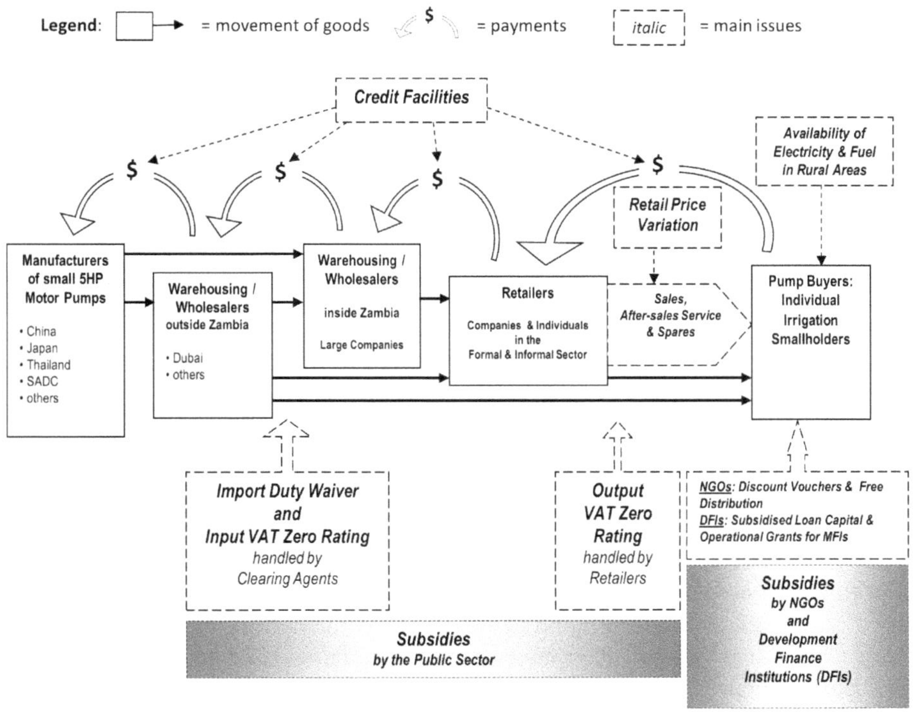

Figure 2. Motor-pump supply chain actors and issues in Zambia.

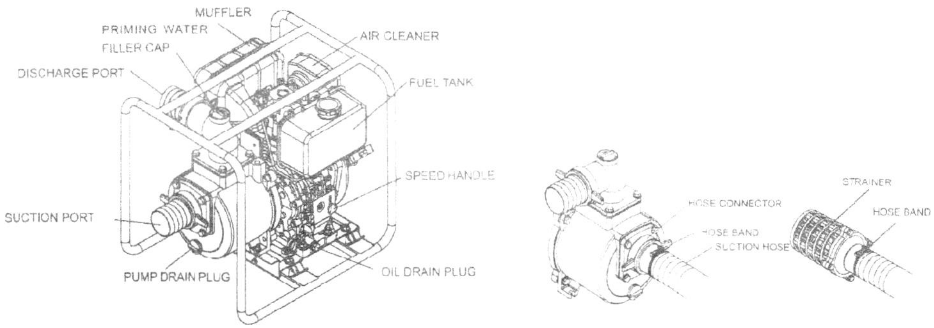

Figure 3. A typical portable motor pump.

imported from Thailand; these are mainly Thai Honda, which are made under license from Honda Japan. Japanese makes are Honda, Koshin and Dayliff. The most common models found in this study are 2-to-3-inch petrol pumps (3 to 5 hp, 30–60 cubic meters per hour), followed by 2-to-4-inch diesel pumps (3.5–9 hp, 20–60 m^3/h). All portable pumps have a suction limit of 8 m. See Figure 3 for a typical portable motor pump.

Import, wholesale, retail and after-sales services

The study identified four groups of actors in the supply-chain, each with specific import, wholesale and retail, and after-sales services arrangements. The first group consists of large importers: those specialized in agricultural and irrigation equipment (of which five were visited); and general retail stores that also sell pumps (of which eight were visited). These bigger companies do sell spares and provide after-sales services in all their shops, but some after-sales services are only provided in their Lusaka headquarters. The bigger stores usually sell a large range of pumps, either retail or wholesale.

The second group consists of smaller importing companies who usually import cheaper Chinese pumps for retail only (of which six were visited). Most of them do not sell spares or provide any after-sales services.

The third group consists of companies that specialize in providing irrigation services to farmers (of which two were visited). They either import themselves or buy locally. They always provide spares and after-sales services to their clients.

The fourth group consists of individuals in the informal sector who import pumps on order from individuals or companies. In this case there is no spare-part provision or after-sales service.

An important obstacle in the supply chain, echoed in the main complaint of smallholder farmers interviewed about the motor-pump supply chain, was the centralization of the companies in the urban centres along Zambia's "line of rail". Farmers complained that there were no nearby, reliable retail shops for pumps and spares, and no repair facilities. The cost of travelling to those urban retail outlets can easily constitute one-third of the price of the pump. For example, motorized-pump irrigators in Livingstone, in the south-west corner of Zambia, said that they buy their motor pumps from individual traders who bring them from shops in South Africa, while others buy pumps from companies in Lusaka, 500 km away. These complaints reflect a demand for pumps in rural areas, which apparently still fails to reach sufficient critical mass to incite formal retailers to open up new sales points there. Purchases over long distances, and informal individual traders and irrigation service providers, currently fill the gap to some extent. The centralized nature of the supply chain also means that farmers in remote rural areas are poorly informed about the range of available pump makes and models and, as elaborated below, their prices.

In order to address the limited availability of spare parts and after-sales services, some retailers mentioned that they are prepared to train mechanics, from areas where there is a market for their pumps, in the maintenance and repair of pumps. These trained mechanics could also order spares from the retailers. A last obstacle observed during the farmer interviews is that farmers lack information and skills on how to handle their pumps properly. For example, some fail to use proper foot-valves, so that the pump sucks water mixed with soil, which causes pump seals to quickly wear out. This indicates that the supply chain is also underdeveloped in terms of after-sales services, information and training.

Fiscal measures

Retail prices can be influenced by governments through fiscal measures. Zambia introduced two fiscal measures with the aim (among others) of increasing the supply of farm equipment to smallholders: an import duty waiver on irrigation equipment and on pumps for agriculture and horticulture (SI No. 13 of 2003); and zero rating of value-added tax (VAT) on the importation of farm equipment, inclusive of pump sets (SI No. 3 of 2009).

Import duty on irrigation equipment would normally be 15%, but since SI No. 13 of 2003, there has been a suspension of duty on irrigation equipment and on pumps for agricultural and horticultural purposes. Duty-free importation of pumps became possible since SI No. 13 of 2003 became effective, but only when an "additional procedure code" (APC 416) could be applied. This was a highly complicated procedure, outside the reach of small farmers. It differentiated between specialized providers of agricultural and irrigation equipment and services, who could bring in pumps duty free, and general retail stores that bring in pumps for general purposes and thus had to pay duty. For small individual importers of pumps who operate in the informal sector, there was no formally registered information on the way they import.

Interviews with nine companies that were importing their own pumps revealed that actors in the supply chain were not always aware of duty and VAT regulations. Categorization of imported items appeared highly complex. ZRA staff confirmed that there is often misunderstanding between the ZRA and importers on classification of goods and that it is advisable to seek regular advice on classification, especially prior to importation. However, since the Customs and Excise Amendment Act of 2011 became effective on 28 December 2011, all portable motor pumps can be imported duty free under Customs Code 8413.19.90.

The second fiscal measure, issued in February 2009 (SI No. 3 of 2009), was the VAT zero rating for importing small-scale agricultural equipment and spares, including portable pumps. The ZRA said that the VAT zero rating will "ensure that unregistered farmers who cannot claim input VAT on such equipment do not incur any VAT. The measure will also reduce input costs and enable farmers to expand production and make farming a viable sector."

The VAT in Zambia is currently 16% and is normally collected at each stage of the supply chain where value is added to goods or services. A business incurs input VAT when purchasing goods or services and charges output VAT when supplying goods or services. At the end of each tax period, the VAT due is arrived at by deducting the total input tax from the total output tax. Any company with an annual turnover of more than USD40,000 is VAT registered, but pumps they import are VAT zero rated. (Such a company pays 35% company tax, as opposed to smaller companies which pay only 3% turnover tax.) VAT zero rating for importing pumps means that the end-buyers (farmers) cannot be charged VAT.

Both fiscal measures are likely to further develop the supply chain. Future research should assess the strategies adopted by importers, wholesalers, retailers and farmers in further detail. They could reduce the price, open up branches in more remote areas, or both.

Retail prices

Prices strongly influence farmers' adoption rates. A comparison was made of the retail prices of different sizes of pumps, as measured by the diameter of the outlet of the pump. The findings are based on a total of 41 different pump makes and models. The analysis highlighted important differences in prices by type of fuel (diesel or petrol), by country of origin and, last but not least, by retailer. Overall, pump prices ranged from the cheapest 2-inch petrol pumps (USD200) to the more expensive 4-inch diesel pumps (USD1000).

Figure 4 shows the comparison between minimum and maximum retail prices of pumps fuelled by petrol or diesel for the same pump diameter. The figure shows that price differences between same-size petrol pumps are much larger than between same-size diesel pumps. Diesel pumps are always more expensive than petrol pumps of the same size, because the diesel engines that drive the pump are more expensive to manufacture. (The pumps that are connected to a diesel or a petrol engine are of the same type and price range.)

The make of the pump also influences the retail price. As shown in Table 1, Thai Honda petrol pumps are 16–39% more expensive than the Chinese makes of pumps. Petrol pumps

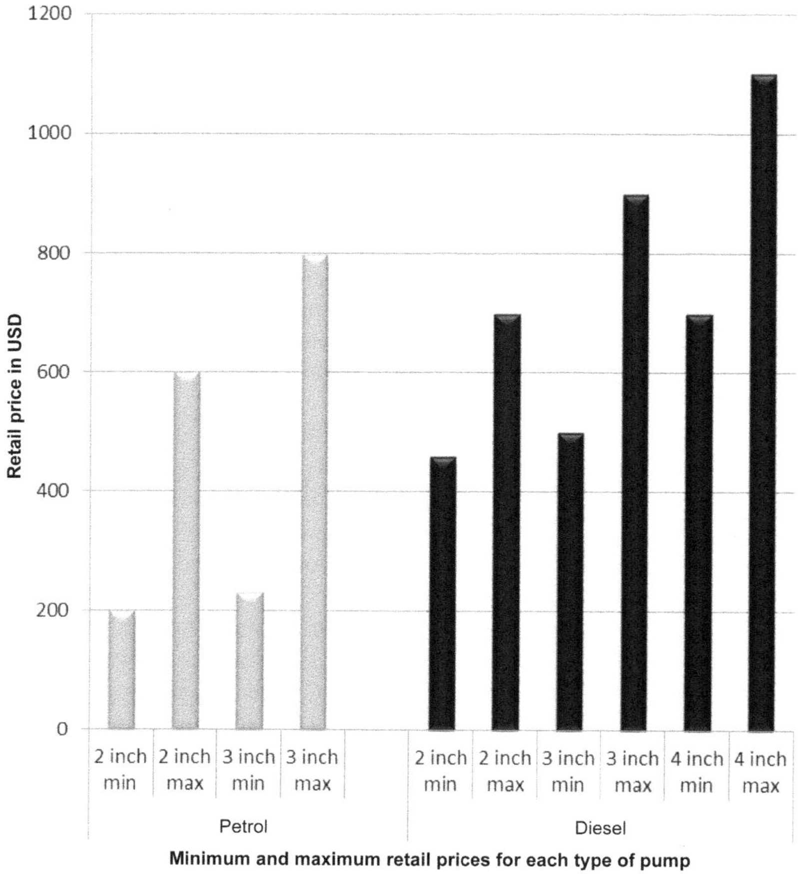

Figure 4. Retail prices by type and size of pump.

of Japanese makes are up to 160% more expensive than the petrol pumps of Chinese makes. Diesel pumps of Japanese makes are 55% more expensive than diesel pumps of Chinese makes. The price difference between Japanese and Chinese makes is much smaller for diesel pumps than for petrol pumps. The cheapest petrol pumps, usually made in China, are of rather poor quality and good parts supply is usually lacking. However, they can still

Table 1. Average retail prices by type and make of pump (n/d: no data available).

	Average retail price (USD)		
Pump country of origin	China	Thailand	Japan
Pump type			
2″ petrol	245	285 (16%)*	600 (145%)*
3″ petrol	270	375 (39%)*	700 (159%)*
2″ diesel	570	n/d	n/d
3″ diesel	580	n/d	900 (55%)*
4″ diesel	800	n/d	n/d

*Retail price premium compared to Chinese pumps.

last for a year or two. There is still demand for these Chinese pump makes, especially from farmers who are changing over for the first time from the bucket or the treadle pump. The expensive pumps are bought by the more experienced farmers who are replacing their cheap start-up pumps.

A third price analysis compared prices of the same models and makes of pumps but from different retail outlets (see Figure 5). There were remarkable differences in retail price for exactly the same make and model of pumps between the different retail outlets. The differences within the capital, Lusaka, were between up-market large agricultural retail shops and the smaller downtown shops in the same markets where smallholders are selling their crops. Pumps were priced lower in the large shops, which also have good after-sales service and spare-part provision. Some of the bigger shops are even giving warranties

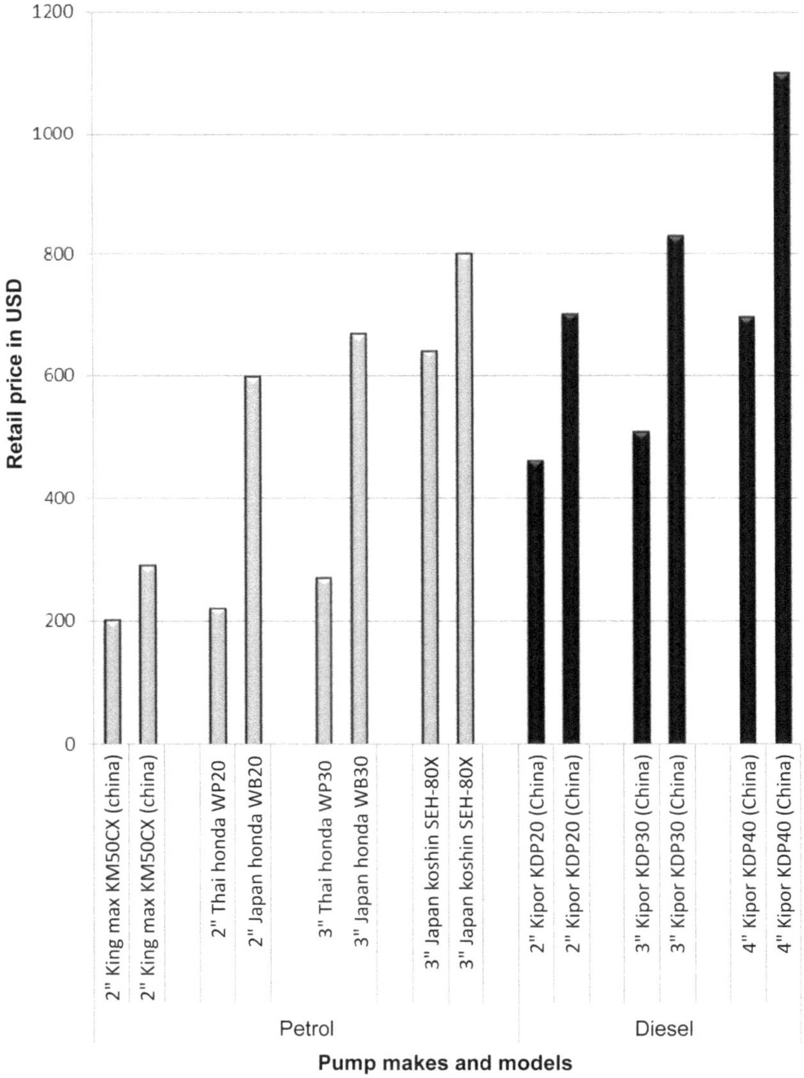

Figure 5. Retail price variation between suppliers of the same pump makes and models.

on portable motor pumps, usually for a period of six months. This points at the diverse marketing strategies of the different retailers. Reasons for this have still to be identified, but one reason could be the economies of scale. The large shops import large quantities at once, just before the irrigation season starts. They are also much more aware of the formal procedure for importing duty free.

Financing facilities: loans and subsidies

Various ways were identified in which NGOs and MFIs currently enable smallholders to purchase pumps, and how facilities can be improved in the future. Only a few MFIs were found to deal specifically with loans to farmers, and only one of them, the Christian Enterprise Trust of Zambia (CETZAM), specializes in loans for irrigation equipment. The few MFIs in rural areas providing credit facilities to smallholders give loans to smallholders who would not qualify for loans from commercial banks because of lack of collateral. This is claimed to justify much higher interest rates than those of commercial banks. Usually the loans are small and for a few months only. When MFIs give loans for pumps, it is at an interest rate of about 4% to 5% per month. They generally recover the loan within six months (three months grace period, three months repayment). Their interest rates are 30–50% per annum. Some of the MFIs buy the pumps on behalf of their client farmers, while others let the farmer buy the pump of their choice within the agreed credit ceiling. Farmers tend to prefer the latter arrangement. The interviewed MFIs were not well informed about where to buy the best pump at the best price.

Discount vouchers support smallholders and the pump supply chain more directly. In Zambia, they have been provided for the promotion of treadle pumps, but not yet for motor pumps. However, a similar system could be possible for motor pumps. For the treadle pumps, a subsidy of 40% of the value of the irrigation equipment is provided by the Mennonite Economic Development Association in the form of a discount voucher whereby the farmer pays the remaining 60%, which means that the farmer buys the equipment at a discount of 40%. The supplier redeems the voucher value from the issuer and so receives the normal retail price for the equipment. Hence, there is no market distortion. The supplier as well as the farmer benefit from this kind of subsidy as it creates a higher demand for the equipment.

Lastly, pumps can be given for free. Such free distribution occurred, for example, in the intensively irrigated areas in the Katuba area, 20 km north of Lusaka. An NGO distributed 30 portable motor pumps free of charge to female, mixed and male groups of smallholders to promote the use of pumps in a particular area. The project's targeting of women's and mixed groups did ensure that more female-headed households got access to the motor pumps than would have if they had had to acquire them privately (Colenbrander et al., 2011). Distribution for free can be justified to promote innovation, create a critical mass, and target specific groups, such as female-headed households. At longer-term or wider scales, free distribution distorts markets.

Discussion and recommendations

Returning to the research question – "What is the nature of the supply chain of portable motor pumps in Zambia, in terms of factors that enable adoption and obstacles that farmers still face?" – the foregoing findings show that the supply chain is still underdeveloped. In no link in the chain is the flow of information, goods or services optimal. This is not surprising given that the supply chain only started developing a decade ago. The supply chain needs to be further developed, and obstacles need to be overcome, so that any smallholder in Zambia who wants to obtain a motorized pump is also able to do that and to use the pump sustainably. This would accelerate the expansion of irrigation in Zambia.

The pump supply chain is primarily underdeveloped in the sense that the provision of goods and services is centralized in Lusaka and few other urban centres, while many smallholders live far from those centres. This applies to retail outlets, to the after-sale services provided by retail outlets, and to the provision of credit facilities. This centralized nature of the supply chain negatively affects smallholders' access to goods and services.

The supply chain is also underdeveloped in that information is concentrated among a few. This is the most striking with regard to pump prices: they vary widely, yet even NGOs and MFIs are unaware of the variations, let alone farmers. The large import and retail companies, and of course the ZRA and the customs clearing agents, are well informed about procedures for duty waiver and VAT zero rating. However, even during this study, it appeared difficult to obtain that information. The smaller retail outlets and the MFIs who deal with agricultural loans are not (or only partly) informed about duty waiver and VAT zero rating.

Smallholder farmers are the least informed. They are in an isolated position in the supply chain: not well informed, and not in a position to influence processes in the chain. They lack information on available pump makes, retail prices and price differences between retail outlets, and the availability of spares and after-sales services. They lack information and training on the proper use and preventive maintenance of the pumps.

So, what is the answer to the second question, "How can farmer organizations, government, donors and NGOs address those obstacles to accelerate pump adoption by smallholders?" It is not feasible for retailers to decentralize and set up outlets in rural areas as long as the effective demand for pumps is limited. Neither is it feasible for MFIs to set up branches in rural areas if their potential client base is small or perceived to be high-risk. The government and NGOs can play supportive roles, but also lack sufficient basis in rural areas to reach all potentially interested smallholders.

This conundrum cannot be solved by isolated individual farmers, but organized farmers could be highly effective. The Zambia National Farmers Union (ZNFU), especially, is well decentralized through its district associations and could unlock each component of the pump supply chain. The ZNFU already submits annual proposals to the Ministry of Finance in which they indicate which issues should be addressed in Zambia's annual budget for the benefit of the farming community. They make proposals for which farming inputs should be free of import duty, to stimulate a particular farming sector. They could also negotiate with a supplier to bring in pumps at a negotiated price favourable for the supplier as well as the farmer (or farmers' associations that have the capacity could import pumps themselves). Further, they could make information available in their magazine, *The Zambian Farmer*, on the range of portable pumps that are available for irrigation in Zambia, their specifications, the sources of supply and the retail prices with the different retailers. They could negotiate the supply of pump spares, so that individual farmers do not have to travel long distances to buy parts. They could explore the observed willingness of retailers to train local mechanics in the maintenance and repair of supplied pumps. Farmers' associations could also approach MFIs and provide them with information useful for the MFIs' market research. This could all be done through advertisements in the printed media and through radio and TV, beside publications in the ZNFU magazine. In that way, smallholders will become the pulling force behind the accelerated mechanization of private irrigation in Zambia.

Acknowledgements

This paper was funded by a grant from the Bill & Melinda Gates Foundation. The findings and conclusions contained within are those of the authors and do not necessarily reflect positions or policies of the Bill & Melinda Gates Foundation.

References

Colenbrander, W., Kabwe, A., & Van Koppen, B.. (2011). *Agricultural water management technology adoption in Zambia, findings of a household survey*. Project Report International Water Management Institute (IWMI). Colombo, Sri Lanka: International Water Management Institute.

Customs and Excise (Amendment) Act No. 13 of 2011; Date of Assent 28th December, 2011. Retrieved from http://www.parliament.gov.zm/index.php?option=com_docman&task=doc_view&gid=921

GoZ. (2004). *Irrigation Policy and Strategy, a Strategy for the Development of Zambia's Irrigation Sector*. Lusaka: Ministry of Agriculture and Cooperatives, Government of the Republic of Zambia.

GoZ. (2010). *Water for agriculture and energy development, national investment profile Zambia*. Draft Final Report of investment in water for agriculture and energy in Zambia. Lusaka: Government of the Republic of Zambia (GoZ) and FAO.

Hichaambwa, M. (2010). *Structure of the Lusaka fresh produce market*. Project Report International Water Management Institute. Colombo, Sri Lanka: IWMI.

Mukherji, A., Facon, T., Burke, J., de Fraiture, C., Faurès, J. -M., Füleki, B., . . . Shah, T. (2009). *Revitalizing Asia's irrigation: To sustainably meet tomorrow's food needs*. Colombo, Sri Lanka: IWMI, FAO, Asia-Pacific Water Forum.

RIU. (2007). *The value chain approach to poverty reduction and development of livelihoods*. Online report, Research Into Use. Retrieved from: http://www.researchintouse.com/nrk/RIUinfo/valuechain/valuechain.htm

Shah, T. (2009). *Taming the anarchy: Groundwater governance in South Asia.Resources for the future*, Washington DC (USA) and IWMI.

Statutory Instrument No. 13 of 2003: The Customs and Excise (Suspension) Regulations. Retrieved from http://zamlaws.zambia.co.zm/si/2003.php

Statutory Instrument No. 3 of 2009: Value Added Tax (Zero Rating Order).

Gender aspects of smallholder private groundwater irrigation in Ghana and Zambia

Barbara Van Koppen, Lesley Hope and Willem Colenbrander

This paper explores gender aspects of smallholders' private technology adoption for groundwater irrigation in Ghana and Zambia. It focuses on two variables of quantitative farm-household surveys: household headship and gendered plot management. The paper compares adoption rates and types of technologies for female- and male-headed households; examines adoption rates when women have their own plots; and compares women's decision making on irrigated plots and rainfed plots. The findings suggest that there are largely untapped synergies between gender-equality and irrigation-policy goals. Systematic gender differentiation in surveys is recommended.

Introduction

Background and aim of the study

Smallholders' private investments in groundwater irrigation are gradually being recognized in public irrigation policies in Sub-Saharan Africa (Villholth, 2013). In the same policies, gender equality figures high as well, as manifest, for example, in the gender strategy of the African Ministers' Council on Water (AMCOW, 2011). However, policy makers have hardly tried to link these two policy goals as yet. It is still unclear whether closing the gender gap would support or hinder achieving water-policy goals to improve adoption rates. If men are better technology adopters, as is often assumed, public efforts to provide equal opportunities to both women and men would reduce the overall impact that could have been achieved by targeting only men. If such were the case, there would have been a trade-off between national gender policies and water policies. On the other hand, if women are also keen adopters, water-policy makers will achieve their own goals better by collaborating with their gender-policy colleagues.

Unfortunately, quantitative empirical evidence on how gender matters in the adoption of private farm technologies in general and groundwater irrigation in particular is scarce. This is partly a methodological issue, further explored in this paper. Household surveys, the main methodology for policy research, typically aggregate information collected for individual male and female household members into a gender-blind unit of analysis: the "household".

The existing literature is mainly qualitative, but allows framing the issue of gender in irrigation-technology adoption more precisely for more quantitative analysis. A key message in the scarce existing literature is that women in rural Africa are as productive as men, provided that they have the same access to land and other productive resources (Quisumbing, 2003). However, access to and control over productive resources and technologies tend to be male-biased (Doss, 1999; FAO, IFAD, & ILO, 2010; Meinzen-Dick et al., 2010; Peterman et al., 2011; Safilios-Rothschild, 1994). Men dominate in technology manufacturing, artisanal repair workshops, hardware stores, formal and informal technical training, and fuel stations and contacts with electricity companies. Access to land for long-term land-bound investments and capital and financing facilities are equally male-biased. Because of culturally defined labour obligations on men's fields, women exert less control over the fruits of investments in those fields. Gender gaps may be exacerbated when men enter what were previously women's domains, such as rainfed high-value horticulture crops. Thus, private irrigation adoption by men for horticulture, the most important irrigated crop, may force women out of their domain. Moreover, women have less time, being disproportionately responsible for domestic chores. Last but not least, public government support for technology promotion, technical training, extension services and input and marketing services is often biased toward men, who are easier to reach and are often believed to better serve the goal of technology adoption for irrigation expansion, precisely because of all the mentioned obstacles that women tend to encounter.

In the face of these obstacles, evidence-based answers are even more important. Are women able to overcome these obstacles, and do they adopt irrigation technologies at any scale at all? If not, the gender-equality goal will remain the main driver behind public support to narrow the gender gaps. However, if women already adopt technologies in spite of these obstacles, synergies can be tapped: technology adoption is promoted by gender equality, and vice versa. This paper seeks to provide some evidence for that question, both substantially and methodologically.

Research design

This work seeks to answer the question, "How does gender matter in the adoption of private groundwater-irrigation technologies in Africa?" through gender-differentiation data collection and analysis in quantitative household surveys as the mainstream method in policy research. The work focuses on two variables:

- *household headship* to compare how female-headed households (FHHs) adopt groundwater irrigation, compared to male-headed households (MHHs)
- *plot management* to compare how women adopt groundwater irrigation on their own plots, compared to male-managed plots

The answers are derived from quantitative household surveys on the adoption of agricultural water management technologies in Ghana and Zambia for the cropping season of 2008/2009 or 2009/2010. In both country studies, several regions (three in Ghana and four in Zambia) were selected where adoption rates of agricultural water management technologies were known to be high. This spread over different regions ensured diversity in ecological and socio-economic contexts. This diversity across countries and regions increases the likelihood that patterns are similar in other sites with high technology development in Sub-Saharan Africa.

In these purposively selected regions, a "hut-to-hut" census was conducted among adopters and non-adopters in selected adjacent areas. This census served as a framework for the selection of representative samples for household surveys, either covering every household in adjacent areas (Ghana) or in randomly selected households from the list (Zambia).

With regard to the technologies surveyed, the focus in all three areas in Ghana was exclusively on groundwater lift irrigation and households' combinations of these technologies. In Zambia, all technologies in the selected region were studied, but only for the main technology per household. In over two-thirds of the lift-irrigation technologies used (i.e. buckets and pumps), the source was groundwater. This allowed assessing the relative importance of smallholders' private groundwater development compared to other public and private technologies. The four regions in Zambia were chosen because they were known to have high adoption rates for one of four specific technologies: river diversions (Mpika), motor pumps (Chibombo), conservation agriculture (Monze) and a public irrigation scheme (Sinazongwe). Table 1 gives an overview of these study designs. No statistical analysis was done on the data presented below, so caution in inference is warranted.

De jure FHHs were defined as households led by unmarried women. In de facto FHHs, women were the main providers due to their husbands being sick or handicapped or otherwise unable to provide. Moreover, they included households in which husbands had migrated. In Ghana, 10% of the households in the sample were found to be headed by females. Between 31% and 47% of these were de facto FHHs in which the men were out-migrants, mainly to neighbouring cities. In Zambia, the proportions of FHHs in three districts were 22–25% (both de jure and de facto FHHs). Mpika was an exception, with only 15% FHHs. In both surveys, the number of unmarried male heads of household was negligible.

The enumerators were instructed to interview the "head of household". They typically interpreted the "head" to be a man. Women were only seen as heads if households had no adult man. In MHHs, the enumerators mainly spoke with men (81% of all respondents in both MHHs and FHHs in Ghana; 95% of all MHHs in Zambia). So, the following findings about MHHs represent men's views. Women's perspectives are likely to be underreported. Alternatives for equal representation would have been interviewing both the most

Table 1. Overview of the study designs.

Country and regions	Focus and selection criterion for the region	Sampling procedure and sample
Ghana Ashanti Greater Accra Volta	Focus on all private water-lifting technologies in the household. Purposive selection of regions with the highest prevalence of private lift irrigation.	Hut-to-hut census among 12,620 households in five regions. Hut-to-hut household survey among 494 households from 44 communities in 17 districts.
Zambia Mpika Chibombo Monze Sinazongwe	Focus on all smallholder technologies but only the main technology per household. Purposive selection of districts with highest prevalence of (respectively) river diversions, motor pumps, conservation agriculture and public irrigation schemes.	Hut-to-hut census among 1935 households. Household survey among 240 representative households, randomly selected from the census.

important man and the most important woman of each household (which renders a household survey much more costly), or the enumerators could have interviewed women in half of the households and men in the other half (at no extra cost).

Technology adoption by household head

Ghana: adoption rates and technologies adopted

As shown in Table 2, adoption rates were higher for MHHs (average 80%) than for FHHs (average 63%). The gap was smallest in the Volta region, where 81% of the adopters were MHHs and 73% were FHHs.

Considering the type of technology among the 397 farm households that had adopted water-lifting devices in Table 3, FHHs adopted a bucket more often (55%) than MHHs (39%). For petrol pumps (without and with buckets), adoption rates were virtually similar. For electric pumps, there was a difference: FHHs hardly adopted electric pumps, while 21% of MHHs did adopt electric lifting devices. Thus, FHHs slightly lagged behind in moving to labour-saving mechanization, but only for electric pumps.

Zambia: adoption rates and technologies adopted

In Zambia, a similar pattern was found. Here, 71% of the MHHs had adopted some agricultural water management technology. Adoption rates for FHHs were 55% (see Table 4). Dis-adoption rates (households that took up irrigation but then abandoned it) were substantial, and similar for MHHs and FHHs (21% and 22%, respectively). Very few MHHs were non-adopters (8%). Among FHHs, this proportion was 22%.

Table 2. Adoption rates, by type of household and region, in Ghana.

	Ashanti		Greater Accra		Volta		Total	
	MHH	FHH	MHH	FHH	MHH	FHH	MHH	FHH
N	156	16	108	18	181	15	445	49
Adopters (%)	71	56	93	61	81	73	80	63

Table 3. Adoption rates, by technology, type of household and region, in Ghana.

	Ashanti		Greater Accra		Volta		Total	
Technology	MHH	FHH	MHH	FHH	MHH	FHH	MHH	FHH
N	111	9	100	18	148	11	359	38
Bucket only (%)	35	33	50	66	39	55	39	55
Treadle pump only (%)	0	11	0	0	0	0	0	3
Bucket and treadle pump (%)	0	0	0	0	1	0	0.8	0
Petrol only (%)	23	11	3	8	16	36	15	16
Petrol and bucket (%)	41	45	41	25	2	0	24	23
Electric only (%)	0	0	1	0	30	0	14	0
Electric and bucket (%)	0	0	5	0	12	9	7	3
Bucket, electric and treadle pump (%)	1	0	0	0	0	0	0.2	0

Table 4. Adoption, dis-adoption and non-adoption rates, by type of household, in Zambia.

	N	Adopters (%)	Dis-adopters (%)	Non-adopters (%)	Total (%)
MHH	191	71	21	8	100
FHH	49	55	22	22	100

Table 5. Technology adopted, by type of household and by district, in Zambia.

			Technology adoption rates (%)								
District	Type of household	*N*	Bucket	River diversion	Public irrigation scheme	Motor pump	Wetland	CA[a]	Treadle pump	Other	Total
Mpika	MHH	256	39	50		0	10	1	0		100
	FHH	31	65	23			10			3	100
Chibombo	MHH	559	64			31			5		100
	FHH	151	81			15			4		100
Monze	MHH	139	81			2		8		9	100
	FHH	37	92					3		6	100
Sinazongwe	MHH	149	62		17		18	1	1	2	100
	FHH	31	55		13		32				100

Note: [a]CA: conservation agriculture. Percentages have been rounded. 0 means there were a few adopters but their percentage has been rounded to 0.

All districts mentioned irrigation with buckets as the main technology adopted, except for Mpika, where 50% of the MHHs used private river diversions, with gravity serving as the cheapest energy source; only 39% of MHHs used buckets. For the FHHs in Mpika, buckets were still more important (65%), and river diversions came in the second place (23%). In Mpika, 10% of both MHHs and FHHs used wetlands (see Table 5).

Motor pumps were primarily found in Chibombo, a district with high groundwater tables just 20 kilometres from the fast-expanding vegetable markets of the capital, Lusaka. After buckets, motor pumps were the most important technology, used by 31% of adopter MHHs. Adoption rates of motorized pumps by FHHs were lower, at 15%. In Monze District, conservation agriculture was used by only 8% of the MHHs and 3% of the FHHs. Monze was selected because of its relatively high adoption rates of conservation agriculture; therefore, adoption rates of this technology elsewhere in Zambia are likely to be even lower. Lastly, the findings in Sinazongwe show the relative importance of private groundwater irrigation compared to the public scheme (lift irrigation from a lake), which served only 17% of the MHHs and 13% of the FHHs. Groundwater irrigation with buckets was again the most frequent, and quite similar for MHHs and FHHs. Wetlands were used more frequently by FHHs (32%) than by MHHs (18%).

Thus, the adoption rates of FHHs were two-third of the rates of MHHs. Their technology choice in most but not all regions is somewhat biased toward buckets and wetlands, while MHHs opt twice as often for labour-saving motor pumps and river diversions. Moreover, groundwater irrigation and wetlands appeared to be by far the most important technology, not only in Chibombo (which was selected for its groundwater motor pumps) but also in all other regions, even though they were selected because of relatively high prevalence of other irrigation technologies.

Discussion of findings by household head

In all regions in Ghana and Zambia, the proportion of FHHs that adopted private irrigation technologies tended to be somewhat lower than MHHs. Further, FHHs' choice of technology was somewhat more biased toward labour-intensive manual lifting and wetland cultivation, instead of labour-saving mechanization and gravity irrigation. However, the differences were modest, and there were exceptions. This evidence suggests that many FHHs overcame the mentioned obstacles for women in accessing productive resources. Removing those obstacles may well lead to higher adoption rates. However, more quantitative research is needed to obtain sufficiently large subsamples for statistical significance. This, plus triangulation with qualitative research, will gradually allow better explanations of the differences and hence the public support required.

A straightforward methodological implication is to systematically include the gender of the household head in quantitative survey data collection and analysis. It requires limited extra effort and concerns a significant proportion of households with different characteristics. However, headship of household is only one way of operationalizing gender. The majority of women live in MHHs. This warrants intra-household gender analysis in both FHHs and MHHs, which is the focus of the next section.

Technology adoption by female- and male-managed plot

The plot as proxy for intra-household production subunits

Global evidence on the conceptualization of the household and intra-household gender relations has shown that the concept of the unitary household, in which all activities, assets and resources are pooled under the custodianship of the male head of the household, fails to match reality (Agarwal, 1994; Meinzen-Dick et al., 2010; Quisumbing, 2003; Safilios-Rothschild, 1994). By now, the household is widely recognized as a domain of cooperation and conflict, with intricate forms of exchange and mutual help in both productive and reproductive spheres. The focus here is on productive spheres in rural family farming. Safilios-Rothschild (1988) conceptualized the farm household as consisting of "intra-household production sub-units", in which adult members manage their own enterprises and, therefore, also decide about investments in technologies for higher productivity of those enterprises. Family members bargain with each other about the allocation of labour, assets and benefits. Each adult member tries to "get the best deal out of it". Ownership of land and other assets, women's cultural labour obligations, marital status, age, stage of the household cycle and other factors determine the outcomes (Safilios-Rothschild, 1988). In her classification, most of rural Africa tends to have a "dual farming system" in which both adults manage their own productive activities, with intricate forms of exchange and bargaining. Rainfed horticulture is one such intra-household production subunit.

This classification aligns with Boserup (1970) and many scholars after her (Doss, 1999; Rogers, 1980; Safilios-Rothschild, 1994), who have underlined the legacy of the ancient "hoe" culture in most parts of rural Sub-Saharan Africa: men and women are both producers and both contribute to providing for their dependents. This differs from ancient rural Europe and Asia, where women in the wealthier, land-owning classes became housewives, while the women in the landless classes became wage workers. This history has even

Table 6. Gendered decision making on the income from sales by the owner of an irrigated plot in Zambia.

		Decision maker on the income gained from sales (% of households)		
Owner of irrigated plot	*N*	Female head or wife	Husband	Others
Female household head	14	93	–	7
Husband in MHH	90	24	57	19
Wife in MHH	13	69	15	15

affected gender relations under urbanization and migration (Holden & Mace, 2003). The above sharpens the question as to how gender matters in irrigation adoption, raising questions such as: To what extent do women and men adopt irrigation in their respective production subunits? Are women more likely to adopt irrigation on their own plots? Is irrigation adoption related to changes in bargaining and decision making compared to traditional rainfed cultivation, either in women's or men's production subunits?

The plot was explored as a proxy for an intra-household production subunit that suits quantitative surveys because a plot is a common unit of data collection and analysis. By establishing whether a plot is male- or female-managed (or, in rare cases, jointly managed), gendered adoption patterns can be established in FHHs and MHHs alike. There is extensive documentation of women's plots in patrilineal and matrilineal societies in Sub-Saharan Africa (Peters, 2010; Van Koppen, 2002; Zwarteveen, 1997). Land and water are linked. Access to land is related to control over the use of and income from irrigation-technology investments. Land tenure is important for land-bound irrigation technologies. Land rights have also long been pivotal in gender-equality policies (Lastarria-Cornhiel, 1997).

Accordingly, the Zambian study labelled plots as either male-managed or female-managed production subunits, based on plot ownership. Plot ownership and decision making are related, as key features of a production subunit. Table 6 illustrates the relationship between decision making over the income gained from sales and irrigated plot ownership (with "ownership" referring to the strongest customary claims to the land).

Almost all female household heads owning plots controlled the income from sales (93%). When wives owned plots, they controlled the money in 69% of the cases. This is higher than when husbands were owners and the husbands only controlled the income (57% of the cases). Remarkably, and for unknown reasons, in 24% of the cases, their wives had the stronger say.

These relationships between gendered farm decision making and plot ownership confirmed the plot as a proxy for an intra-household production subunit. Below are presented two applications of plot-related gendered decision making and technology adoption.

Adoption rates on female-managed plots

For comparing irrigation adoption rates on male- and female-managed plots, data were first taken from the Zambian census on the households with women-managed plots, whether irrigated, rainfed or both, in three of the four districts. As shown in Table 7, in 70% to 95% of the FHHs, women had their own plot(s). That percentage is lower in MHHs. In patrilineal

Table 7. Proportion of households with women-managed plots (rainfed, irrigated or both), by type of household and district, in Zambia.

District	Type of household	*N*	Proportion of households with women-managed plots (rainfed, irrigated or both) (%)
Mpika	MHH	346	34
	FHH	61	87
Monze	MHH	205	44
	FHH	57	95
Sinazongwe	MHH	191	47
	FHH	56	70
Total		916	100

Monze and Sinazongwe, almost half of the MHHs (44–47%) have women managing their own plots. However, in Mpika, women had their own plot in only 34% of the MHHs. It is noted that plot ownership is a complex concept in this matrilineal society. If land is being cleared for shifting cultivation, it is referred to by the name of the man. During cultivation and with regard to the produce, the plot is referred to by the name of the woman. As most respondents were men, there may have been a bias towards ownership claims by men. On the other hand, if a household claims to have a plot owned by a woman, it may actually refer to the main household field.

For the question on the relative adoption rates on female-managed plots, only the households with women's plots were considered. Table 8 shows the percentage of households with female-managed plots in which that plot was irrigated, and compares that adoption rate with the overall adoption rates, by household headship and by district, as found in the census.

Table 8 shows that the adoption rates by female heads of households, managing their own plot, were higher than the average adoption rates in FHHs in that region. So, self-management of a plot is positively related to adoption of irrigation in FHHs.

For MHHs, an even stronger relationship between women's self-management and adoption of AWM technologies was found in Sinazongwe. When wives in MHHs had their own plot, they irrigated in 85% of the cases. This is considerably higher than the average of

Table 8. Comparison between irrigation adoption rates on women-managed plots and overall adoption rates, by type of household and district, in Zambia.

		Households with female-managed plots		All households	
District	Type of household	*N*	Irrigation adoption rate (%)	*N*	Irrigation adoption rate (%)
Mpika	MHH	118	49	346	63
	FHH	53	42	61	41
Monze	MHH	90	29	210	53
	FHH	54	52	58	45
Sinazongwe	MHH	85	85	216	54
	FHH	39	69	61	38
Total		**439**		**952**	

MHHs adopting irrigation, which is 54%. Women with their own plots in MHHs irrigated even more often than women with their own plots in FHHs, who irrigated their own plots in 69% of the cases. In contrast, in the MHHs in Monze and Mpika, adoption on women's self-managed plots was less than average overall rates.

In sum, women managed their own plots in between one-third and half of the MHHs, and in 70% to 95% of the FHHs, in the Zambian regions. In FHHs, such self-management was related to higher-than-average adoption rates. For MHHs, there was no clear pattern. In two regions, women-owned plots in MHHs were irrigated less often. However, women with their own plots in Sinazongwe had the highest adoption rates of all households interviewed in the census. Although the samples are too small for any generalized conclusion, one contributing reason could be that women with stronger land rights and a stronger say over the fruits of their labour and investments are keener to adopt irrigation. If so, public interventions for gender equality in land tenure may well boost irrigation adoption also. Gender equality and agricultural growth would go together.

Changes in gendered decision making

Differentiation by plot in the Zambian household survey also shed light on gendered decision making for the new irrigated plots (for any crop) as compared to rainfed-plot adopters and non-adopters (for any crop). The decision studied was the use of the produce of a plot, i.e. whether to sell it or use it for household consumption. Table 9 gives an indication.

Table 9. Gendered decision-making on the use of produce by plot ownership for rainfed and irrigated plots among adopters and non-adopters, and by type of household.

			Decision-maker on the use of the produce (%)				
Owner of plot	N=240	Type of plot	Husband	Wife	Female household head	Others or not applicable	
MHH: Husband	N=41	NON-ADOPTER rainfed plot	71	4	–	24	100
	N=103	ADOPTER rainfed plot	71	6	–	24	100
	N=109	ADOPTER irrigation plot	69	13	–	19	100
MHH: Wife	N=7	NON-ADOPTER rainfed plot	43	43	–	14	100
	N=6	ADOPTER rainfed plot	33	50	–	17	100
	N=15	ADOPTER irrigation plot	20	60	–	20	100
FHH: Female household-head	N=16	NON-ADOPTER rainfed plot	-	–	94	6	100
	N=21	ADOPTER rainfed plot	-	–	100	0	100
	N=20	ADOPTER irrigation plot	-	–	80	20	100

Source: This study.
Note: Non-adopters include dis-adopters.

In MHHs, the decision making of married women on how to use the produce of their own and their husbands' irrigated plots tended to be slightly stronger than for their husbands' and own rainfed plots. Husbands decided in 70% of the cases for their own male-managed plots, both rainfed and irrigated. Wives hardly decided about the use of the produce of their husbands' rainfed plots: just 4% to 6%. However, this was 13% for the irrigated plots managed by their husbands.

On their own plots, wives were also decision makers for irrigated plots more often than for rainfed plots. The proportion of wives' irrigated plots with husbands deciding on the produce was only 20%. Among adopters, husbands decided about the produce of their wives' rainfed plots in 33% of the cases. Among non-adopters, this proportion was even higher: husbands decided for 43% of their wives' rainfed plots. Overall, men had a stronger say over the use of the produce of women-owned plots than women had over the produce of their husbands' plots. However, the difference was smaller for irrigated plots than for rainfed plots. When female household heads were owners, they decided in 80% to 100% of the cases. However, the lowest say of female heads (80% of the cases) was for their irrigated plots. "Others" had somewhat more influence than women for irrigated plots compared to rainfed plots.

These findings suggest that innovation through irrigation is related to women's somewhat stronger voices, either as a causal factor or as an effect. More quantitative research on decision making (differentiated by plot, with larger subsamples) and more qualitative research are needed, both to corroborate the methodology and to generate evidence.

Discussion of findings by female- and male-managed plots

The conceptualization of gender-specific intra-household production subunits as male- or female-managed plots offered opportunities for gender-differentiated data collection and analysis in both MHHs and FHHs. Irrigation adoption rates for women with their own plots depended on headship of household and site. Women's decision making for irrigated plots, either their own or their husbands', seemed somewhat stronger than for rainfed plots.

Conclusions and recommendations

The research findings in this paper confirm, in general, the vital role of smallholders' investments in groundwater for irrigation expansion in Sub-Saharan Africa. Across the seven regions in Ghana and Zambia, from more than half to virtually all smallholders had privately invested in groundwater-irrigation technologies. Even where a public scheme existed, there were six times as many private smallholder investors in groundwater irrigation as public-scheme beneficiaries. Of smallholders, a significant proportion (Ghana) or the majority (Zambia) use buckets. This forms an important potential for further mechanization and expansion of irrigation, a trend already well underway. Further, this study filled methodological and substantive gaps in the study of how gender matters in these developments.

Methodologically, gender-differentiated data collection and analysis were undertaken in household surveys, avoiding the aggregation of gender-specific data collected for smaller units of analysis into a gender-blind "household". Adding the variable of gender of the household head required limited extra efforts in data collection and analysis. This appeared relevant for the up to a quarter of households which are female-headed in both countries. Therefore, headship of household should systematically be included in farm-household surveys. The study in Zambia also included plot-level data collection and analysis. The plot was taken as a proxy for an intra-household production subunit, in which adults negotiate mutual labour allocations and benefits, but in which either a man or a woman was the ultimate decision maker. Establishing decision making for a female- or male-managed plot required more effort, but allowed studying intra-household gendered production relations both in FHHs and in the majority of households: MHHs. Further corroboration of this method, also through qualitative research, for standard inclusion in mainstream farm households is recommended.

Again, these substantive insights from the studies are indicative only, because statistical significance was not assessed, the subsamples were often small, and the issues are complex, with many factors that could be either cause or effect. Nevertheless, the findings were quite similar across all regions (though with local exceptions). Overall, women seem to be proactive irrigation adopters, in spite of the many obstacles they face in accessing productive resources. In terms of headship of household, adoption rates for FHHs were some two-thirds of the adoption rates for MHHs. FHHs tended to adopt labour-intensive buckets somewhat more often than MHHs, while MHHs favoured labour-saving mechanized pumps (and in Zambia, river diversions).

The analysis of intra-household production subunits with plots as a proxy confirmed women's irrigation adoption, also for women in MHHs. The census conducted in Zambia found that women had their own plots in well over two-thirds of FHHs and between one-third and one-half of the MHHs. On their own plots, women in FHHs had consistently higher-than-average adoption rates. In one site (Sinazongwe), women in MHHs with their own plots had the highest adoption rates of all respondents in Zambia. However, adoption rates of married women on their own plots were lower than average in other sites. The comparison between women's farm decision making on irrigated plots compared with rainfed plots showed that women had a somewhat stronger say on irrigated plots. This may be a cause or an effect.

In summary, the study provides some evidence of a link between irrigation and gender policies. Women smallholders in Ghana and Zambia, and probably also elsewhere in Sub-Saharan Africa, are already overcoming the obstacles they face in accessing productive resources, at significant scales. By removing these gender-specific obstacles, water-policy makers will further accelerate adoption rates, and serve their countries' gender-policy goals at the same time. These are not trade-offs; rather, they are synergies between the two policy aims that can be tapped. More gender equality in public support serves irrigation expansion in Africa's dual farming system. As a farm leader in Chibombo said, capturing the benefits of gender equality for his household: "My wife has her plot and I have my plot. Together we prosper."

Acknowledgements

This research was conducted under the AgWater Solutions Project, led by the International Water Management Institute (IWMI), in collaboration with the Food and Agriculture Organization of the United Nations, iDE, the International Food Policy Research Institute and the Stockholm Environment Institute. We gratefully acknowledge the contributions made by Regassa Namara (IWMI), Eric Sarpong Owusu (Food Research Institute, Council for Scientific and Industrial Research, Accra, Ghana, formerly IWMI), Andrew Kabwe (project manager for the Livingstone Green Initiative), Vincent Akamandisa (consultant at Mutakamwa Productions), Mukelabai Ndioyi (Farming Systems Association of Zambia [FASAZ]), Kondwani Gondwe (FASAZ), Lewis Bangwe (FASAZ), field enumerators, and staff of the ministries of agriculture in Ghana and Zambia. We especially thank Meredith Giordano (IWMI) and Charlotte de Fraiture (UNESCO-IHE Institute for Water Education, formerly IWMI) for their support of this study. The Rockefeller Foundation facilitated a highly effective convention in Bellagio on groundwater irrigation in Africa. The field research was funded by the Bill & Melinda Gates Foundation. The findings and conclusions contained within are those of the authors and do not necessarily reflect the positions or policies of the Bill & Melinda Gates Foundation.

References

Agarwal, B. (1994). *A field of one's own: Gender and land rights in South Asia. Cambridge South Asian Studies* (Vol. 58). Cambridge: Cambridge University Press.

AMCOW (African Ministers' Council on Water). (2011). *Policy and strategy for mainstreaming gender in Africa's water sector.* Tunisia: African Ministers' Council on Water (AMCOW). Retrieved from http://www.amcow-online.org/index.php?option=com_content&view=article&id=274&Itemid=143&lang=en

Boserup, E. (1970). *Women's role in economic development.* New York, NY: St. Martin's Press.

Doss, C. R. (1999). *Twenty-five years of research on women in Africa: Lessons and implications for agricultural research institutions - With an annotated bibliography.* CIMMYT Economics Program Paper 99–02. Mexico, DE: International Maize and Wheat Improvement Center (CIMMYT).

FAO (Food and Agriculture Organization of the United Nations), IFAD (International Fund for Agricultural Development) and ILO (International Labour Organization). (2010). *Gender dimensions of agricultural and rural employment: Differentiated pathways out of poverty. Status, trends and gaps.* Rome: FAO, IFAD and ILO.

Holden, C. J., & Mace, R. (2003). Spread of cattle led to the loss of matrilineal descent in Africa: A coevolutionary analysis. *Proceedings of the Royal Society B: Biological Sciences, 270,* 2425–2433.

Lastarria-Cornhiel, S. (1997). Impact of privatization on gender and property rights in Africa. *World Development, 28* (8), 1317–1334.

Meinzen-Dick, R., Quisumbing, A., Behrman, J., Biermayr-Jenzano, P., Wilde, V., Noordeloos, & Beintema, N. (2010). *Engendering agricultural research. IFPRI Discussion Paper 00973. Environment and Production Technology Division.* Washington, DC: International Food Policy Research Institute.

Peterman, A., Behrman, J., & Quisumbing, A. (2011). *A review of empirical evidence on gender differences in non-land agricultural inputs, technology, and services in developing countries. ESA Working Paper No. 11–11. Agricultural Development Economics Division.* Rome: Food and Agriculture Organization of the United Nations (FAO) and International Food Policy Research Institute (IFPRI). Retrieved from www.fao.org/economic/esa

Peters, P. E. (2010). Our daughters inherit our land, but our sons use their wives' fields. Matrilineal-matrilocal land tenure and the New Land Policy in Malawi. *Journal of Eastern African Studies, 4* (1), 179–199.

Quisumbing, A. R. (Ed.). (2003). *Household decisions, gender, and development: Asynthesis of recent research.* Washington, DC: International Food Policy Research Institute (IFPRI).

Rogers, B. (1980). *The domestication of women: Discrimination in developing societies.* New York, NY: Tavistock Publications.

Safilios-Rothschild, C. (1988). *Farming systems and gender issues: Implications for agricultural training and projects* (Unpublished manuscript). Wageningen: Ministry of Agriculture and Fisheries of the Netherlands and the International Agricultural Centre.

Safilios-Rothschild, C. (1994). Agricultural policies and women producers. In A. Adepoju, & C. Oppong(Eds.), *Gender, work and population in Sub-Saharan Africa* (pp. 54–63). London: James Currey.

Van Koppen, B. (2002). *A gender performance indicator for irrigation: Concepts, tools and applications. IWMI Research Report 59.* Colombo: International Water Management Institute (IWMI).

Villholth, K. G. (2013). Groundwater irrigation for smallholders in Sub-Saharan Africa: A synthesis of current knowledge to guide sustainable outcomes. *Water International 38*(4): 369–391.

Zwarteveen, M. Z. (1997). *A plot of one's own: Gender relations and irrigated land allocation policies in Burkina Faso. IWMI Research Report 10.* Colombo: International Irrigation Management Institute.

Integrated assessment of hydrogeology and water quality for groundwater-based irrigation development in the Raya Valley, northern Ethiopia

Tenalem Ayenew, Merhawi GebreEgziabher, Seifu Kebede and Sileshi Mamo

Ambitious plans for expanding the area under irrigation in Ethiopia have not been adequately underpinned by assessments of the actual potential. A detailed hydrogeological study was conducted in one of the focal areas for expansion, in the northern part of the country. The study revealed that there are large groundwater reserves within the valley floor's Quaternary alluvial sediments and underlying Tertiary fractured volcanic rocks. Levels of salinity and sodicity do not pose constraints for agricultural water use. In the valley-floor aquifers there would appear to be sufficient water to support large- and small-scale irrigation development.

Introduction

In many countries groundwater has begun to play an important role in changing the livelihoods of rural communities through groundwater-based irrigation development. Within Sub-Saharan Africa, groundwater use is relatively high in absolute and percentage terms, driven by a supportive policy regime (Giordano, 2006; Kebede, 2013). Ethiopia is lagging, however. At present, only in very few places are farmers using groundwater for irrigation. One such area is the Raya Valley, located in the mountainous regions of northern Ethiopia.

In this study an attempt is made to evaluate the groundwater potential of the Raya Valley through an examination of groundwater quality and subsurface hydrodynamics. Groundwater potential is assessed by detailed hydrogeological mapping, combined with groundwater flow dynamics based on a water-level inventory. Water quality is assessed from the point of view of its use for irrigation purposes, as that is its primary prospective use. Environmental isotopes and hydrochemistry provide ancillary information to understand the groundwater flow system and recharge mechanisms. The focus is on the valley fill aquifers situated within the intermountain graben.

Overview of the study area

The Raya Valley basin is located in the southern part of the Tigray regional state in northern Ethiopia (Figure 1). The basin is bounded by the North-Western Ethiopian Plateau in the

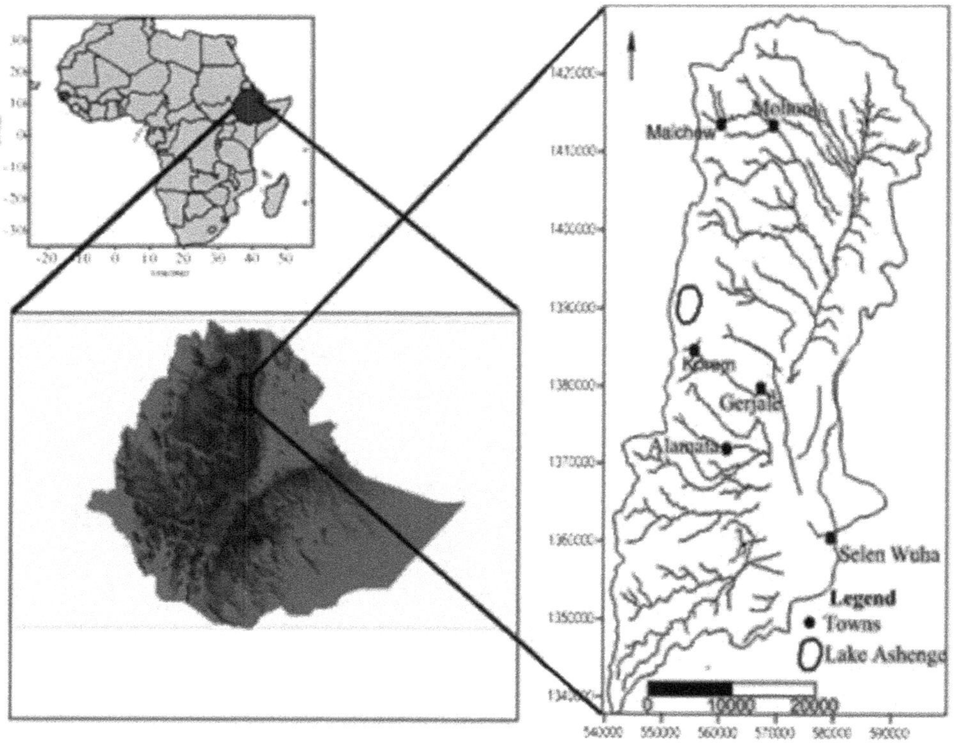

Figure 1. Location map of the Raya Valley basin.

west and the Afar Rift in the east. The focal area is an elongated intermountain graben filled with Quaternary sediments, bordered to the east and west by rugged volcanic mountains with relatively high elevation, ranging from 3820 m asl (above sea level) in the southwestern mountains to around 1310 m asl within the valley floors. North-south-oriented mountains in the west, a steep fault scarp in the east, a major graben and isolated hills within the graben constitute the main physiographic units.

The western mountainous area features sharp peaks and steep gorges highly dissected by streams. The steep slope of the western highlands favours high runoff, and as a result the valley floor is seasonally inundated.

The total land in the project area is estimated to be 2480 km^2. The largest portion (48.2%) is covered by potential crop-production areas, of which 706 km^2 (28.4%) is currently under cultivation. Forest and pasture lands cover 977 km^2 (39.3%) and 323 km^2 (13%), respectively. Water bodies and developed area cover 0.6% and 1.3%, respectively (REST [Relief Society of Tigray], 1998).

Bushes cover the western mountains and the escarpment in the east. In the valley floor there are patchy cactus and acacia trees. Eucalyptus trees are common in settlement areas. Relatively dense acacia trees are common in major settlement areas and towns.

The mean annual precipitation, potential evapotranspiration, actual evapotranspiration and temperature of the area are 813 mm, 1065 mm, 733 mm and 23.3 °C, respectively (Ayenew et al., 2011). The study area is characterized by an erratic, bimodal rainfall pattern. The main rainy season lasts from late June to early September, with the highest rainfall recorded in July and August, whereas the short spring rainy season (*belg*) extends from February to March.

The valley floor is characterized by a semi-arid climate with seasonal scarcity of surface water. Local farming is largely rainfed. In recent years farmers have started to irrigate small farm plots by diverting flash floods from the western side of the mountain. This has changed the lives of people and reduced the effect of the recurrent droughts that are common in the region (REST, 1998). In recent years the government of Ethiopia has drilled many boreholes for irrigation. Most of the boreholes considered in this study were drilled for furrow irrigation.

Geology and hydrostratigraphic units

Geology

The study area comprises a wide range of rocks, ranging in age from Precambrian to Quaternary. The Precambrian rocks include meta-dunite and meta-pyroxinite. These outcrops are situated around the Sulula River outlet. Meta-pyroxinite with pyroxene, olivine, serpentine and opaque minerals and meta-dunite are present below small outcrops of Lower Sandstone in the Michare area of the north-eastern escarpment. There are also a few meta-granite outcrops. From a hydrogeological viewpoint, the Precambrian rocks offer limited development potential.

The Mesozoic sedimentary rocks include the Antalo Limestone and Amba Aradam Formation. The Antalo Limestone is highly jointed and occupies the lower slope of a fault scarp around Gaduwo Stream (Michare area) up to the Daleti area, underlying the Upper Sandstone unit. The Amba Aradom Formation or Upper Sandstone is light pink, medium-to-coarse-grained and friable. Highly weathered beds are evident at the bottom of this unit. A conglomeratic layer toward the base, and sandy and crystalline limestone interbeds are also present. This sandstone is underlain by Antalo Limestone in the eastern escarpment and overlain by Tertiary basalts. This sandstone is also found at the foot of the western mountains, underlying the basalt.

The Tertiary volcanics are predominantly basalt. The basalts are fine-grained and light to dark greenish-grey in colour. They occupy significant parts of the western mountain and escarpment areas. They are slightly to moderately fractured and highly weathered in the upper part. The basalts in the western mountains and eastern escarpment overlie the Upper Sandstone. In the southern adjacent Kobo Valley, Tertiary rhyolites form ridge features.

Quaternary sediments fill the valley floor and form important aquifers. Recent sediments consist of fluvio-lacustrine deposits, which are at places covered with channel and fan deposits, and by colluvial deposits at the foot of the eastern escarpment. The borehole logs clearly display that the Quaternary sediments vary widely in grain size and are characterized by inter-fingering in the valley floor, dominantly deposited by fluvial processes.

The fluvio-lacustrine deposits seem to have been deposited due to the damming of the proto-river during the faulting that gave rise to the eastern escarpment. These sediments are sandy toward the western and eastern sides, and deep (see the geophysics logs in Ayenew et al., 2011), and become silty and clayey toward the centre.

The different rocks are extremely faulted. The faults are prominent in the eastern and western escarpments, forming the normal marginal faults that form the graben. The graben floor is filled with thick Quaternary sediments. The prominent geological structures in the study area are normal faults trending in the NW-SE, NE-SW and N-S directions, and a few E-W faults. Fractures, joints and dykes are common along major tectonic trends.

Hydrostratigraphic units

In the Raya Valley basin, three hydrostratigraphic units can be identified: Quaternary alluvial aquifers; western and eastern Tertiary volcanic aquifers; and basement aquicludes.

Quaternary alluvial aquifers. This hydrostratigraphic unit is localized within the valley floor. The aquifers are composed of intercalating layers of gravels, gravelly sand, silty sand, silty clay and clay. The dominant coarse sand, gravel and boulders are found in the western part of the alluvial deposit, where the energy of the transporting water agents was highest. Therefore, the most productive aquifers are localized in these areas. The grain size of the quaternary deposits decreases from the foot of the western highlands to the eastern part of the valley. The maximum depth is located at the central part of the valley.

Tertiary volcanic aquifers. The Tertiary volcanic basaltic aquifer covers the western part of the study area. The Ashenge Formation is characterized by fissural flood basalts, moderately to highly weathered and fractured, and tilted at angles of 30–40° towards the east (Raya graben). Significant recharge comes to these aquifers both as mountain-front recharge and seasonal floods. More than 85% of the boreholes are drilled in this aquifer. Springs emerging from the western highlands provide moderately high yield during rainy seasons in the escarpment areas.

Aquicludes. The underlying basement complex rocks are considered to be aquicludes as they are not significantly fractured and weathered, with the exception of the highly fractured metamorphic rocks around Selen Wuha (in the graben floor), which is highly affected by a normal fault trending E-W that may localize groundwater at greater depth. No wells are drilled in this unit (Figure 2).

Aquifer hydraulic parameters

The yields of wells drilled in the valley floor are highly variable (REST, 1998). This is in part related to differences in aquifer parameters. Table 1 summarizes the transmissivity and hydraulic conductivity of the different aquifers, mostly drilled in the valley floor. Data from more than 150 boreholes are used here, from previous studies (Hagos, 2005; Hailu, 2010; Nadew, 2003; REST, 1998; Water Work Design and Supervision Enterprise WWDSE, 2008) and from the primary data collected in this study. Hydraulic conductivity and transmissivity decrease in general towards the valley floor as the grain size of the Quaternary sediments decreases. The most productive wells are confined to the western margin of the valley floor, and the yield decreases towards the centre.

Groundwater balance

An attempt was made to assess the water balance of the basin using groundwater flow modelling. The aerial distribution of rainfall was estimated based on long-term (15 years on average) meteorological records (both in the highlands and lowlands) from 11 stations. The Thiessen polygon method was used to estimate the areal depth of rainfall. Evapotranspiration and groundwater recharge were estimated using the WATBAL soil water balance model (Ayenew et al., 2011). The regional semi-distributed hydrometeorological data were used as input to a steady-state groundwater flow model.

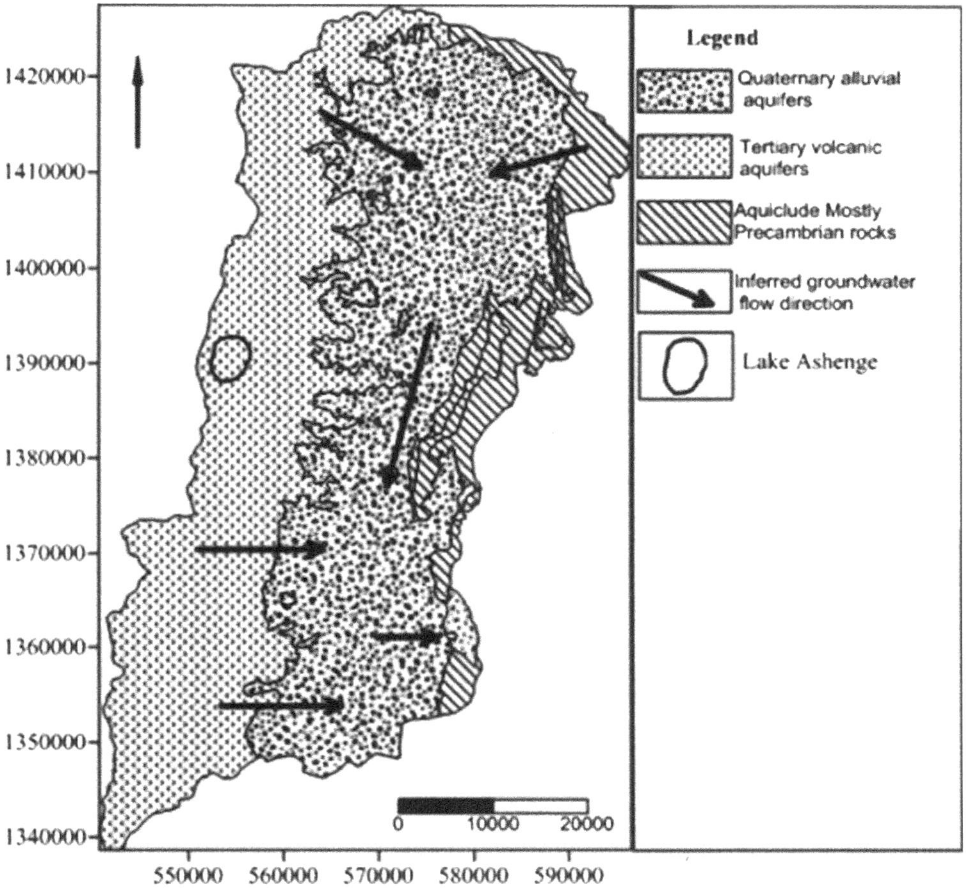

Figure 2. Simplified hydrogeological map of the basin.

Table 1. Summary aquifer parameters and elevation of the static water level for the valley-floor aquifers.

Statistical variables	Standing water level (m asl)	Transmissivity (m^2/day)	Hydraulic conductivity (m/day)
Minimum	1363.6	2	0.01
Maximum	2375.7	5210	124
Mean	1483.5	799	12.6
Standard deviation	96.5	845	15.1

MODFLOW was used to run a steady-state model to quantify important components of the water balance of the basin on a daily basis. MODFLOW is a three-dimensional finite-difference groundwater flow model (McDonald & Harbaugh, 1988). A steady-state model run was made for the whole basin. The distributed recharge for the model was derived from hydrometeorological data, the hydraulic conductivity of the rocks and the topographic factors that influence the recharge process. The pumping test data from over

Table 2. Water balance of the groundwater model (all figures in 10^6 m^3/y).

Water budget components	In	Out	Difference (in – out)
Groundwater recharge	213	0	213
Outflow through Selen Wuha area (GHB)	1551	1752	−201
Outflow through Lake Ashenge	0.21	1.83	−2
Wells (community supply and irrigation wells)	0	5	−5
Springs (Waja and Gerjale)	0	5	−5
Evapotranspiration	0	1	−1

100 wells was used as input for hydraulic conductivity mapping and model calibration. The calibration between the observed and simulated head was found to be very good, with a linear correlation coefficient of 0.99. Table 2 shows the steady-state water balance of the basin.

The most important direct recharge is from rainfall in the highlands. Rainfall is transported to the valley floor as mountain-front recharge and overland flow. The distributed annual recharge in the entire basin is estimated at 213×10^6 m^3. Over 70% of this goes to the valley floor as mountain-front recharge. Direct recharge from rainfall in the valley floor is insignificant.

The major outflow in the modelled area is around the Selen Wuha area in the south. The outflow calculated by the general head boundary package of the flow model is around Selen Wuha and accounts for 201×10^6 m^3 per year. This is expected to go to the southern adjacent Kobo Basin. The groundwater discharge from the basin through the drain package for springs and the well package for pumping wells accounts for 5×10^6 m^3 in total per year. A small swampy area in the valley floor was simulated using evapotranspiration package of the well. The loss from this accounts for 10^6 m^3 per year.

Hydrochemistry

A total of 119 water samples were taken from various groundwater points for hydrochemical and isotopic analysis. A summary of the data is given in Table 3. The results show significant variations in total ionic concentration. This is illustrated with the help of a Piper diagram in Figure 3.

Based on the Piper (trilinear) plot, different water types can be identified: Mg-Ca-HCO_3, Ca-Mg-HCO_3, Ca-HCO_3, Mg-Ca-Na-HCO_3, Ca-Mg-Na-HCO_3, Ca-Na-Mg-HCO_3, Ca-Na-HCO_3 and Na-Mg-Ca-HCO_3. These types of water are found in the western, central and eastern part of the study area. The dominance of calcium and magnesium is mainly associated with the basic volcanic (basaltic) rocks. Sodium is more dominant in the

Table 3. Summary of hydrochemical and isotope data (TDS in mg/L, EC in µS/cm, temperature in °C, anions and cations in mg/L, $\delta^{18}O$ in ‰).

Parameter	pH	EC	Temp.	TDS	Na	Ca	Mg	F	HCO_3	NO_3	SO_4	$\delta^{18}O$
Min	6.44	249	19	265	13	5	6	0.0	131	0	0.2	−65.20
Max	8.78	2730	35	2413	180	210	220	3.83	641	113	975	7.13
Mean	7.54	734	27	583	43	70	36	0.47	368	18	44	−5.16
Standard deviation	0.39	365	3.5	304	27	28	27	0.53	89	22	107	−5.16

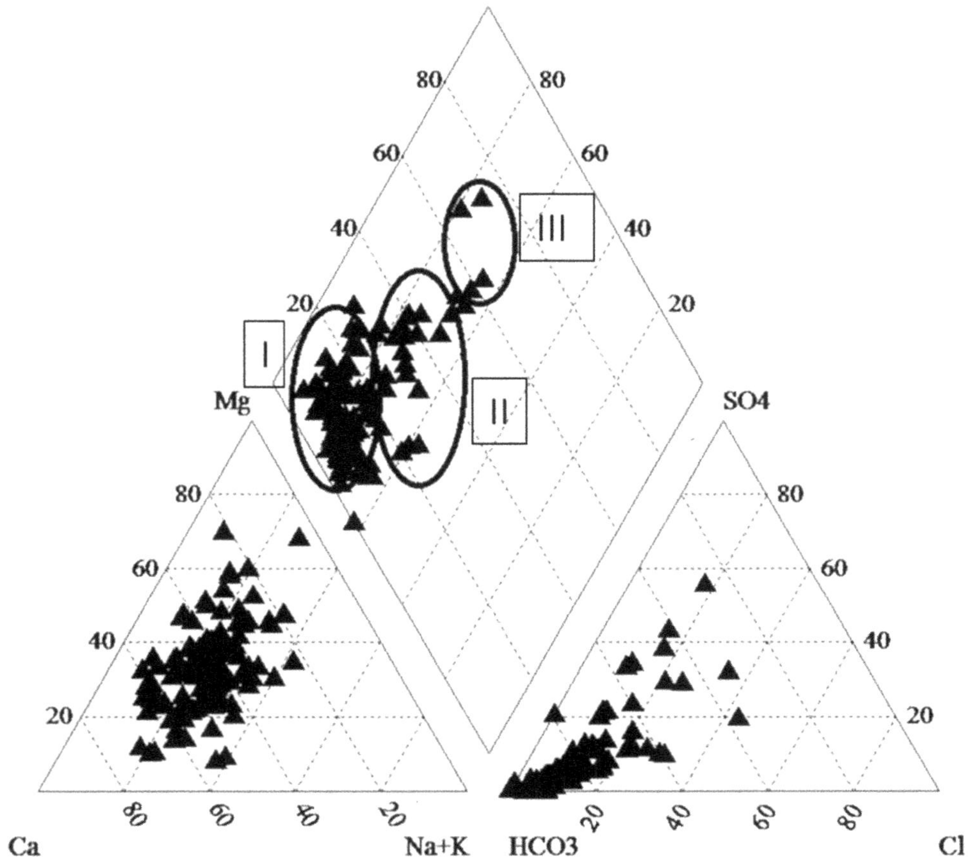

Figure 3. Piper diagram showing the three distinct hydrochemical regimes in the groundwater. I: water from the western highland, characterized by low TDS and EC. II: water from western escarpment and western and central alluvial deposit, characterized by medium TDS and EC. III: water from the eastern alluvial deposit and eastern highlands, characterized by high TDS and EC.

valley floor. Total ionic concentration increases towards the valley floor. The concentration of sodium in the groundwater, especially in the central graben and groundwater outflow regime, is approximately 5–11% of the total cations in the groundwater samples (Ayenew, et al., 2011; GebreEgziabher, 2012).

The majority of the water samples of the Mg-Ca-HCO_3 type, with relatively low total dissolved solids (TDS), are in the highlands and escarpment volcanic terrain. In this area the majority of the groundwater undergoes low rock–water interactions, with low residence times and high throughflow. This water type is found mainly around the areas where recharge takes the form of mountain-front recharge to the valley floors via fractures, dominantly on the western part (Ayenew, Kebede, & Alemyahu, 2007; GebreEgziabher, 2012; Kebede et al., 2008). Some of the geochemical data for Mohoni Basin, in the north-eastern part of the study area, shows Mg-Ca-HCO_3 types of water, indicating that this area is also a recharge area.

The hydrochemistry of the groundwater changes from domination by Mg-Ca-HCO_3 in the recharge area to Ca-Mg-Na-HCO_3 and Ca-Na-Mg-HCO_3 in the eastern escarpment and in the centre and western side of the alluvial deposits. In the Mohoni sub-basin, the

Mg-Ca-HCO_3 water type evolves to Ca-Na-Mg-HCO_3. When it joins the Alamata sub-basin, in the valley floor around the south-eastern part of the study area, the hydrochemistry of the water changes to Mg-Ca-Na-HCO_3. Generally, the ionic concentration, expressed as electrical conductivity (EC) or TDS, increases towards the valley floor.

The hydrochemistry of the Alamata sub-basin changes from Mg-Ca-HCO_3 in the western highland to Ca-Na-Mg-HCO_3 + Ca-Mg-HCO_3 towards the valley floor. This shows that hydrolysis of silicate minerals in volcanic rock with the presence of CO_2 from different sources contributes free sodium in addition to bicarbonate (HCO_3^-) in both sub-basins. Around Selen Wuha, the geochemistry of the water changes from Ca-Na-Mg-HCO_3 + Ca-Mg-HCO_3 to Mg-Ca-Na-HCO_3 and is characterized by high concentration of sulphate (SO_4^{-2}) and chloride (Cl^-). A water sample at Selen Wuha shows a signature of Mg-Ca-Na-SO_4-HCO_3, with very high EC and TDS values of 2730 μS/cm and 2412 mg/L, respectively. This in part indicates that the groundwater outflow from the basin is at Selen Wuha, as also demonstrated in the groundwater flow model. Despite the high EC value, the groundwater around Selen Wuha is fresh, possibly due to flushing by the high rainfall (800 mm/y).

The hydrolysis of silicate minerals in volcanic rocks in the presence of CO_2 from the atmosphere, soil zone and deep sources has resulted in high amounts of sodium (Na^+) and bicarbonate (HCO_3^-) in the groundwater systems in the region (Ayalew, 2009; UNDP, 1973). The high sodium content in the valley floor is likely to be due in part to the reaction:

$$CO_2 + H_2O + \text{Na, K silicates} \rightarrow HCO_3^- + \text{Na, K, H silicates}$$

The leaching of solutes from rocks (sodium) due to water–rock interactions has important implications in irrigation practice, as discussed below.

Environmental isotopes

Environmental isotopes were used together with hydrochemical data to assess the groundwater dynamics within the basin. A total of 110 samples were collected for isotopic analysis from Lake Ashenge (3 samples), springs in the western highlands, escarpment and in the valley floor (11), boreholes and hand-dug wells (64), river (2), swamps (1), flood water (1) and rainfall (28). The result shows significant spatial variations in isotopic signature (Figure 4).

The climate of the region ranges from humid in the highlands to semi-arid in the valley floor, which is characterized by very high evaporation; therefore, secondary evaporation is expected during rainfall. When rain is falling through a dry air column, some molecules of water can evaporate (Yin et al., 2010). Friedman, Machta, and Soller (1962) first showed that evaporation during rainfall would shift the water away from the global meteoric water line (GMWL) and the local meteoric water line (LMWL). The LMWL is not significantly different from the GMWL (Ayenew et al., 2011).

The isotopic composition of precipitation in the area is within the range of –6.03‰ to 1.92‰ for $\delta^{18}O$ (oxygen-18) and –26.4‰ to 27.2‰ for δ^2H (deuterium). The average values of δ^2H and $\delta^{18}O$ in precipitation are –2.18‰ and –1.78‰, respectively. The deuterium values are linearly and positively correlated with the $\delta^{18}O$ value and plot very close to the GMWL.

The value of $\delta^{18}O$ in the groundwater samples ranges from –4.93‰ to 0.6‰, and δ^2H ranges from –24.6‰ to 5.6‰. The mean values of δ^2H and $\delta^{18}O$ are –7.9‰ and –2.22‰,

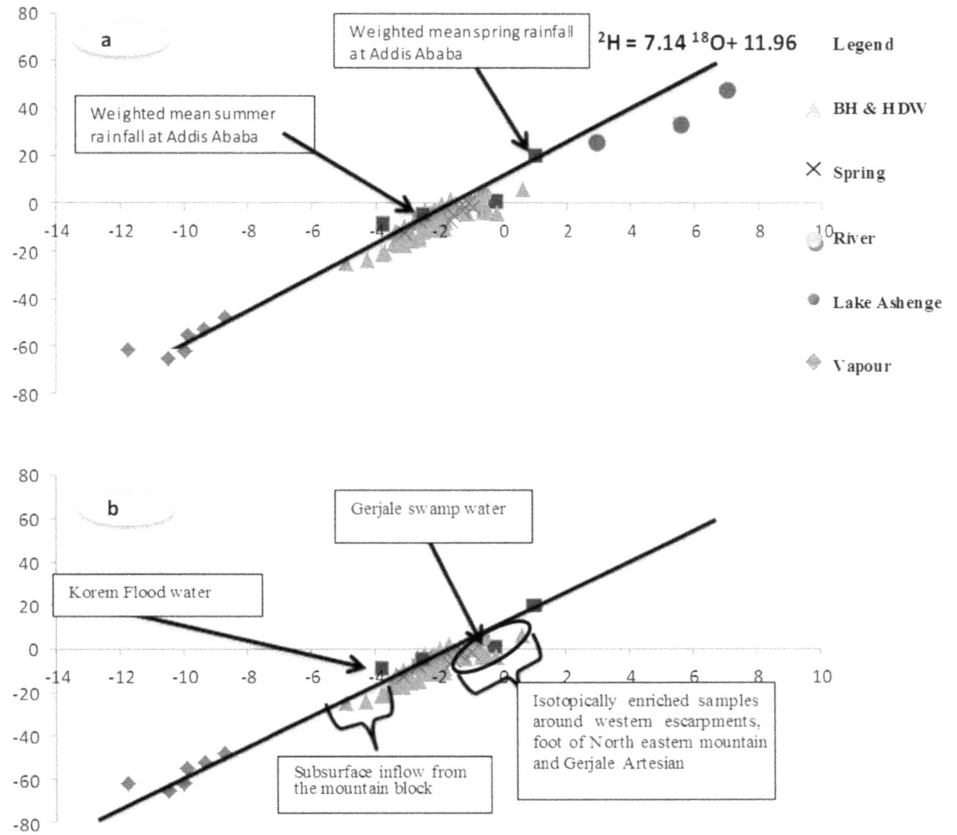

Figure 4. (a) Isotopic compositions of all water samples with weighted mean spring and summer rainfall. (b) Plot of concentrations of deuterium (δ^2H) and oxygen-18 ($\delta^{18}O$) isotopes for groundwater samples classified based on isotopic value.

respectively. δ^2H is linearly and positively correlated with $\delta^{18}O$. All groundwater samples are plotted below and close to the LMWL, which implies the groundwater is of meteoric origin. But the plotting position varies widely, indicating evaporative fractionation prior to recharge, mainly in the valley floor.

Three major groundwater types are identified based on their δ^2H and $\delta^{18}O$ values. Group A groundwater is plotted on the upper right side, showing enrichment. Group B groundwater samples are highly depleted compared to the weighted-mean modern summer rainfall of Addis Ababa and the mean summer rainy season of the Raya Valley. Group C groundwater samples are plotted between Group A (isotopically enriched) and Group B (isotopically depleted).

Group A groundwater is plotted and a bit shifted in the upper-right corner. These samples are dominantly found in the western escarpment, with only one case from the eastern part, near the foot of the mountain, and another from an artesian well. This result can be interpreted as flash-flood water generated from the western highland during the rainy season that undergoes evaporative fractionation prior to recharging the aquifer.

Group B waters are the most isotopically depleted. They are dominantly found in the western mountain block, although some highly depleted samples are found in the north-eastern part and the valley floor. Group B groundwater could be a result of recharge to the groundwater during heavy rainfall events and mountain-block recharge from the highland toward the alluvial aquifer in the form of mountain-front recharge in summer.

The lake water is highly enriched, as is usually the case. It is clearly evident that the groundwater is meteoric in origin. However, a few groundwater samples are highly depleted, probably due to both recharge from non-evaporated flash floods and slight rock–water interactions. Most of the groundwater samples in the western mountain block are isotopically depleted. This indicates that the western mountain block is the principal groundwater recharge zone. There is no clear evidence of old isotopically fractionated groundwater in the valley-floor aquifers.

Groundwater quality for irrigation

Determining water-quality criteria for irrigation water use is not straightforward, as it is dependent on several factors. Among the most important are the chemical composition of the water, the nature and composition of the soil and subsoil, the topography of the land, the amount of water used for (and the methods of) irrigation, the kinds of crops grown, the climate of the region (especially the amount and distribution of rainfall), and the nature of the groundwater and surface-water drainage system (Hoffman, Irmak, Kranz, Martinm, & van Donk, 2010).

Understanding the quality of groundwater for irrigation is critical for managing the groundwater resource and for the long-term agricultural productivity of smallholder farmers in the Raya Valley. Here an attempt is made to assess the quality of the groundwater for irrigation based on a few water-quality variables. The most important variables considered are salinity, sodicity and specific ion toxicities. The water-balance study indicates that there is a good groundwater reserve. A brief account of the groundwater quality is given below.

Salinity

Based on the salinity classification standard of the US Department of Agriculture diagram or Wilcox plots the hydrochemical samples ranged from low to very high salinity (Wilcox, 1955) (Figure 5).

Salinity increases systematically from the highland area towards the lowland area. Very-high-salinity groundwater is situated around the groundwater outflow region in the southern part of the valley floor. However, most of the groundwater from the highland region is characterized by medium to low salinity. In the central graben floor, the salinity generally ranges from medium to high levels. One sample from the groundwater outflow region exhibited very high salinity.

The salinity of groundwater in the south-eastern part of the valley floor limits the mode and extent of irrigation. In particular, those waters in the higher salinity range should not be used for flood or furrow irrigation. In the valley floor close to the escarpment, the salinity is lower.

Sodicity

Sodicity is the amount of sodium relative to that of calcium and magnesium. While high salt content in water leads to the formation of saline soil, high sodium content leads to the development of alkaline soil. Irrigation with sodium-enriched water results in ion-exchange reactions, which cause aggregates of some soils to break down or disperse, leading to sealing of soil pores and a reduction in the permeability of the soil.

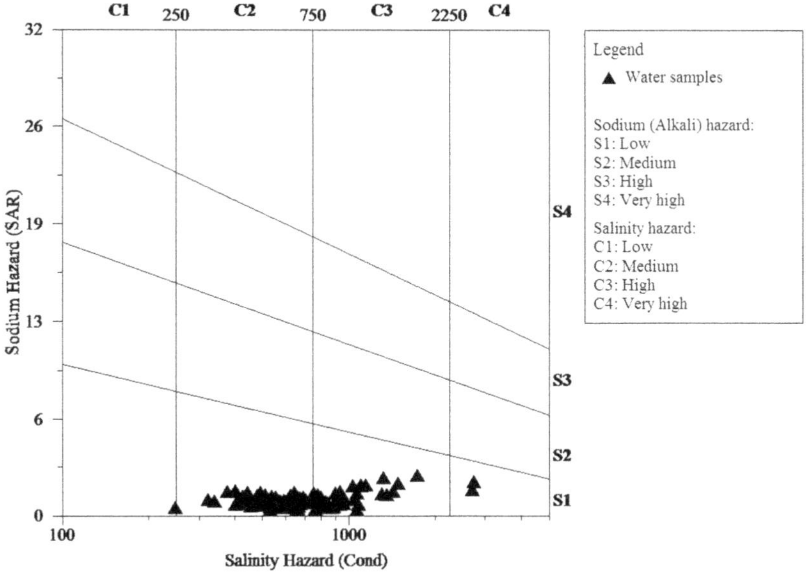

Figure 5. Wilcox diagram classification of salinity and sodicity levels to determine water quality for irrigation.

The sodium level in water used in irrigation is determined by the absolute and relative concentrations of sodium, calcium and magnesium (Hoffman et al., 2010). The sodium absorption ratio (SAR) is given by

$$\mathrm{SAR} = \frac{\mathrm{Na}^2}{\sqrt{\frac{\mathrm{Ca}^2 + \mathrm{Mg}^2}{2}}} \tag{1}$$

Ionic concentration is given in milliequivalents per liter (meq/L).

There is a strong positive correlation between SAR values of irrigation water and the extent to which sodium is absorbed by the soils. Continued use of water with a high SAR value leads to a breakdown in the physical structure of the soil caused by excessive amounts of collodially absorbed sodium. This breakdown results in the dispersion of soil clay, causing the soil to become hard and compact when dry and increasingly impervious to water penetration (due to dispersion and swelling) when wet. Fine-textured soils such as clays are especially vulnerable. The central part of the valley floor has more clay-dominated soils. Therefore, this needs special attention during irrigation.

The water samples are grouped based on SAR and EC (Figure 5). In general, most of the water samples fall in the low-sodium zone. There is a slight increment in the amount of sodium in the groundwater along the groundwater flow direction, i.e. in the southern valley floor. This is due to interaction with weathered basaltic rocks in the presence of free carbon dioxide (UNDP, 1973). In general no high sodium levels have been identified in the limited amount of groundwater currently used for irrigation in the basin.

Conclusions

This comprehensive and integrated detailed hydrogeological investigation in the Raya Valley basin enabled an understanding of the groundwater potential, the groundwater flow dynamics and the quality of the groundwater. Hydrochemical and isotope techniques enable the understanding of the groundwater flow regime and source of recharge.

Conventional hydrogeological mapping and pumping test analysis indicates that there are two important exploitable hydrostratigraphic units: (1) the valley floor aquifers, which are dominantly Quaternary sediments; and (2) the fractured volcanic aquifers in the highlands and underlying the valley floor. The aquifer is predominantly unconfined, so it can easily be used by smallholder farmers through hand-dug wells at depths of less than 50 metres. Deep wells can also be developed.

The simple steady-state groundwater flow model simulation indicates that the groundwater recharge of the entire basin is 213×10^6 m^3 per year. There are strong interactions among local precipitation, Lake Ashenge, local intermittent rivers and groundwater. The swampy area is believed to discharge 10^6 m^3 per year from groundwater as evapotranspiration.

The hydrochemical signature displays large spatial variability. In general, the total ionic concentration increases towards the valley floor. The hydrochemistry of the groundwater changes from Mg-Ca-HCO_3 dominance in the recharge area to Ca-Mg-Na-HCO_3 and Ca-Na-Mg-HCO_3 in the eastern escarpment and central and western side of the alluvial deposits. At Selen Wuha the hydrochemistry of the groundwater is Mg-Ca-Na-SO_4-HCO_3 and the EC and the TDS values are quite high. The highland groundwater is of high quality, of Mg-Ca-HCO_3 type with relatively low TDS. This indicates that the majority of the groundwater is flushed and young and undergoes limited rock–water interactions.

All groundwater samples are plotted just below the LMWL. This indicates that the groundwater's origin is meteoric and that it undergoes slight evaporative fractionation prior to recharge. Based on the Wilcox diagram classification, most of the highland and central graben groundwater is moderately saline, which is more or less suitable for careful furrow irrigation. However, groundwater from the eastern escarpment is characterized by high salinity and therefore not suitable, especially for furrow (flood) irrigation. There is however no significant sodicity in the study area. With these qualifications, this study reveals high groundwater potential, basically derived from highland rainfall.

Acknowledgments

The authors are grateful for the support provided by the International Water Management Institute and research funding made available from the Rockefeller Foundation through project number 2008 AGR 305. Numerous institutions helped by providing data and analyzing samples. They include Addis Ababa University, University of Avignon (Laboratory of Hydrogeology), International Atomic Energy Agency, Water Works Design and Supervision Enterprise, Geological Survey of Ethiopia and Bahir Dar University.

References

Ayalew, E. (2009). *Growing lake with growing problems: Integrated hydrogeological Investigation on Lake Beseka, Ethiopia*. (Published Doctoral Dissertation). University of Bonn, Germany.

Ayenew, T., Kebede, S., & Alemyahu, T. (2007). Environmental isotopes and hydrochemical study as applied to surface water and groundwater interaction in the Awash River basin. *Hydrological Processes*. doi:10.1007/s00254-007-0914-4

Ayenew, T., Mamo, S., Ali, S., Abdu, A., GebreEgziabher, M., & Desalegn, A. (2011). *Hydrogeological and groundwater potential evaluation, water quality assessment and socioeconomic surveying for small-scale irrigation development in the Raya-Kobo Valleys, Northern Ethiopia*. Report to the International Water Management Institute, Ethiopia, September 2011.

Friedman, O., Machta, L., & Soller, R. (1962). Water vapour exchange between a water droplet and its environment. *J Geophys Res*, *67*, 2761–2766.

GebreEgziabher, M. (2012). *An Integrated Hydrogeological study to understand the groundwater flow dynamics of Raya Valley Basin, Northern Ethiopia. Hydrochemistry, Isotope Hydrology, and Flow Modeling Approaches* (Unpublished Master's Thesis). Addis Ababa University, Ethiopia.

Giordano, M. (2006). Agricultural groundwater use and rural livelihoods in sub-Saharan Africa: A first-cut assessment. *Hydrogeology Journal*, *14*, 310–318.

Hagos, E. (2005). *Hydrogeology of Mehoni sub-basin and Lake Ashenge catchment in the Raya Valley Northern Ethiopia* (Unpublished Master's Thesis). Addis Ababa University, Ethiopia.

Hailu, A. (2010). *Aquifer Characterization of the Raya Valley based on pumping test and geophysical data analysis, Northern Ethiopia* (Unpublished Master's Thesis). Addis Ababa University, Ethiopia.

Hoffman, G. J., Irmak, S., Kranz, W. L., Martinm, D. L., & van Donk, S. J. (2010). *Water quality criteria for irrigation. University of Nebraska-Lincoln Extension*. Retrieved from http://ianrpubs.unl.edu/epublic/live/ec782/build/ec782.pdf

Kebede, S. (2013). *Groundwater in Ethiopia: Features, vital numbers and opportunities*, 297 pp., Springer Hydrogeology Series, ISBN: 978-3-642-30390-6 (Print) 978-3-642-30391-3 (Online).

Kebede, S., Travi, Y., Asrat, A., Alemayehu, T., Ayenew, T., & Tessema, Z. (2008). Groundwater origin and flow along selected transects in Ethiopian rift volcanic aquifers. *Hydrogeology Journal*, *16*, 55–73.

McDonald, M. G., & Harbaugh, A. W. (1988). *A modular three-dimensional finite-difference groundwater flow model*. U.S. Geological Survey Techniques of Water-Resources Investigations Book 6, Chapter A1, 586 p.

Nadew, D. (2003). *Aquifer characterization and hydrochemical investigation in Raya Valley* (Unpublished Doctoral Dissertation). University of Natural and Applied Life Sciences (Boko) Vienna, Austria.

Relief Society of Tigray (REST). (1998). *Raya Valley Development Study Project: Feasibility study report, Hydrogeology*. Mekelle, Ethiopia.

UNDP (1973). *Geology, geochemistry and hydrology of hot springs of the East African rift system within Ethiopia*. New York: United Nations.

Water Work Design and Supervision Enterprise (WWDSE). (2008). *Geological investigation report on Raya valley basin, Northern Ethiopia*. Unpublished report.

Wilcox, L. V. (1955). *Classification and use of irrigation water*. Circular 969. US Department of Agriculture.

Yin, L., Hou, G., Su, X. S., Wang, D., Dong, J., Hao, Y., & Wang, X. (2010). Isotopes (dD and d18O) in precipitation, groundwater and surface water in the Ordos Plateau, China: Implications with respect to groundwater recharge and circulation. *Hydrogeology Journal*. doi:10.1007/s10040-010-0671-4

Cost-benefit analysis and ideas for cost sharing of groundwater irrigation: evidence from north-eastern Ethiopia

Gebrehaweria Gebregziabher, Karen G. Villholth, Munir A. Hanjra, Muleta Yirga and Regassa E. Namara

The government of Ethiopia has invested in groundwater development for smallholder irrigation in the Raya Valley and Kobo Valley, north-eastern Ethiopia, where the hydrogeological potential is large but not fully developed. A cost-benefit analysis shows that investment in deep groundwater irrigation development is viable at a 9.5% discount rate in 75% of the wells. Assuming full cost recovery of capital investment, the annual payment rates (annuity) that irrigation users should pay over the wells' service life (25 years) were estimated. It is recommended that future investment be based on cost sharing rather than full cost recovery to facilitate uptake and address financial realities.

Introduction

Agriculture is the main sector of the Ethiopian economy. It contributes approximately 42% of the GDP, generates more than 85% of the foreign exchange earnings and employs over 80% of the population (CSA, 2004; MoFED, 2010). The government of Ethiopia is committed to rapid agricultural growth as a means of accelerating economic growth and reducing poverty (Hanjra, Ferede, & Gutta, 2009). Despite impressive achievements over the last two decades, Ethiopia remains one of the poorest countries in the world, with over 12 million people still experiencing food insecurity. Agricultural productivity is constrained by high climate variability (Ferede, Ayenew, & Hanjra, 2013) rather than low water availability. Rainfall distribution is extremely uneven, both spatially and temporally, which has negative implications for the livelihood of the population (FAO, 2005). Drought frequently results in crop failure, low agricultural productivity and high levels of food insecurity (Deressa, 2007; Lautze et al., 2003).

The problem is particularly serious in the moisture-stressed northern highlands (Hagos, Pender, & Gebreselassie, 1999; Pender & Gebremedhin, 2004, 2008). As part of a strategy to address these problems, groundwater irrigation has been given greater emphasis in the Raya Valley and Kobo Valley (here after Raya-Kobo Valley), which crosses the Tigray and Amhara Regions. For example, the Golgol Raya Development Project and the Kobo-Girana

Valley Development Project were established in the 1990s to develop groundwater irrigation in the Raya-Kobo Valley. In line with this, a number of deep wells were drilled and equipped with drip or sprinkler systems and electric pumps. While the capital investment was financed by the government (both federal and regional governments), the end users were smallholder farmers. Irrigation infrastructure was transferred to smallholder farmer groups through water user associations (WUAs). In many public-funded irrigation projects, capital investment, operation and maintenance costs are high, while beneficiaries' willingness and ability to pay these costs are low (Easter & Liu, 2005). This poses large risks for the financial feasibility and sustainability of such projects.

Investment costs include borehole drilling, casing, electric power connection and power house, and installation of drip and sprinkler distribution systems. Although the cost of well drilling showed a declining trend between 1998 and 2004 (Moges, 2011), this was to the detriment of quality, mainly due to poor steel casing and lack of standard design and supervision guidelines. Despite some discrepancies in estimates (Awulachew et al., 2007; Kebede, 2010; Kobo-Girana Valley Development Project (KGVDP), 2009; Moges, 2011; Relief Society of Tigray (REST), 1996), the groundwater potential in the Raya-Kobo Valley is substantial, and sufficient to irrigate about 20% (35,000 ha) of the valley, implying that the current irrigation use of less than 1000 ha is far below the potential.

The Ethiopian Water Resources Management Proclamation (FDRE, 2000) stipulates that farmers have to pay fees for water use and share part of the investment costs. However, the implementation of this is in the early stages. The idea of cost sharing emerges for two reasons. First, beneficiaries have to cover part of the capital investments in addition to the full cost of operation and maintenance so that the project is financially sustainable. Second, the idea of cost sharing emerges from a water use efficiency perspective, because when farmers pay for the assets they use (groundwater resource and irrigation infrastructure), ideally, they value them and try to use them efficiently, potentially maximizing the return per unit of water and per unit of investment. This is in line with the literature. For example, a study from Kenya (Marks & Davis, 2012) reported that rural households' contributions for the construction of a rural water supply system were associated with a twofold increase in their sense of ownership. Moreover, Easter and Hearne (1993) demonstrated that user participation in irrigation management and financing established a sense of ownership among farmers and increased transparency as well as accountability in decision making. Since cost sharing is a missing element so far in the Ethiopian irrigation subsector, the contribution of this study is largely in relation to proposing strategies to inform policy making.

To the best of our knowledge, no study has been made to assess the costs and benefits of groundwater irrigation investments in the Raya-Kobo Valley. However, Godswill and Krishna (2006) found that gross income per hectare of borehole users in the Alamata *woreda*[1] within the Raya-Kobo Valley was higher than that of non-users.

In this context, the study has two main objectives. The first objective is to evaluate the viability of investments in groundwater-based irrigation in the Raya-Kobo Valley in north-eastern Ethiopia. The second, given that cost-benefit analysis shows the viability of such investment, is to estimate annual instalment (annuity) rates that users can afford to pay as a part of a cost-sharing plan.

Data and methodology

The study area (the Raya-Kobo Valley) is located in north-eastern Ethiopia. It covers two *woreda*s (Alamata and Raya-Azebo) in the lowland part of the southern Tigray Region and one (Kobo) in the Amhara Region (Figure 1).

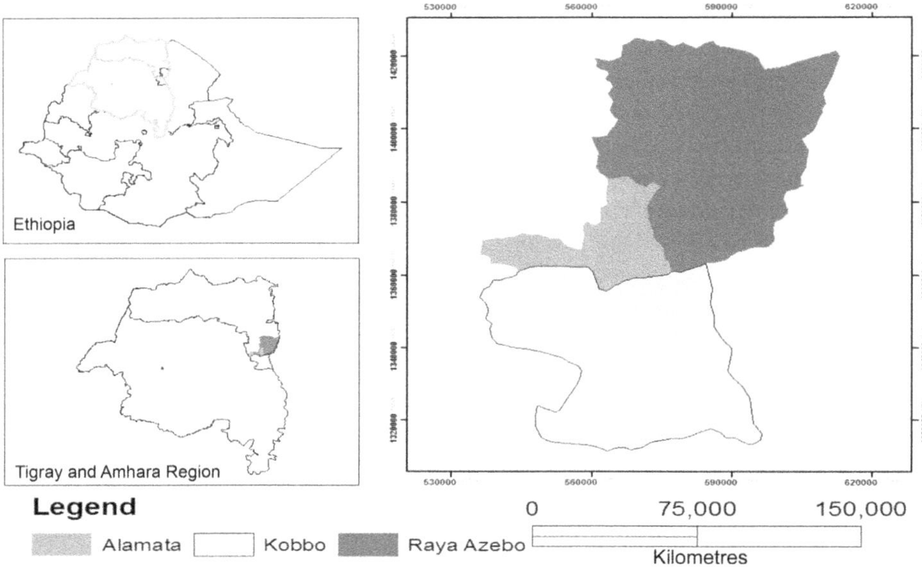

Figure 1. Map of the study area: Raya-Kobo Valley, consisting of the three *woreda*s Alamata, Raya Azebo (Raya Valley) and Kobo (Kobo Valley) in north-eastern Ethiopia.

The study area is located within UTM 37P (11°56′ N 39°16′ E to 12°45′ N 39°54′ E) and covers an area of 3640 km^2. As in other parts of rural Ethiopia, the livelihood of farmers in the study area is based on traditional smallholder farming systems.

The data used for the cost-benefit analysis were acquired from the Kobo-Giranna Valley Development Project office, the Relief Society of Tigray, the agriculture offices of the Alamata and Raya Azebo *woreda*s, and the Tekeze deep well drilling company. These organizations are actively involved in groundwater development, irrigation management, and extension and marketing services. The study focuses on 20 deep wells (9 in the Raya and 11 in the Kobo Valley) installed between 1997 and 2000. These wells are used to irrigate 292 ha in Raya and 460 ha in Kobo, benefitting 1044 and 1540 farm households, respectively. Table 1 describes the wells. The well depth and the area that each well can irrigate varies between 91 and 130 metres and between 30 and 50 ha, respectively. In the Raya Valley, the average investment for a well (average depth 129 m, irrigating about 32 ha) was ETB 3.7 million (equivalent to USD 202,000 at the current exchange rate[2]), implying an average capital investment per hectare of ETB 115,000 (USD 6250) (Table 1). Similarly, in the Kobo Valley, the average investment costs per well (average depth 108 m) and per hectare were found to be ETB 3.0 million (USD 163,000) and ETB 71,500 (USD 3900), respectively.

Ownership of the irrigation infrastructure is partially public, because the initial capital investment was made by the public sector, while the actual users are smallholder farmers who have access to land in the command area. Management of the irrigation systems is transferred to the users, who are expected to contribute, based on the size of their land, to cover maintenance and operation costs. The size of irrigated landholding per household depends on the size of the command area that each well can irrigate and the number of beneficiary households, and varies between 0.21 and 0.60 ha. The main irrigated crops produced are vegetables, while maize, pulses and fruits are grown on a limited scale.

Table 1. Description of all the wells used in this analysis.

Borehole name/site	Management	Average depth (m)	Average area of irrigated land (ha)	Number of beneficiary households	Average area of irrigated land per household (ha)	Total investment cost (ETB/well)	Average annual gross income per household per year (ETB)
Raya Valley							
Alamata (WF 0 BH-6)–S. Bikalsi	WUA	128.5	36.0	160	0.23	4,889,254	12,620
Alamata (WF 0 BH-5)–S. Bikalsi	WUA	128.5	36.0	165	0.22	4,709,040	12,237
Alamata (WF 0 BH-7)–S. Bikalsi	WUA	128.5	36.0	160	0.23	4,726,183	12,620
Raya-Azebo (WF 13 BH-1)–Wargba	WUA	128.5	30.7	111	0.25	4,683,641	32,911
Raya-Azebo (WF 13 BH-2)–Wargba	WUA	128.5	30.7	122	0.31	4,769,521	29,944
Raya-Azebo (WF 9 BH-1)–Warabaye	WUA	128.5	30.7	99	0.26	4,656,599	36,900
Raya -Azebo (WF 9 BH-21)–Karra Adishuhu	WUA	128.5	30.7	119	0.60	1,832,582	23,877
Raya -Azebo (WF 9 BH-3)–Karra Adishuhu	WUA	128.5	30.7	51	0.54	1,570,785	47,753
Raya -Azebo (WF 9 BH-4)–Karra Adishuhu	WUA	128.5	30.7	57	0.42	1,570,785	42,727
Average (Raya Valley)		**128.5**	**32.4**	**116**	**0.32**	**3,712,043**	**23,687**
Kobo Valley							
HG1	WUA & KGVDP	112.0	34.0	167	0.20	2,093,542	3,217
HG2	WUA & KGVDP	91.0	40.0	169	0.24	2,772,958	3,380
HG4	WUA & KGVDP	109.0	50.0	167	0.30	3,995,330	4,622
HG6	WUA & KGVDP	101.0	48.0	168	0.29	3,580,965	8,371
HG7	WUA & KGVDP	105.5	48.0	119	0.40	3,564,686	10,976
HG8	WUA & KGVDP	110.0	42.0	200	0.21	2,666,378	6,558
HG10	WUA & KGVDP	100.0	30.0	72	0.42	2,182,225	10,462
WG1	WUA & KGVDP	110.4	41.0	154	0.27	2,798,879	4,190
WG2	WUA & KGVDP	130.0	31.0	63	0.49	2,294,318	15,602
WG4	WUA & KGVDP	111.0	48.0	167	0.29	3,341,970	5,469
WG5	WUA & KGVDP	111.0	48.0	94	0.51	3,606,309	7,508
Average (Kobo Valley)		**108.3**	**41.8**	**140**	**0.33**	**2,990,687**	**6,432**
Average (total)		**117.4**	**37.6**	**129**	**0.32**	**3,210,620**	**13,403**

The UNIDO (1972) approach, based upon determining the net present value and internal rate of return as specified below, was used for the cost-benefit analysis. The UNIDO approach is the one commonly used for project cost-benefit analysis, especially in developing countries.

$$\mathrm{NPV} = \sum_{t=1}^{n} \frac{(B_t - C_t)}{(1+r)^t} \tag{1}$$

In Equation (1), NPV stands for net present value; n is the time period (25 years), B_t is benefits at time t (measured in years); C_t is cost at time t; and r is the real discount rate. The NPV is the present value of net cash inflows generated by a project. It is a reliable measure used in capital budgeting because it accounts for temporal changes in the value of money by using discounted cash inflows.

The internal rate of return (IRR) is defined as the rate of return on an investment which will equate the net present value of benefits and costs (such that the net present value of the investment becomes zero). The IRR is found by an iterative process and is equivalent to the discount rate (r) that satisfies the following equation (Dixon, Carpenter, Fallon, Sherman, & Manipomoke, 1986):

$$\sum_{t=1}^{n} \frac{B_t}{(1+r)^t} = \sum_{t=1}^{n} \frac{C_t}{(1+r)^t} \tag{2}$$

Strictly speaking, the IRR does not give one the discount rate; it merely finds the value of r that meets the above condition of zero net present value. For convenience, the ETB was used as the common unit for the cost-benefit analysis.

Costs considered in the analysis include: fixed investment costs (including well drilling, pump and installation, electric power-line installation, drip or sprinkler system installation, power house, and other costs related to design and study); operating and maintenance costs; and production (variable) costs, such as for seed, fertilizer, pesticide, labour and energy (Figure 2).

Due to lack of data, costs related to supervision and extension support by the *woreda* agriculture offices were not considered in the cost-benefit analysis. Market value of irrigated crops as reported by respondents was used to estimate benefits. Official interest rates from the National Bank of Ethiopia were used for discount rates. Effective December 2010, the Development Bank of Ethiopia (DBE, 2010) increased the lending interest rate from 8% to 8.5% for priority investment sectors and to 9.5% for non-priority sectors. Investment in agriculture in general and irrigation in particular is among the priority investments. In line with this, discount rates of 8%, 8.5% and 9.5% were applied to compare and perform sensitivity analysis related to financial viability. The sensitivity analysis was important to test and understand the robustness of the feasibility analysis in the face of dynamic interest rate regimes. The service life of the deep wells is assumed to be 25 years, as adopted by the Relief Society of Tigray (Tesfay, 2008).

Once the feasibility of groundwater irrigation investments was assessed in terms of NPV and IRR, the average annual instalment (annuity) that beneficiaries should pay to recover the full cost of the capital investment was estimated. The estimation procedure is specified as follows:

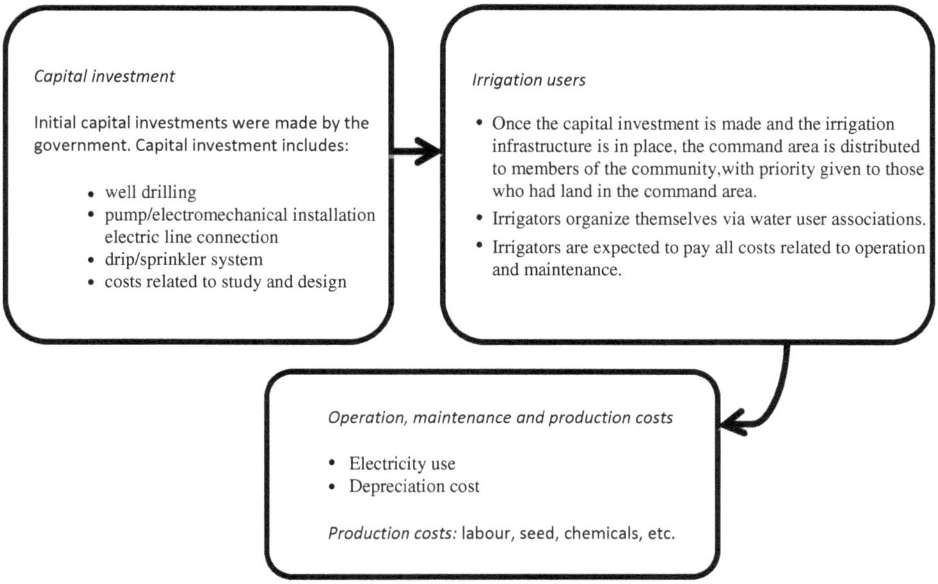

Figure 2. Implementation and operation components and costs for the Raya-Kobo groundwater irrigation system, north-eastern Ethiopia.

$$A = \frac{NPV}{\frac{1-(1+i)^{-n}}{i}} = NPV\frac{i}{1-(1+i)^{-n}} \quad (3)$$

where A is the annual instalment payment (annuity), NPV is the net present value of capital investment, i is the interest rate per time period, and n is the time period (25 years). In the cost recovery analysis (using the annual instalment (annuity)), only the capital investment costs are included, as farmers have, at least in principle, already covered operational costs.

Because establishment cost and productivity are not uniform across wells, meaning that the cost to irrigate a given area of land and the area of land that each well can irrigate vary, annuity was estimated both per well and per hectare. Using a quarter-hectare as the average irrigated landholding per household, the annual instalment (annuity) per household was also estimated.

Finally, a detailed analysis and recommendations for the implementation of full cost recovery and partial cost sharing are presented.

Results and discussion

Financial feasibility and cost recovery analysis

Table 2 presents the financial feasibility analysis at the well level using the three discount rates (8%, 8.5% and 9.5%). Except for five wells (HG1, HG2, HG4, WG1 and WG5) in the project area, the NPV of the groundwater irrigation wells was found to be positive at a 9.5% discount rate, implying that investment in groundwater irrigation in the study area is generally profitable even at the highest discount rate. Similarly, except for the same five wells mentioned above, the IRR was found to be higher than the highest lending interest rate (9.5%), also indicating that investment in groundwater irrigation is feasible even at interest rates higher than 9.5%.

Table 2. Sensitivity analysis of financial feasibility at the well level (in ETB).

Project name	NPV at 8%	NPV at 8.5%	NPV at 9.5%	IRR
P1 Selam Bikalsi (WF 0 BH-6)	11,750,110	11,063,345	9,821,620	0.32
P2 Selam Bikalsi (WF 0 BH-5)	11,830,503	11,147,858	9,913,582	0.33
P3 Selam Bikalsi (WF 0 BH-7)	11,913,181	11,226,417	9,984,692	0.33
P4 Wargaba (WF 13 BH-1)	17,441,385	16,528,208	14,877,111	0.44
P5 Wargaba (WF 13 BH-2)	17,135,898	16,231,785	14,597,077	0.43
P6 Warabaye (WF 9 BH-1)	17,708,002	16,784,937	15,115,962	0.45
P7 Karra Adishuhu (WF 9 BH-21)	11,466,875	10,917,960	9,925,478	0.68
P8 Karra Adishuhu (WF 9 BH-3)	8,753,316	8,327,204	7,556,760	0.62
P9 Karra Adishuhu (WF 9 BH-4)	8,633,530	8,212,362	7,450,858	0.61
Average (Raya Valley)	**12,959,200**	**12,271,120**	**11,027,016**	**0.47**
HG1	−1,337,921	−1,369,108	−1,425,497	0.01
HG2	−1,960,855	−1,994,373	−2,054,977	0.03
HG4	−1,264,623	−1,377,329	−1,581,110	0.04
HG6	5,783,384	5,396,884	4,698,063	0.24
HG7	5,709,181	5,326,416	4,634,347	0.24
HG8	5,526,063	5,187,932	4,576,565	0.29
HG10	3,051,582	2,835,565	2,444,988	0.22
WG1	−922,468	−999,914	−1,139,943	0.04
WG2	5,458,977	5,138,972	4,560,376	0.32
WG4	796,248	625,449	316,631	0.11
WG5	−231,055	−370,364	−622,245	0.07
Average (Kobo Valley)	**1,873,501**	**1,672,739**	**1,309,745**	**0.14**
Average (total)	6,862,066	6,442,010	5,682,517	0.29

When comparing the IRRs of wells in the Raya and Kobo Valleys (Table 2), those in the Raya Valley generally show better performance (0.47 on average, compared to 0.14). While all wells considered in the cost-benefit analysis operate in adjacent *woreda*s using similar water sources, irrigation technology, markets, etc., this may indicate a difference at the organizational level leading to differing performance between the *woreda*s. Further research is needed to pinpoint the precise reasons.

Cost-sharing scheme

Despite the feasibility of irrigation investment and the untapped groundwater-irrigation potential, the current scale of irrigation development is far below the potential. The major question is, however, how the investment cost required to develop the potential can be financed, as it requires large capital investments. Because the smallholder farmers are unlikely to be in a position to finance the full cost of capital investment, a cost-sharing system that can be implemented between the public sector and the smallholders was also analyzed.

First, annual instalments (annuity) that users should pay assuming a full cost recovery plan were estimated. Considering the maximum interest rate (9.5%), the estimated annual payment (annuity) rate per well ranges between ETB 166,439 and ETB 518,062 (USD 9051 and USD 28,171) in Raya and between ETB 221,830 and ETB 423,343 (USD 12,063 and USD 23,020) in the Kobo Valley (Table 3). This translates to annual instalments per hectare of between ETB 5427 and ETB 16,478 (USD 295 and USD 896) in Raya and ETB 6524 and ETB 8467 (USD 355 and USD 460) in the Kobo Valley. Furthermore, assuming an average of 0.25 ha of irrigated landholding per household, the average annual

Table 3. Sensitivity analysis of annual instalment (annuity) per well (in ETB).

Project (well) name	8%	8.5%	9.5%
P1 Selam Bikalsi (WF 0 BH-6)	458,019	477,737	518,062
P2 Selam Bikalsi (WF 0 BH-5)	441,137	460,128	498,967
P3 Selam Bikalsi (WF 0 BH-7)	442,743	461,803	500,783
P4 Wargaba (WF 13 BH-1)	438758	457,646	496,276
P5 Wargaba (WF 13 BH-2)	446,803	466,038	505,375
P6 Warabaye (WF 9 BH-1)	436,224	455,004	493,410
P7 Karra Adishuhu (WF 9 BH-21)	171,674	179,065	194,179
P8 Karra Adishuhu (WF 9 BH-3)	147,149	153,484	166,439
P9 Karra Adishuhu (WF 9 BH-4)	147,149	153,484	166,439
Average (Raya Valley)	**347,740**	**362,710**	**393,326**
HG1	196,120	204,564	221,830
HG2	259,767	270,950	293,821
HG4	374,278	390,390	423,343
HG6	335,460	349,902	379,437
HG7	333,935	348,311	377,712
HG8	249,783	260,536	282,528
HG10	204,428	213,229	231,227
WG1	262,196	273,483	296,567
WG2	214,929	224,182	243,105
WG4	313,072	326,550	354,113
WG5	337,835	352,378	382,122
Average (Kobo Valley)	**280,164**	**292,225**	**316,891**
Average (total)	**310,573**	**323,943**	**351,287**

instalment (annuity) per household is ETB 3003 (USD 163) in Raya and ETB 1885 (USD 102) in the Kobo Valley (Table 4).

For full cost recovery, the estimated annual instalment per household (quarter-hectare, 9.5% interest rate) constitutes on average 13% and 28% of the estimated annual gross income of households due to irrigated crops in Raya and Kobo, respectively. This is unrealistically high for the smallholders to pay, implying that there is a need to come up with more realistic rates and recommendations. In this regard, two potential cost-sharing scenarios are proposed: the first, with smallholders paying 25% of the capital investment over 10 years; the second, with smallholders paying 5% of their annual gross income until 25% of capital costs are covered.

Scenario 1 assumes that every household equally shares only 25% (as opposed to 100% in the full cost recovery case) of the investment cost of their well over a period of 10 years. Accordingly, the annual payment per household ranges between ETB 385 and ETB 1176 (USD 21 and USD 64) in Raya and between ETB 313 and ETB 959 (USD 17 and USD 52) in Kobo (Table 5). Interestingly, although in absolute terms the values for Kobo are lower, the required proportion of annual household income is much higher than in Raya (as seen from Columns 2 and 3 in Table 5). The annual payment expected from smallholder farmers in Kobo constitutes between 5% and 13% of their estimated annual income, while in Raya it is only between 2% and 6%. The limitation of Scenario 1 is the large disparity between required payments in the two valleys and even between individual households within the valleys.

Scenario 2 provides an alternative to the high disparity associated with Scenario 1 (Table 6). This scenario seems more just, as all farmers pay a proportion relative to their income. Due to the large variability in costs and incomes across the project areas and

Table 4. Sensitivity analysis of annual instalments (annuity) per hectare and per quarter-hectare (in ETB).

	Annual instalment (annuity) per ha			Annual instalment (annuity) per 0.25 ha		
Project (well) name	8%	8.5%	9.5%	8%	8.5%	9.5%
P1 Selam Bikalsi (WF 0 BH-6)	12,723	13,270	14,391	3,181	3,318	3,598
P2 Selam Bikalsi (WF 0 BH-5)	12,254	12,781	13,860	3,063	3,195	3,465
P3 Selam Bikalsi (WF 0 BH-7)	12,298	12,828	13,911	3,075	3,207	3,478
P4 Wargaba (WF 13 BH-1)	14,306	14,922	16,181	3,576	3,730	4,045
P5 Wargaba (WF 13 BH-2)	14,568	15,195	16,478	3,642	3,799	4,119
P6 Warabaye (WF 9 BH-1)	14,223	14,835	16,088	3,556	3,709	4,022
P7 Karra Adishuhu (WF 9 BH-21)	5,597	5,838	6,331	1,399	1,460	1,583
P8 Karra Adishuhu (WF 9 BH-3)	4,798	5,004	5,427	1,199	1,251	1,357
P9 Karra Adishuhu (WF 9 BH-4)	4,801	5,008	5,430	1,200	1,252	1,358
Average (Raya Valley)	**10,619**	**11,076**	**12,011**	**2,655**	**2,769**	**3,003**
HG1	5,768	6,017	6,524	1,442	1,504	1,631
HG2	6,494	6,774	7,346	1,624	1,693	1,836
HG4	7,486	7,808	8,467	1,871	1,952	2,117
HG6	6,989	7,290	7,905	1,747	1,822	1,976
HG7	6,957	7,256	7,869	1,739	1,814	1,967
HG8	5,947	6,203	6,727	1,487	1,551	1,682
HG10	6,814	7,108	7,708	1,704	1,777	1,927
WG1	6,395	6,670	7,233	1,599	1,668	1,808
WG2	6,933	7,232	7,842	1,733	1,808	1,961
WG4	6,522	6,803	7,377	1,631	1,701	1,844
WG5	7,038	7,341	7,961	1,760	1,835	1,990
Average (Kobo Valley)	**6,668**	**6,955**	**7,542**	**1,667**	**1,739**	**1,885**
Average (total)	**8,446**	**8,809**	**9,553**	**2,111**	**2,202**	**2,388**

wells (Table 1), the ultimate model may be one of sharing the partial cost (say 25%) of all wells in the project (rather than individual wells) among all the farmers, but proportionally to acreage (as farmers with larger land are expected to earn proportionally more) and with possible differentiation in payback time, if individual farmers apply for extension. For a 10-year pay-back period, this would amount to annual instalments varying from ETB 161 to ETB 2388 (USD 9 to USD 130) for the various farmers. The upper figure is comparable to the annuity for Scenario 1.

Conclusions and recommendations

With only a few exceptions, the empirical results show that investment in groundwater irrigation is profitable, suggesting that more investment in groundwater irrigation is worthwhile (sufficient groundwater resources appear to be available to support further development). In addition, an average annual instalment (annuity) that a household irrigating the average quarter-hectare holding should pay was estimated, assuming full recovery of all capital investment costs and a payback period of 25 years for the expected life of the infrastructure. The results show that, on average, a farm household should pay ETB 3003 (USD 163) per year in the Raya Valley and ETB 1885 (USD 102) in the Kobo Valley. Due to the unrealistically high payments that would be expected relative to the annual incomes of the farmers, an alternative cost-sharing model was proposed whereby farmers annually

Table 5. Cost-Sharing Scenario 1: user households share and pay 25% of the capital investment in their respective well over 10 years.

Well	Estimated annual instalment per household (ETB)	Proportion of household income (%)	Total annual payments (ETB)	Payback period (years)
P1 Selam Bikalsi (WF 0 BH-6)	764	6	118,279	10
P2 Selam Bikalsi (WF 0 BH-5)	713	6	113,919	10
P3 Selam Bikalsi (WF 0 BH-7)	738	6	114,334	10
P4 Wargaba (WF 13 BH-1)	1,055	3	113,305	10
P5 Wargaba (WF 13 BH-2)	977	3	115,383	10
P6 Warabaye (WF 9 BH-1)	1,176	3	112,651	10
P7 Karra Adishuhu (WF 9 BH-21)	385	2	44,333	10
P8 Karra Adishuhu (WF 9 BH-3)	770	2	38,000	10
P9 Karra Adishuhu (WF 9 BH-4)	689	2	38,000	10
Average (Raya Valley)	**808**	**4**	**89,800**	**10**
HG1	313	9	50,646	10
HG2	410	12	67,082	10
HG4	598	13	96,654	10
HG6	533	6	86,629	10
HG7	749	7	86,236	10
HG8	333	5	64,504	10
HG10	758	7	52,792	10
WG1	454	10	67,709	10
WG2	910	6	55,503	10
WG4	500	9	80,848	10
WG5	959	12	87,243	10
Average (Kobo Valley)	**593**	**9**	**72,350**	**10**
Average (total)	**689**	**6**	**80,203**	**10**

pay 5% of their income towards partial coverage of infrastructure capital costs, until 25% of the investment costs are recovered. Finally, a more pragmatic and workable model of partial cost-sharing across all wells in the project, according to household irrigated area and individually negotiated pay-back times, provided a better feasible alternative.

However, the institutionalization of cost sharing has challenges and carries associated transaction costs. Since it is an emerging proposition in Ethiopia, one has to expect reluctance from both farmers and implementers. Likewise, formalizing and implementing a cost-sharing scheme would require a legal framework for how to collect and administer the funds raised.

Institutionalizing cost sharing effectively requires that policy makers and researchers also give emphasis to: (1) improving on-farm production efficiency, which presents opportunities for increasing income and hence farmers' capacity to share part of the initial investment costs; (2) strengthening the capacity of the WUAs to manage and maintain the systems and foster the active participation of farmers in promoting irrigation and the cost-sharing scheme; (3) expanding rural electrification, which would benefit the rural community as a whole; and (4) generally improving the boundary conditions for groundwater irrigation. In terms of those boundaries, it is well accepted that investment in groundwater irrigation and access to irrigation in general can play a great role in improving people's livelihoods and transforming agriculture. But its effect could be higher in the face of enhanced supply of (and access to) improved seeds, farm inputs and better output market

Table 6. Cost-Sharing Scenario 2: users pay 5% of their annual gross income until 25% of the capital investment is covered.

Well	Average annual gross income per well per year (ETB)	Estimated annual instalment per household (ETB)	Total annual payment (ETB)	Payback period (years)
P1 Selam Bikalsi (WF 0 BH-6)	2,019,140	631	100,957	12
P2 Selam Bikalsi (WF 0 BH-5)	2,019,140	612	100,957	12
P3 Selam Bikalsi (WF 0 BH-7)	2,019,140	631	100,957	12
P4 Wargaba (WF 13 BH-1)	3,653,131	1,646	182,657	6
P5 Wargaba (WF 13 BH-2)	3,653,131	1,497	182,657	7
P6 Warabaye (WF 9 BH-1)	3,653,131	1,845	182,657	6
P7 Karra Adishuhu (WF 9 BH-21)	2,841,324	1,194	142,066	3
P8 Karra Adishuhu (WF 9 BH-3)	2,435,421	2,388	121,771	3
P9 Karra Adishuhu (WF 9 BH-4)	2,435,421	2,136	121,771	3
Average (Raya Valley)	**2,747,664**	**1,398**	**137,383**	**7.11**
HG1	537,228	161	26,861	19
HG2	571,199	169	28,560	24
HG4	771,811	231	38,590	26
HG6	1,406,303	419	70,315	13
HG7	1,306,186	549	65,309	14
HG8	1,311,609	328	65,580	10
HG10	753,275	523	37,664	14
WG1	645,236	209	32,262	22
WG2	982,910	780	49,146	12
WG4	913,363	273	45,668	18
WG5	705,736	375	35,287	26
Average (Kobo Valley)	**900,441**	**365**	**45,022**	**18**
Average (total)	**1,731,692**	**830**	**86,585**	**13.1**

conditions (Hanjra et al., 2009). Both input and output markets are highly fragmented and inefficient, because informal brokers currently exercise excessive power in setting market prices, usually against the interest of farmers. Hence, by improving the extension and marketing system, farm households' technology adoption, income and willingness to contribute to investment cost could be improved. Addressing these issues through integrated and participation-based measures would capitalize on the willingness of farmers to adopt new approaches that provide evidence of increased yields and higher sustained incomes.

Notes

1. The *woreda* (district) is the smallest administrative unit in Ethiopian government structure.
2. Using the February 2013 exchange rate of ETB 18.39 = USD 1, due to a lack of information on the exchange rate at the time of the investments in the mid-1990s.

References

Awulachew, S. B., Yilma, A. D., Loulseged, M., Loiskandl, W., Ayana, M., & Alamirew, T. (2007). *Water resources and irrigation development in Ethiopia* (78p). Colombo, Sri Lanka: International Water Management Institute. Working Paper 123.

Central Statistics Authority (CSA). (2004). *The Federal Democratic Republic of Ethiopia Statistical abstract for 2003*. Addis Ababa, Ethiopia: CSA.

Deressa, T. (2007). *Measuring the economic impact of climate change on Ethiopian agriculture: Ricardian approach: World Bank Policy Research Paper No. 4342*. Washington, DC: The World Bank.

Development Bank of Ethiopia. (2010). Development Bank first to increase lending interest. Accessed September, 13, 2013. Retrieved from http://www.ethiopiainvestor.com/index.php?option=com_content&task=view&id=2101&Itemid=88

Dixon, J. A., Carpenter, R. A., Fallon, L. A., Sherman, P. B., & Manipomoke, S. (1986). *Economic analysis of the environmental impacts of development projects*. London, UK: Asian Development Bank, Manilla, the Philippines & Earthscan.

Easter, K. W., & Hearne, R. R. (1993). *Decentralizing resource management: Economic incentives, accountability and assurance*. Policy Research Working Paper 1219. Washington, DC: The World Bank.

Easter, K. W., & Liu, Y. (2005). *Cost recovery and water pricing for irrigation and drainage projects: Agriculture and rural development*. Washington, DC: The World Bank.

FAO (Food and Agriculture Organization of the United Nations). (2005). *Irrigation in Africa in figures: AQUASTAT survey – 2005*. FAO Water Report No. 29. Rome: FAO. , Accessed August 13, 2012. Retrieved from ftp://ftp.fao.org/agl/aglw/docs/wr29_eng.pdf

FDRE. (2000). *Ethiopian Water Resources Management Proclamation*. Proclamation No. 197/2000. Federal Negarit Gazeta, 6th Year No. 25. Addis Ababa, Ethiopia.

Ferede, T., Ayenew, A. B., & Hanjra, M. A. (2013). Agroecology matters: Impacts of climate change on agriculture and its implications for food security in Ethiopia. In: M. A. Hanjra (Eds.), *Global food security: Emerging issues and economic implications* (Chapter 3, pp. 71–112). New York, NY: Nova Science Publishers.

Godswill, M., & Krishna, C. P. (2006). *Addressing irrigation needs of Alamata farmers: Options and scope.* International Water Management (IWMI), Draft Report. Retrieved on September 13, 2013 from http://www.ipms-ethiopia.org/content/files/Documents/publications/IPMS%20Other%20Studies/Addressing%20Irrigation%20Needs%20of%20Alamata%20Farmers.pdf

Hagos, F., Pender, J., & Gebreselassie, N. (1999). *Land degradation in the highlands of Tigray and strategies for sustainable land management*. Socioeconomic and Policy Research Working Paper No. 25. Ethiopia: International Livestock Research Institute.

Hanjra, M. A., Ferede, T., & Gutta, D. G. (2009). Pathways to breaking the poverty trap in Ethiopia: Investments in agricultural water, education, and markets. *Agricultural Water Management, 96*(11), 1596–1604.

Kebede, S. (2010). *Groundwater in Ethiopia: Occurrence, drought proofing and technology options*. Addis Ababa University, Department of Earth Sciences. Retrieved on July 10, 2013 from http://ellasprings.files.wordpress.com/2010/09/gw-ethiopia-hq.pdf

Kobo-Girana Valley Development Project (KGVDP). (2009). Performance evaluation report (Amharic Version).Addis Ababa, Ethiopia.

Lautze, S., Aklilu, Y., Raven-Roberts, A., Young, H., Kebede, G., & Learning, J. (2003). *Risk and vulnerability in Ethiopia: Learning from the past, responding to the present, preparing for the future*. Addis Ababa, Ethiopia: Report for the U.S. Agency for International Development.

Marks, S. J., & Davis, J. (2012). Does user participation lead to sense of ownership for rural water systems? Evidence from Kenya. *World Development, 40*(8): 1569–1576.

MoFED. (2010). *Annual report on macroeconomic development*. Retrieved on July 10, 2013 from http://www.mofed.gov.et/English/Resources/Documents/2002MEDRAnnual.pdf

Moges, S. (2011). *Agricultural use of ground water in Ethiopia: assessment of potentials, analysis of economics, policies, constraints and opportunities (final draft)*. Addis Ababa, Ethiopia.

Pender, J., & Gebremedhin, B. (2004). Impacts of policies and technologies in dryland agriculture: Evidence from Northern Ethiopia. In S.C. Rao and J. Ryan (Eds.), *Challenges and strategies for dryland agriculture*. Crop Science Society of America (CSSA) Special Publication no. 32. Addis Ababa, Ethiopia.

Pender, J., & Gebremedhin, B. (2008). Determinants of agricultural and land management practices and impacts on crop production and household income in the highlands of Tigray, Ethiopia. *Journal of African Economies, 17*(3), 395–450.

Relief Society of Tigray (REST). (2011). *Raya valley development study project: Phase I reconnaissance phase report volume I, Main Report*. Alamata, Ethiopia.

Tesfay, H. (2008). *Impact of irrigation development on poverty reduction in Northern Ethiopia* (unpublished doctoral thesis). Cork: National University of Ireland.

United Nations Industrial Development Organization (UNIDO). (1972). *Guidelines for project evaluation*. New York.

Predicting groundwater recharge in Ghana by estimating evapotranspiration

Fathi M. Anayah, Jagath J. Kaluarachchi, Paul Pavelic and Vladimir Smakhtin

This study uses a modified Granger and Gray model to estimate evapotranspiration and then groundwater recharge in Ghana. The overall results show that the model is capable of reliably predicting regional evapotranspiration using a small number of monitoring stations with meteorological data only. This information allows the estimation of groundwater recharge via the water balance equation. The results indicate that the aquifer system is sufficiently recharged, especially in northern Ghana, where dry conditions prevail, to allow the development of groundwater resources to satisfy increasing water demands.

Introduction

The focus of this study is on estimating groundwater recharge in the Volta River basin of Ghana through evapotranspiration (ET) calculations. In the water balance equation, recharge is calculated as the residual of rainfall, ET and surface runoff. In arid and semi-arid regions such as the Volta River basin (Carrier, 2008; Obuobie, 2008), runoff accounts for an insignificant portion of the rainfall. Therefore, ET estimation, combined with rainfall measurements, can be used to derive groundwater recharge.

Description of the study area

The Volta River basin in Ghana was selected for this study for several reasons. Ghana was the focus of several important studies conducted more than 50 years ago by pioneering scientists such as Penman (as cited in Rietveld, 1978). Field studies are conducted there by the International Water Management Institute (IWMI), the GLOWA Volta Project funded by the German Technical Cooperation (GTZ), and the Canadian International Development Agency (CIDA).

The Volta River basin covers about 70% of the total area of Ghana (or 238,538 km^2), mainly in the northern and central parts of the country (Figure 1). The basin comprises four principal sub-basins: the Black, White, Oti, and Lower Volta. The agricultural land

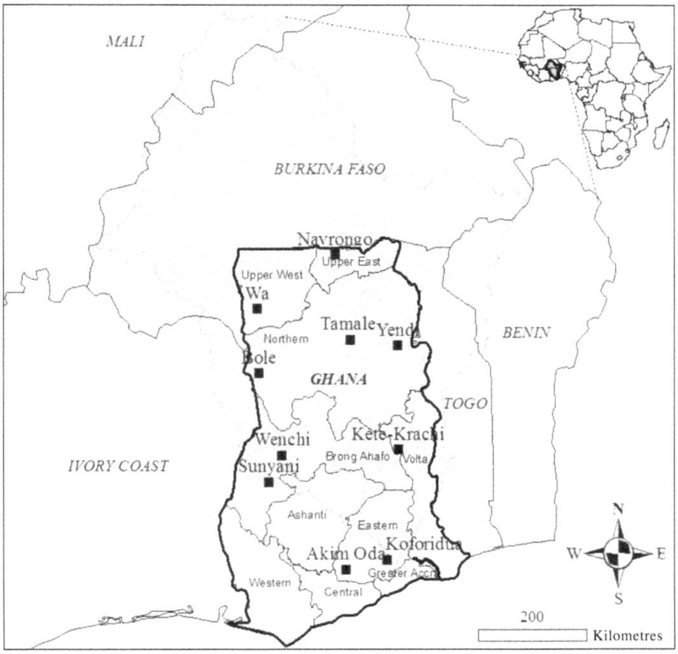

Figure 1. Map of Ghana (within dark solid line) and the Volta River basin (hatched area). The dark squares show the synoptic stations used in the study; the interior divisions are administrative regions of Ghana.

of Ghana is about 57% of the total area (Statistics, Research and Information Directorate [SRID], 2010). Ghana is administratively divided into 10 regions. The population of Ghana in 2009 was 24 million, with a growth rate of 2.1%, and 68% of the population live in rural areas (SRID, 2010). Agriculture is the major source of income and employment for rural communities.

Ghana lies within 4 major agro-ecological zones, and the 10 synoptic stations fall across all 4: the Guinea Savannah Zone (includes Navrongo, Wa, Tamale, Yendi and Bole), the Transitional Zone (Kete-Krachi and Wenchi), the Rain Forest Zone (Sunyani and Akim Oda) and the Coastal Savannah Zone (Koforidua). The climates of Ghana vary from semi-arid in the north, with a monomodal rainfall pattern, to subtropical humid in the south, with a bimodal rainfall pattern. The average annual rainfall varies between 963 mm and 1432 mm from north to south, respectively. At Akim Oda in the south, for instance, the major rainfall season is from March to August, while the minor season is from September to October. In contrast, there is only one rainfall season, from May to September, at Navrongo in northern Ghana (i.e. summer rainfall). Rainfall is high during these months, and ET is low due to increased humidity. In this humid climate, recharge is likely to be high. As in most semi-arid regions, surface-water resources are considerably unreliable due to the high inter- and intra-annual variability in rainfall in the north. The average daily temperature varies from 26.4 °C to 29.3 °C, while the average daily relative humidity fluctuates considerably between 53.8% and 79.2%; the same is true with sunshine hours and average wind speed. Topography is relatively undulating, with gentle slopes, and the average elevation is about 190 m above mean sea level. The topsoil is mostly sandy loam and loam, and the gravel content in the sandy soils increases with depth, indicating lower soil

Table 1. Details of the 10 synoptic stations in Ghana.

Station	Lat. (°)	Long. (°)	Elev. (m)	AI*	Rainfall (mm/y)	T (°C)	RH (%)	S (h)	U (m/s)	Land cover
Navrongo	10.90	−1.10	201	24.5	963	29.3	53.8	8.0	1.5	Savanna woodland
Wa	10.05	−2.50	323	27.0	1040	28.5	58.3	7.8	1.1	Savanna woodland
Tamale	9.50	−0.85	168	26.8	1040	28.8	60.3	7.5	1.8	Savanna woodland
Yendi	9.45	−0.02	195	33.6	1291	28.4	62.7	7.7	0.8	Savanna woodland
Bole	9.03	−2.48	299	31.2	1157	27.1	66.9	6.9	1.3	Savanna woodland
Kete-Krachi	7.82	−0.03	122	35.4	1353	28.3	73.2	7.2	0.8	Savanna woodland
Wenchi	7.75	−2.10	339	34.5	1258	26.5	73.2	6.4	1.0	Herb/bush
Sunyani	7.33	2.33	309	32.9	1197	26.4	75.3	5.4	2.2	Settlement
Koforidua	6.08	−0.25	166	37.1	1384	27.3	77.6	6.1	0.7	Herb/bush
Akim Oda	5.93	−0.98	139	38.4	1432	27.3	79.2	5.3	0.6	Herb/bush

*AI is defined as very humid for higher than 35, humid between 28 and 35, subhumid between 24 and 28, Mediterranean between 20 and 24, semi-arid between 10 and 20, and arid when less than 10 (de Martonne, 1925).

stability, but the drainage characteristics are better during the rainy season (O'Driscoll, Clinton, Jefferson, Manda, & McMillan, 2010).

All the data used in this work were gathered from the Meteorological Services Department of Ghana, the Water Resources Commission of Ghana, CIDA, and the IWMI. Information related to the 10 synoptic stations used here is shown in Table 1 and Figure 1. Daily values of minimum air temperature (T_{min}), maximum air temperature (T_{max}), minimum relative humidity (RH_{min}), maximum relative humidity (RH_{max}), sunshine hours (S), and average wind speed (U) are available from 2000 to 2005.

Methodology

Complementary methods, including the complementary relationship areal ET (CRAE) method of Morton (1983), the advection-aridity (AA) method of Brutsaert and Stricker (1979) and the method of Granger and Gray (1989) (GG), offer viable alternatives to classical methods for calculating ET using meteorological data only under diverse physical and climatic conditions (see the Appendix). According to Morton (1983), complementary methods offer simplicity and practicality, and provide reliable, physically based operational ET estimates at regional scales. The methods also consider the surrounding climatic conditions regardless of the underlying soil-plant system, avoid locally calibrated coefficients, and require no more than the minimal meteorological data typically monitored in most rural areas.

In the complementary methods, wet environment ET (ETW) is the ET that occurs if the soil-plant system is saturated and ET approaches its potential rate, ETP. In spite of the fact that ET is negatively correlated with ETP (Figure 2), there is no defined shape of that relation (see Morton, 1983). The complementary theory posits a complementary relationship between ET and point ETP estimates in a given area. Accordingly, the relationship between ET and ETP under limited water supply is not necessarily linear and does not follow a given law. The earliest complementary method states:

$$\mathrm{ET} + \mathrm{ETP} = 2\mathrm{ETW} \tag{1}$$

where ET, ETW and ETP are in mm/day. Equation (1) indicates that an increase in ET is accompanied by an equivalent decrease in ETP, i.e. $\delta\mathrm{ET} = -\delta\mathrm{ETP}$. In other words, as the

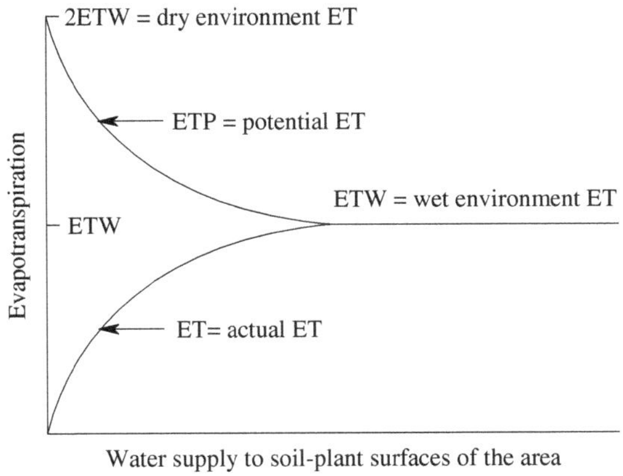

Figure 2. Schematic representation of the complementary relationship between ET and ETP (after Morton, 1983).

surface dries, areal ET decreases, causing a decrease in humidity and an increase in the temperature of the surrounding air, and as a result, ETP increases. This means that ETP is a function of ET, in contrast to the model suggested by Penman (1948), where ETP is independent of ET. This latter claim is true for large moist areas where the effect of ET on temperature and humidity is fully developed and ET and ETP are equal.

The CRAE and AA methods use the same complementary relationship originally developed and shown in Equation (1). However, Granger and Gray (1989) modified the relationship as follows:

$$\mathrm{ET} + \frac{\gamma}{\Delta}\mathrm{ETP} = \left(1 + \frac{\gamma}{\Delta}\right)\mathrm{ETW} \tag{2}$$

where γ is the psychrometric constant (in kPa/°C) and Δ is the rate of change of saturation vapour pressure with temperature (in kPa/°C). Hence, Equation (2) can be reduced to Equation (1) only when $\gamma = \Delta$.

For several decades, the complementary methods have been studied and compared to other methods over different spatial and temporal scales and climates. Oguntunde, van de Giesen, and Andreini (2005) applied the AA method in semi-arid to subhumid tropical regions of West Africa and noted that the complementary relationship is valid across the Volta Basin, but the ET estimates have to be verified. It was found that in cold semi-arid regions of Canada, the CRAE method is not suitable for short periods or field-size areas (Granger & Gray, 1990). However, Doyle (1990) compared the CRAE method and the Thornthwaite soil moisture model on an Irish humid basin to water balance data and found the CRAE method to be valid. The CRAE and AA methods produced ET values at 139 minimally impacted basins in the US (Hobbins, Ramirez, Brown, & Claessens, 2001), and the results showed that as aridity increases, ET is overestimated by the CRAE method and underestimated by the AA method. Still, the CRAE method predicted monthly ET over the climate spectrum. A more detailed discussion of the complementary methods and their applicability and limitations is provided by Anayah (2012).

In a recent study, Anayah (2012) developed a modified complementary model based on the GG model that was validated using measured ET data from 34 global sites with contrasting climatic and physical conditions. The climates of these global sites vary from tropical to subarctic, while the average ET values range from 10.5 mm/month to 134.3 mm/month. The results showed that the average root mean square error, mean absolute bias and R^2 across the 34 global sites were 20.57 mm/month, 10.55 mm/month and 0.64, respectively. It was found that the ET estimates of the proposed model, which requires meteorological data only, outperform those of the most recent ET studies. Given the validity of the proposed modified GG model by Anayah (2012), the purpose of this study is to extend the work to estimate ET and groundwater recharge in Ghana.

Prior studies, for example by Carrier (2008), Friesen, Andreini, Andah, Amisigo, and van de Giesen (2005) and Wagner (2008), showed that there is substantial uncertainty and discrepancy between different studies estimating recharge in Ghana. Given the fact that the aquifer systems of Ghana are not yet fully delineated (Enoch Asare, personal communication, 25 July 2009), regional-scale recharge estimation using the classical methods is difficult.

Carrier (2008) estimated groundwater recharge in northern Ghana using soil moisture and chloride mass balance approaches. Recharge rates from soil moisture balance were found to vary from 1% to 16% of the average annual rainfall. Martin (2006) and Obuobie (2008) used chloride mass balance to estimate recharge in the Upper East Region of Ghana; the estimates of the latter study were considerably higher than those of the former. These findings indicate the uncertainty in the method and suggest that there is no single comprehensive estimation method applicable at regional scales. In general, it is not possible to directly measure fluxes into and out of groundwater over large spatial scales, and therefore indirect methods must be used. The water balance method will offer a good alternative once the ET fluxes are accurately estimated in the highly diverse environments of Ghana.

The modified complementary method

The modified GG model proposed by Anayah (2012) is an enhanced version of the original GG method. Anayah (2012) found that the complementary relationship shown in Figure 2 and Equation (1) can describe the behaviour of ET fluxes more precisely than the more generic expression derived in the original work by Granger and Gray (1989), i.e. Equation (2). More importantly, the predictive power of the GG method is improved when the Priestley and Taylor (1972) (P-T) equation shown in Equation (3) is used to calculate ETW instead of the Penman (1948) equation used in the original GG method.

$$\mathrm{ETW} = \alpha \frac{\Delta}{\gamma + \Delta}(R_n - G_{soil}) \tag{3}$$

Here, α is a coefficient typically equal to 1.26, R_n is the net radiation in mm/d, and G_{soil} is the soil heat flux density in mm/d. The procedure described by Allen et al. (2005) is used to calculate daily net radiation for both actual and crop ET estimates. The daily data required to calculate R_n are simply T_{min}, T_{max}, RH_{min}, RH_{max}, and S. It is noticed that soil heat flux density (G_{soil}) is negligible for daily time steps, especially when compared to R_n.

In the GG method, two new parameters were proposed and empirically correlated together: relative drying power D and relative evaporation G, shown in Equations (4) and (5), respectively. Parameter D indicates the dryness of the surface, i.e. higher D indicates a drier surface. Relative evaopration (parameter G) is assumed to occur under wind and humidity conditions of a saturated surface at its actual temperature (Granger & Gray, 1989).

$$D = \frac{E_{\mathrm{a}}}{E_{\mathrm{a}} + R_{\mathrm{n}}} \tag{4}$$

$$G = \frac{\mathrm{ET}}{\mathrm{ETP}} \tag{5}$$

$$G = \frac{1}{1 + 0.028\mathrm{e}^{8.045D}} \tag{6}$$

In Equation (4), E_{a} is the drying power of air in mm per day. The average daily values of U are required to calculate E_{a}. Solving Equation (5) for ETP, substituting this into Equation (1), and solving for ET provides an estimate of ET in the modified GG model, as shown in Equation (7).

$$\mathrm{ET} = \frac{2G}{G+1}\mathrm{ETW} \tag{7}$$

Classical methods

In this work, five additional classical methods of estimating ET are used for comparison purposes; the ASCE method with two reference crops (Allen et al., 2005), the Hargreaves method (Hargreaves, Hargreaves, & Riley, 1985), Turc's method (Turc, 1961), and the Jensen and Haise method (Jensen-Haise, 1963). These classical methods require the calculation of ET_{o} first, followed by the use of the crop coefficient K_{c} to calculate crop ET. The Hargreaves method was developed using the conditions of West Africa, while Turc's method is based on Western Europe under humid conditions (Jensen, Burman, & Allen, 1990). The Jensen-Haise method was developed for well-watered crops in the western United States (Jensen et al., 1990). Hereafter, Jensen-Haise, Hargreaves, Turc, ASCE with grass reference and ASCE with alfalfa reference are referred to as JH, HV, TC, GR and AL, respectively.

Groundwater recharge

Groundwater recharge will be computed using the water balance equation, as shown in Equation (8).

$$\mathrm{Rech} = P - Q - \mathrm{ET} - \Delta S \tag{8}$$

where Rech is the groundwater recharge, P is the rainfall, Q is the surface runoff, and ΔS is the change in soil water storage. All terms of the water balance equation should have the same unit, usually given as a rate (unit depth per unit time) such as mm/month or mm/year. It should be noted that ΔS approaches zero if calculations are made on an annual basis.

Results and discussion

Climate class

The climatic condition of each station was defined using the aridity index (AI) defined by de Martonne (1925). In this method, AI is based on the average annual rainfall and average annual temperature. The reasons for using the AI proposed by de Martonne (1925) compared to other, newer models are its ability to capture the climatic variability across Ghana and the simplicity of data required. In Table 1, AI indicates contrasting climates, varying from subhumid in three stations in the north to very humid conditions towards the south.

Potential evapotranspiration

ETP or ET_0 is the first step in calculating crop ET with the classical methods, while the proposed modified GG model estimates ETP and ETW; the results are shown in Figure 3. It is observed that ETP from the modified GG model is high compared to other estimates. The pan evaporation measurements of Acheampong (1986) at Navrongo, Tamale and Wenchi, and the corresponding ETP predictions from the modified GG model, are close and comparable. Similarly, Chiew and Leahy (2003) and Hobbins et al. (2001) found that ETP from the CRAE method is high enough to be a good estimator of pan evaporation that considers ideal conditions where neither energy nor water is limited. ETP is calculated using ET from Equation (7) and the parameter G in Equation (5). G indicates the dryness of the surface and is empirically correlated to D in Equation (6), which represents the convective properties of the environment. High values of ETP indicate convective environments (high D values) under dry surface conditions (low G values). Apparently, ETP values in the north are considerably higher than those in the south.

In contrast, ETW predicted by the modified GG model is comparable to other ET_0 estimates. For example, the mean values of ETP predicted by the modified GG model range from 157 to 222 mm/month, whereas mean ETW values from the same model vary between 116 and 138 mm/month. The corresponding mean values of ET_0 for grass and alfalfa range from 100 to 155 mm/month and from 109 to 199 mm/month, respectively. These results suggest the possibility of comparing the ETW of the modified GG model with the ET_0 of the GR method. Compared to the ETW estimates, the GR method slightly overestimates ET_0 under arid climates and underestimates it under humid climates (see Figure 3 and Table 2). The variability of ET_0 from the GR method is slightly higher than ETW with the modified GG model. It is noticed that the AL method overestimates ET_0 compared to the GR method; however, the difference gradually diminishes as humidity increases.

Table 2 shows the annual estimates of ETP, ETW and ET_0 predicted by the different methods. The ETW estimates are consistent, and the variation among the stations is not substantial. The relative absolute difference (RAD) is simply defined as the absolute value of the difference between ETW from the modified GG model and ET_0 from the classical method, divided by ETW. It is noticed however that RAD_{GR} ranges from zero at Bole to 16% at Akim Oda, with an average of 9% and no obvious trend. Basically, this is a comparison between the Penman-Monteith (P-M) and P-T equations. The P-M equation is a combination of the radiative and convective energy demands of the climate, while the P-T equation consists of the radiative demand multiplied by the coefficient α which accounts for the convective demand. Although advection energy is lumped into the radiation energy of the P-T equation in the modified GG model, the P-T equation is still a good substitute for the P-M equation. The results of the JH method are close to those of the AL method. The performance of the HV method improves with aridity, whereas the TC method moves

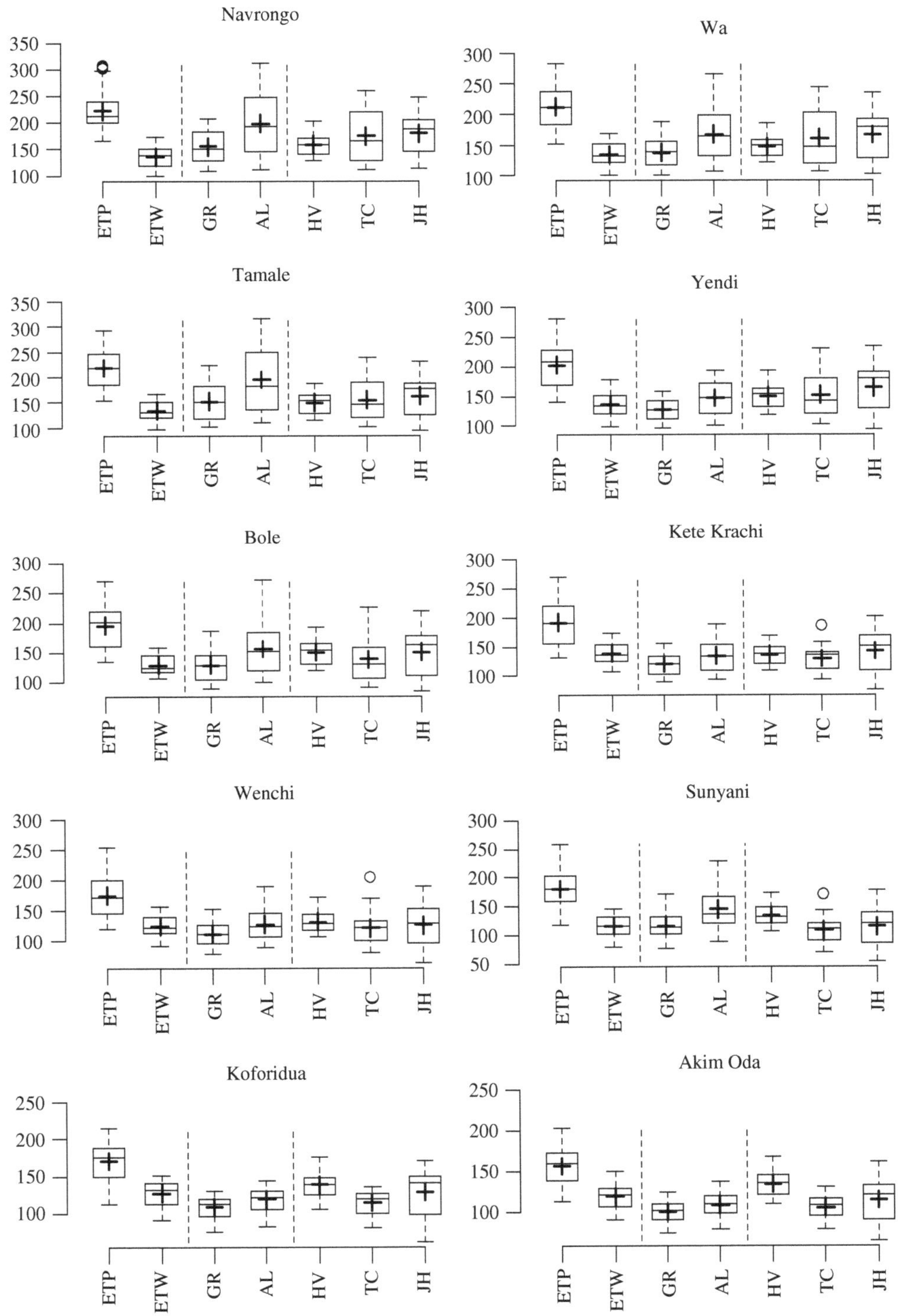

Figure 3. Boxplots for monthly potential evapotranspiration, in mm. (+: average value. ETP: GG model. ETW: GG model. GR: grass ET_o. AL: alfalfa ET_o. HV: Hargreaves ET_o. TC: Turc ET_o. JH: Jensen-Haise ET_o.)

Table 2. Average annual ETP estimates (mm) computed by different methods and their relative absolute difference (RAD) as a percentage.

	[1]	[2]	[3]	[4]	[5]	[6]	[7]	[8]	[9]	[10]
Method	Complementary		Classical						RAD_{GR}	RAD_{AVG}
	Modified GG		ASCE					Avg.	$\frac{\lvert[2]-[3]\rvert}{[2]}$	$\frac{\lvert[2]-[8]\rvert}{[2]}$
Station	ETP	ETW	GR	AL	HV	TC	JH	[3] to [7]		
Navrongo	2662	1623	1862	2386	1905	2102	2161	2083	15	28
Wa	2547	1616	1651	2008	1767	1928	2001	1871	2	16
Tamale	2614	1609	1824	2343	1793	1856	1943	1952	13	21
Yendi	2432	1635	1527	1766	1801	1826	1964	1777	7	9
Bole	2341	1540	1540	1864	1794	1648	1801	1729	0	12
Kete-Krachi	2308	1657	1446	1609	1642	1549	1713	1592	13	4
Wenchi	2105	1497	1341	1530	1576	1460	1514	1484	10	1
Sunyani	2165	1389	1398	1736	1608	1302	1385	1486	1	7
Koforidua	2044	1523	1296	1420	1654	1364	1533	1453	15	5
Akim Oda	1882	1426	1202	1305	1614	1269	1374	1353	16	5
Min	1882	1389	1202	1305	1576	1269	1374	1353	0	1
Avg.	2310	1551	1509	1797	1716	1630	1739	1678	9	11
Max	2662	1657	1862	2386	1905	2102	2161	2083	16	28

in the opposite direction. One criticism of these two methods is the opposite behaviour of the results. A possible reason for this behaviour may be that the parameters used are locally developed under specific conditions and may not be applicable elsewhere. The HV method was constructed to calculate the water requirements for crops under dry conditions in Senegal, Mauritania and Mali. The TC method was exclusively developed for Western Europe under more humid conditions. Nevertheless, an interesting trend is observed when ETW and the average of the five ET_0 estimates from the classical methods are compared by RAD_{AVG} values, as shown in Column 10 of Table 2. The closure between these two estimates significantly increases with humidity.

Acheampong (1986) estimated pan evaporation at Navrongo, Tamale and Wenchi to be 2550, 3085 and 1985 mm/y, respectively. These estimates are close to the ETP values from the modified GG model, except at Tamale, where the difference is about 18%. The estimates from the other three methods are in agreement with the estimates from the modified GG method. At Navrongo, for instance, Compaore et al. (2008) stated that average annual ETP from 1961 to 2003 was 2423 mm. Carrier (2008) studied ET in northern Ghana, including eight stations discussed in this work. It is no surprise that ET_0 from the GR method agreed at around 90% to 95% with the estimates of Carrier (2008), who used the P-M equation. In a total of seven previous studies (e.g. Acheampong, 1986; Carrier, 2008; Compaore et al., 2008), ET_0 was estimated using both Penman and Thornthwaite methods at Navrongo, and the average annual ET_0 ranged from 1647 mm to 2423 mm. In Table 2, the values of ETW and ETP at Navrongo are 1623 mm/y and 2662 mm/y, corresponding to the lower and upper bounds, respectively, of the ET_0 range of the seven previous studies.

Actual ET

Actual ET is the total water loss from the land surface, which may or may not include vegetation, and includes evaporation from water bodies and soil moisture. Since most synoptic stations used in this study are located within agricultural areas, crop ET is comparable to actual ET.

It is worthwhile to add that the data used by Anayah (2012) are at monthly time steps from global sites, while the data used from Ghana are daily. The question is whether this difference in temporal resolution may affect the results of the modified GG model. For this purpose, the time series of monthly meteorological data was made available by the Meteorological Services Department of Ghana from 1961 to 2005 for five northern stations, namely Navrongo, Tamale, Wa, Bole and Yendi. Monthly data from these five stations from 2000 to 2005 were used for comparison with the daily estimates of the modified GG model. The root mean square errors (RMSEs) of the two estimates ranged from 2.8 mm/month at Wa to 5.9 mm/month at Tamale. The bias of the mean values varied between −1.6 mm/month at Bole and 2.6 mm/month at Tamale, and the lowest coefficient of determination (R^2) was 0.96, at Bole. These results suggest that the modified GG model can accommodate both daily and monthly time series to produce consistent results.

Crop ET

The ASCE method is used to calculate crop ET, and the crop coefficients used are adjusted to suit the existing climatic conditions of each station. Since the growing stages of each crop vary by region and climate, the data provided by Allen, Pereira, Raes, and Smith (1998) cannot be used. Therefore, the detailed data for Ghana published by the FAO (n.d.) were used. For example, the growing seasons for 29 crops are locally adapted to suit the four major agro-ecological zones of Ghana. The other key information needed was the cropping patterns across Ghana. The distribution of each crop in all 138 districts in 2008 was available from the Ghanaian Ministry of Food and Agriculture.[1] For this calculation, the representative area of each station was determined using the Thiessen polygon method. These calculations were conducted on the basis that "agriculture is predominantly on a smallholder basis in Ghana" and "about 90% of farm holdings are less than two hectares in size" (SRID, 2010). Although the details of crop coefficients are not given here, the data are readily available from the authors.

Comparison of actual and crop ET

Figure 4 shows the average monthly actual ET computed from the modified GG model and crop ET computed by the GR method. The actual ET estimates in most stations are close to those of crop ET, in particular during the growing seasons. There is one major exception at Sunyani and a minor discrepancy at Tamale. Kete-Krachi shows a unique behaviour that does not exist in other stations. Sunyani is the only station where crop ET is always higher than actual ET. This station exists in a residential area (see Table 1), and typically residential areas have less ET than agricultural areas. This is similar to the urbanization (or "deforestation") effect explained by Morton (1983), which causes increased runoff and decreased ET. This argument is proven here too, where the modified GG model captures the land cover/land use effect using climatic variables only. In essence, the complementary method represented here using the modified GG model recognizes the effect of land use/cover on the adjacent atmospheric layer without the details of soil–plant interactions.

A similar effect is noticed at Tamale, where the crop ET is higher than the actual ET in the last half of the year during the growing cycle. This could be simply referred to as an urbanization effect, similar to in Sunyani, because Tamale is the capital city and the economic hub of northern Ghana. Urban areas with a high proportion of impervious

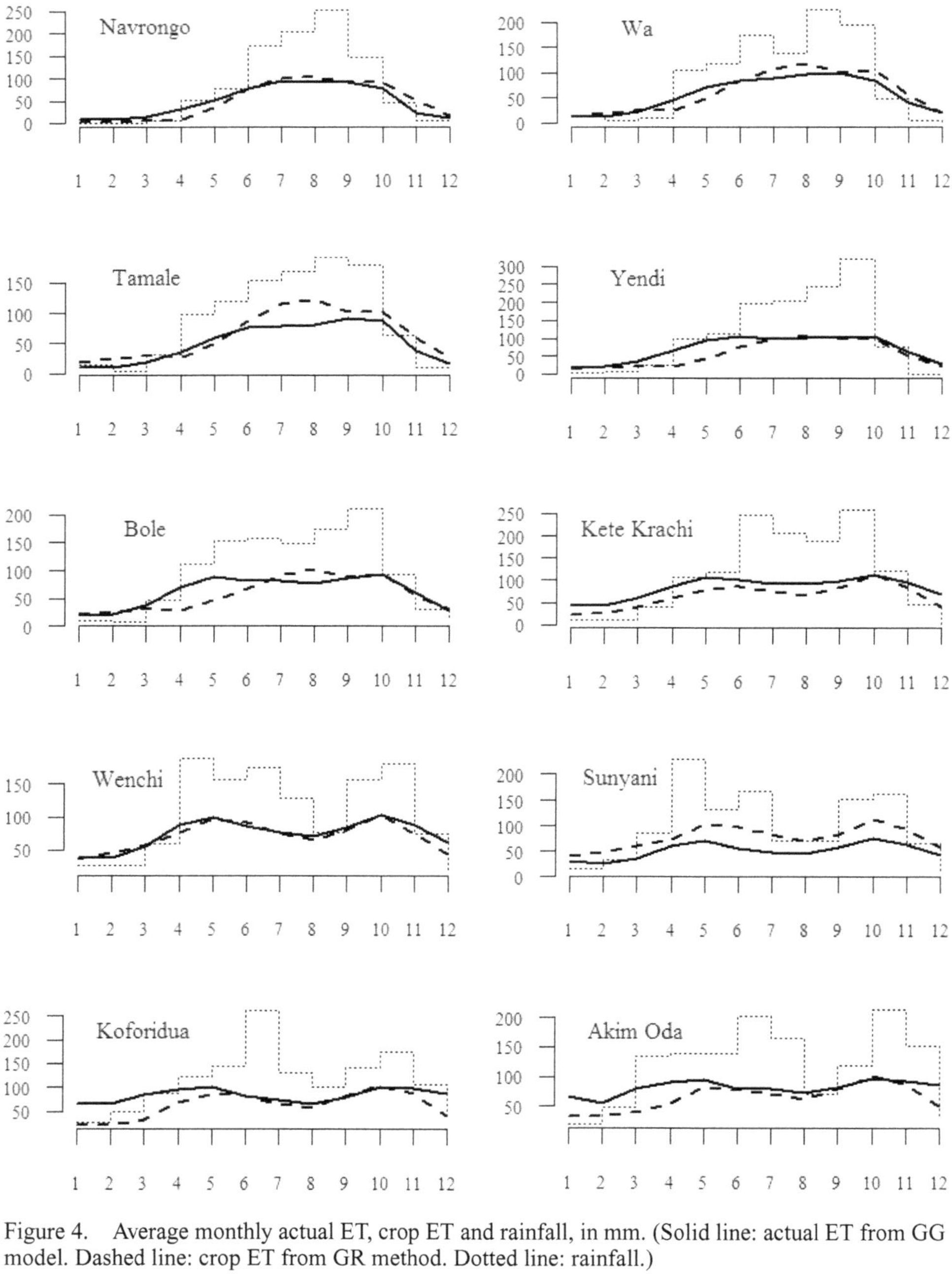

Figure 4. Average monthly actual ET, crop ET and rainfall, in mm. (Solid line: actual ET from GG model. Dashed line: crop ET from GR method. Dotted line: rainfall.)

surfaces "are drier and release more heat than surrounding rural areas", which is the "urban heat island effect" explained by O'Driscoll et al. (2010). Compared to vegetated surfaces, impervious surfaces have lower ET rates, and therefore rainfall is more likely to produce runoff or infiltration (O'Driscoll et al., 2010).

Crop ET is marginally higher than actual ET during the growing season in other semi-arid stations, Navrongo and Wa, indicating that a smaller number of crops may actually exist. Kete-Krachi is uniquely different from all other stations. Kete-Krachi is situated at

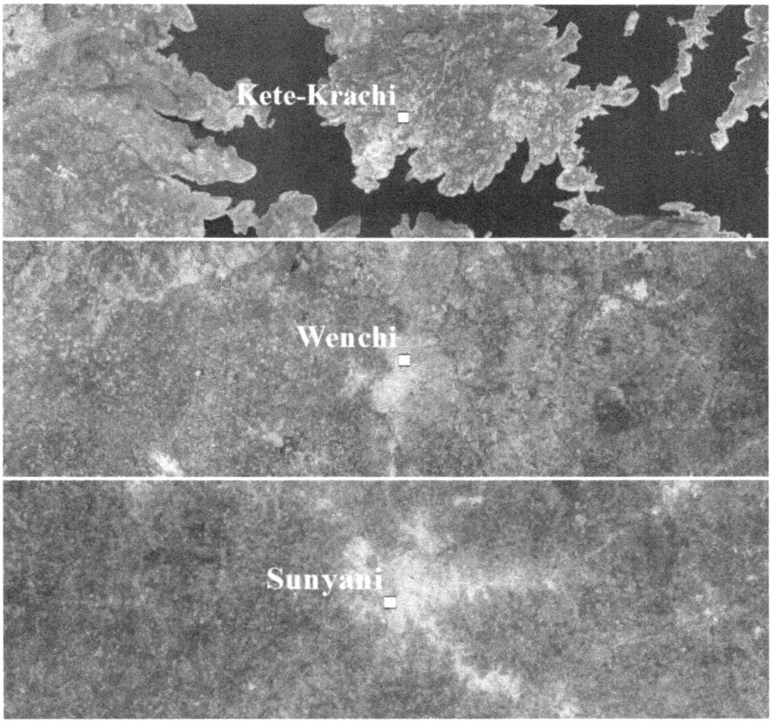

Figure 5. Land use in areas surrounding the Kete-Krachi, Wenchi and Sunyani stations, using the areal basemap feature of ArcMap 10 at 1:250000 scale.

the confluence of the river discharging from the Oti sub-basin and the river discharging from the Black and White sub-basins. It therefore acts as an "oasis", because the open water body surrounding the station causes ET to elevate beyond typical values of crop ET. The excess water vapour in the low atmospheric layer indicates higher relative humidity and lower temperatures in the surrounding air, which could be related to cooling caused by higher evaporation. The single exception occurs when the peak of the growing season is reached in September/October and water is abundant on the land surface. Hence, the difference between actual ET and crop ET at this station could be justified by the additional evaporation from the water bodies in the surrounding area.

Figure 5 shows the areal basemap generated by ArcMap 10 showing the land use pattern of each station. The maps clearly show that Kete-Krachi is surrounded by large water bodies whereas Wenchi and Sunyani are not. The signals from the meteorological variables (i.e. temperature, relative humidity, etc.) are induced by the interactions developed between the atmosphere and the underlying soil-plant system. These signals are processed and translated by the modified GG model to produce the regional estimated ET. Based on the previous discussion and the images shown in Figure 5, it can be concluded that the modified GG model is capable of predicting ET or total water loss from the land surface using meteorological data only and independent of land cover/land use information.

It is noticed that actual ET is higher than crop ET in late spring and early summer. This is the onset of the rainfall season, when the crops are in the initial stages of growing with a high water demand (high R_n and therefore high ETP). Therefore, ET is high during this period due to the high rates of evaporation. However, this is not the case during the minor rainfall season since soil moisture is still high enough to accommodate the water demand; therefore actual ET and crop ET are almost same during this period. In the fall season, actual ET in many stations is again higher than crop ET due to the end of the growing cycle when the soil is moderately moist.

Annual ET estimates

To better understand ET distribution in Ghana, annual ET estimates and ratios of ET to rainfall are shown in Table 3. Generally, there is a trend of ET decreasing with latitude. At Navrongo, the actual ET estimates are in good agreement with all crop ET estimates, but not at Sunyani. In general, the agreement between the actual ET and the crop ET estimates increases with latitude. The crop ET estimates from the GR and AL methods are almost identical, as expected.

Comparing actual ET from the modified GG model and crop ET from the GR method, the absolute difference between the two estimates (as a percentage of rainfall) ranges from zero at Navrongo to 27% at Sunyani (Table 3, Column 13). The average value of that absolute difference is 12%. This value is still close to the typical measurement uncertainty of about 10%. This finding also indicates that the modified GG model can reliably estimate ET using minimal data and unequivocally represent areal ET. The actual ET estimates from the modified GG model vary between 51% and 74% of the average annual rainfall, and the average ET/P value is 65%. This value is in excellent agreement with the globally observed average percentage of ET/P of 60% to 65% (Brutsaert, 1982) and is exactly equal to the mean value of the average percentages from the five classical methods as well. The crop ET estimates from the GR method range from 52% to 78% of the average annual rainfall, with an average of 62%. The metrics used here to compare the monthly ET estimates from the modified GG model and the GR method are root mean square error (RMSE), the bias of the mean values, and the coefficient of determination (R^2). The RMSE ranges from 11 mm/month at Wenchi to 32 mm/month at the neighbouring station, Sunyani. The

Table 3. Computed average annual actual or crop ET (mm) and corresponding ET/P ratios as a percentage. Diff. is the absolute difference between the percentages in Columns 2 and 4.

		[1]	[2]	[3]	[4]	[5]	[6]	[7]	[8]	[9]	[10]	[11]	[12]	[13]
Method		Complementary		Classical										Diff.
		Modified GG		ASCE										
Station	Rainfall	ET	%	GR	%	AL	%	HV	%	TC	%	JH	%	|[2] - [4]|
Navrongo	963	584	61	585	61	586	61	647	67	618	64	592	61	0
Wa	1040	686	66	723	69	713	69	808	78	787	76	735	71	3
Tamale	1040	603	58	765	74	784	75	809	78	770	74	702	67	16
Yendi	1291	838	65	670	52	647	50	797	62	746	58	708	55	13
Bole	1157	739	64	683	59	684	59	831	72	714	62	668	58	5
Kete-Krachi	1353	1006	74	776	57	747	55	875	65	830	61	783	58	17
Wenchi	1258	889	71	848	67	836	66	998	79	906	72	810	64	3
Sunyani	1197	612	51	930	78	974	81	1085	91	874	73	773	65	27
Koforidua	1384	1001	72	749	54	724	52	950	69	791	57	763	55	18
Akim Oda	1432	970	68	765	53	737	51	1021	71	814	57	766	53	14
Min	963	584	51	585	52	586	50	647	62	618	57	592	53	0
Avg.	1212	793	65	749	62	743	62	882	73	785	65	730	61	12
Max	1432	1006	74	930	78	974	81	1085	91	906	76	810	71	27

average RMSE is 22 mm/month, which is acceptable, given the uncertainty of the methods and the diversity of the environments. Knowing that the average bias is 3.7 mm/month, Sunyani has the minimum bias of –26 mm/month while Koforidua has the maximum of 21 mm/month. R^2 values vary between 0.34 at Koforidua and 0.83 at Wa, with an average of 0.65. As a result, it can be concluded that the modified GG model produces good regional estimates of ET.

The HV method tends to overestimate crop ET compared to the GR method, specifically under humid climates. As mentioned earlier, the reason could be the arid conditions under which the HV method was developed. The TC method simulates the ET estimates from the GR and AL methods perfectly among all different climates. With the exception of Sunyani, the JH method shares the same behaviour with the TC method. This information indicates that ET in the region is driven by radiation fluxes more than temperature. Hence, the radiation methods, developed under humid conditions, outperform the temperature method, which was implicitly developed for the study region. Therefore, the use of simple methods that require minimal data can be a good alternative to the more complex methods, such as the P-M equation, that are data intensive. Yet, estimates are sometimes incorrect and misleading, as the simple methods are unable to explain the dynamic interactions between the soil-plant system and the surrounding climate from the temperature signal only. The HV method, developed exclusively for West Africa, cannot be readily used in the same area under different hydro-climatic conditions. Each method constructed for a specific region under specific hydro-climatic conditions is subject to verification and perhaps calibration when applied elsewhere.

Comparison with prior studies

Acheampong (1986) computed ETP by three methods at six synoptic stations across Ghana, whereas Asare, Banini, Ayeh, and Amenorpe (2011) recently estimated ETP at the Atomic-Kwabenya site, 20 km north of Accra, using six models. Such comparative studies do not exist for actual ET estimates in Ghana. Given the fact that actual ET varies significantly, regional studies are important in managing the transboundary water resources of the Volta Basin. Oguntunde et al. (2005) verified the complementary relationship between ETP and ET for the Volta Basin, but they also found that actual ET estimates were unacceptably high (even higher than rainfall), and the values have still to be validated by further studies.

Remote sensing provides a good alternative for regional studies that can demonstrate the spatial distribution of ET. Remote sensing was applied to estimate actual ET using Makkink's equation and compared to scintillometer observations from August to December 2002 at three stations, namely Ejura (in the humid tropical region), Navrongo and Tamale (Schuttemeyer, Schillings, Moene, & de Bruin, 2007). In Upper East Ghana, Compaore et al. (2008) used remote sensing during the dry season of 2002/2003 at two spatial scales: MODIS (1000 m) and Landsat (30 m). In effect, prior remote-sensing studies conducted in Ghana are spatially and temporally confined and, more importantly, required large amounts of data and information.

The average annual actual and crop ET estimates are compared to the estimates from prior studies in the region, and the results are shown in Table 4. The comparison to prior studies conducted in different time periods should be made carefully, as hydro-climatic conditions may vary significantly. Rainfall, as an example, could play a pivotal role in explaining the uncertainties in the ET estimates from the different studies shown in Table 4. This is particularly true for Ghana, where rainfall varies significantly from north to south.

Table 4. Comparison of annual ET estimates (mm) and percentage of annual ET to annual rainfall (%) from this study with results from prior studies conducted in the region.

		Present study				[1][a]		[2][a]		[3][a]		[4][a]		[5][a]		[6][a]		[7][a]	
Method		Modified GG		Classical methods		Rushton		K-J/Mike SHE		Rushton and WASIM		WASIM		Ritchie/SWAT		Water balance		Thornthwaite–Mather	
Study area		Ghana		Ghana		Ghana		Densu River basin		Upper East		White Volta		White Volta		Volta Basin		Volta Basin	
Period		2000–05		2000–05		2000–05		2007–08		2003–04		1961–00		1986–99		1936–98		1931–95	
ET and ET/P		mm	%	mm	%	mm	%	mm	%	mm	%	mm	%	mm	%	mm	%	mm	%
Station	Navrongo	584	61	606	63	758	79			762	76								
	Tamale	603	58	766	74	862	83					739	75	603	74				
	Wa	686	66	753	72	891	86												
	Bole	739	64	716	62	970	84												
	Yendi	838	65	714	55	919	71												
	Kete-Krachi	1006	74	802	59	915	68												
	Wenchi	889	71	880	70	1071	85									910	91	892	89
	Sunyani	612	51	927	77	1006	84												
	Koforidua	1001	72	795	57				71										
	Akim Oda	970	68	821	57														
Average		793	65	778	65	886	80												

[a] [1] Carrier (2008), [2] Alfa et al. (2011), [3] Martin (2006), [4] Wagner (2008), [5] Obuobie (2008), [6] Andreini et al. (2000), and [7] Friesen et al. (2005).

Thus, ET/*P* ratios are better indicators than the absolute ET rates. It should be mentioned here that average annual crop ET represents the average value of the five crop ET estimates from the classical methods. First, comparing the estimates of the modified GG method with those of the classical methods, both sets of results have similar average values (793 and 778 mm/year) and percentage of rainfall (65%). Second, areal averages (not shown here) were calculated for both estimates using Thiessen polygons, and these areal average values are similar to the simple average values shown in Table 4. However, the areal average values cannot be used for comparison with estimates from the prior studies that only used simple average values.

The results of the classical methods show that the estimates from the proposed modified GG model are closer to those of both Obuobie (2008) and Wagner (2008), who worked in the White Volta River basin. However, Carrier (2008) and Martin (2006) overestimated actual ET, and both studies used the same method. Alfa, Hasholt, Jørgensen, and Banoeng-Yakubo (2011) used the MIKE SHE hydrologic model, which uses the Kristensen-Jensen (K-J) method, to predict ET over the Densu River basin, and their ET/*P* percentage (no mention of actual values) is close to that of the modified GG model at Koforidua, which lies within the same basin. The remaining two studies, by Andreini, van de Giesen, van Edig, Fosu, and Andah (2000) and Friesen et al. (2005), are generic studies in the upper basin, where climate is semi-arid to hyper-arid. The high value of ET/*P* may be due to the areas of high aridity, where most rainfall is lost as ET. Each of these studies produced an overall average ET that is close to that of Carrier (2008), whose area of study is much wetter (due to relatively high rainfall) compared to the upper basin. This observation is also shown by the difference in ET/*P* ratios, since the upper basin has much less water available for ET. A study in the upper basin, specifically south-eastern Burkina Faso, produced an ET/*P* ratio range of only 28% to 32% (Bagayoko, 2006). In contrast, the ratios from Andreini et al. (2000) and Friesen et al. (2005) are much higher than those found in similar regions of Africa (Brutsaert, 1982). Therefore, the ratios cannot be compared to the typical values of Ghana, in which the climate varies from semi-arid in the north to subtropical humid in the south. In addition, the tropical evergreen forests of south-western Ghana have higher rates of rainfall and very humid climatic conditions. In summary, the estimates of the modified GG model are comparable and similar to the average from most prior studies conducted in the same region. More noteworthy is that the ET/*P* ratio of the modified GG model agrees well with the globally known ET/*P* ratio.

Spatial distribution of ET

Figure 6 shows the spatial distribution of ET across Ghana using the simple kernel smooth method in ArcMap 10. ET decreases from south-east to north-west, but the trend changes to north-east in northern Ghana. This trend is similar to the distribution of rainfall, which is apparently the main driver of ET. Volta Lake contributes significantly to the water loss in the area which was evident at Koforidua and Kete-Krachi. Carrier (2008) conducted a study in northern Ghana and found that the trend is from south-south-west to north-north-east, consistent with the findings of this study.

The polygons in Figure 6 are the representative areas of synoptic stations used in this study; the numbers inside each polygon provide the total water loss as ET using actual ET from the modified GG model. The total volume of annual rainfall is approximately 297 km^3, while the actual ET represents 66% of the rainfall volume, or about 195 km^3. The difference between these two values is the water excess which is available as surface runoff and groundwater recharge.

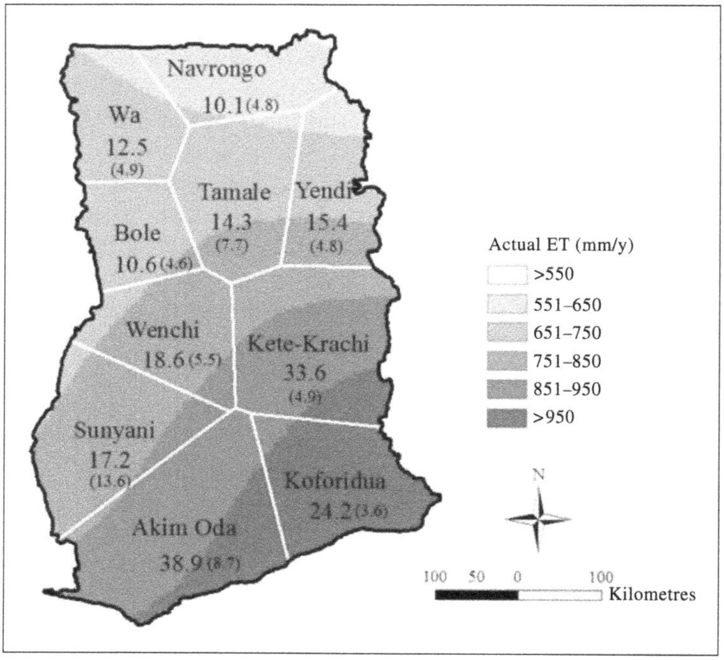

Figure 6. Spatial distribution of actual ET, where the polygons show the representative areas falling within each synoptic station. The number inside each Thiessen polygon is the volume of annual water loss from each area in km^3, while the number in parentheses is the volume of annual groundwater recharge in km^3.

Andreini et al. (2000) performed a water budget analysis for the Volta Basin and found the average total rainfall to be 400 km^3, of which 195 km^3 (or 48.8%) falls in Ghana. An additional 100 km^3 falls in southern Ghana, outside the Volta River basin (see Figure 1). Given the high humidity and prevailing tropical climate, rainfall in this southern part is high. Friesen et al. (2005) performed a similar water balance in the Volta Basin and found that the average annual rainfall is 401 km^3, while the average ET accounts to 357 km^3, or 89% of the rainfall, which is much higher than predicted by this study. The model used by the other study was a modified Thornthwaite-Mather, and no details of the model were provided. There is no obvious reason for this extremely high value, but compared to all other studies, the ET$/P$ ratio of 89% is too high.

Groundwater recharge

Normally, surface runoff data are critical to determining groundwater recharge, but this information is not readily available in the Volta River basin (Carrier, 2008; Obuobie, 2008). Yet runoff can be estimated, since it accounts for a relatively small portion of the rainfall in this region. Most studies, including Andreini et al. (2000), Barry, Obuobie, Andreini, Andah, and Pluquet (2005), Carrier (2008), Friesen et al. (2005) and others, used simple methods such as a constant runoff coefficient to estimate runoff. In this method, runoff is computed as a fixed percentage (constant coefficient) of the rainfall for a given catchment; these coefficients are derived from historic observations and previous knowledge of hydrology of the region. Runoff derived by this method may include contributions from baseflow or interflow. Therefore, runoff estimates may be overestimated, especially in humid regions. Runoff estimation adds more uncertainty to the recharge predictions.

The Nawuni station, which has the longest streamflow time series, is considered the outflow to the White Volta River (see Obuobie, 2008; Wagner, 2008). Daily streamflow data (in m^3/s) are sporadically available from 1954 to 2005. Although not shown here, plotted streamflow showed interesting and different patterns from 1954 to 1994 and from 1995 to 2005. Before 1995, the average streamflow volumes were almost negligible in the dry season of January to April. However, these volumes increased significantly (4–9 times higher) during the dry season in 1995, after the construction of the Bagre Dam in Burkina Faso in the same year. Given these uncertainties and variability, runoff is simply computed using the constant coefficient method, assuming that the contribution of baseflow is insignificant, especially in dry areas.

In this study, the runoff coefficients developed by Barry et al. (2005) are used for two reasons. First, coefficients are given for the four sub-basins of the Volta River basin; and second, the coefficients are adjusted to suit the Volta River basin of Ghana, which is wetter than the conditions of neighbouring Burkina Faso. Barry et al. (2005) found runoff coefficients (as a portion of the annual average rainfall) of 8.3% for Black Volta, 10.8% for White Volta, 14.8% for Oti River and 17% for Lower Volta. One may note that Akim Oda lies outside the boundaries of the Volta River basin. Therefore, the runoff coefficient of the closest station (Koforidua) is used, and this value conforms with the data of Alfa et al. (2011). Eventually, the average annual recharge rates are determined from Equation (8) and shown in Table 5.

According to Table 5, groundwater recharge ranges from 11% to 41% of the average annual rainfall. An obvious trend is that groundwater recharge rates (absolute values) generally increase with latitude in Ghana. Gumma and Pavelic (2012) delineated zones of high groundwater potential using many techniques and found the same trend with latitude (with minor exceptions, mostly in south-western Ghana). Forkuor, Pavelic, Asare, and Obuobie (2013) developed a groundwater development potential map for the northern regions, namely the Upper East, the Upper West, and the Northern.

The upper limit may be questionable; Sunyani, which has a unique behaviour as per this study, should have generated higher rates of runoff rather than recharge. Sunyani also

Table 5. Average annual values of water budget components (mm) and the corresponding percentages of the average annual rainfall rate.

		Rainfall	Evapotranspiration		Runoff		Recharge	
Station	Sub-basin	P	ET	%	Q	%	Rech.	%
Navrongo	White	963	584	61	104	11	275	29
Wa	Black	1040	686	66	86	8	267	26
Tamale	White	1040	604	58	112	11	324	31
Yendi	Oti	1291	838	65	191	15	262	20
Bole	Black	1157	739	64	96	8	322	28
Kete-Krachi	Oti	1353	1006	74	200	15	147	11
Wenchi	Black	1258	889	71	104	8	265	21
Sunyani	Black	1197	612	51	99	8	485	41
Koforidua	Lower	1384	1001	72	235	17	147	11
Akim Oda	Lower	1432	970	68	243	17	218	15
Average		1211	793	65	147	12	271	23

produced a lower percentage of ET/*P*. Sunyani is the capital of the Brong-Ahafo Region, which is a part of the forest zone of Ghana. Typically, forest ecosystems have the probability of decreasing runoff and increasing ET. Such trends, however, are more prominent at the neighbouring station, Wenchi (see Table 5). Gumma and Pavelic (2012) found that the groundwater potential is higher at Wenchi than at Sunyani. The authors believe that further data and information are required from this region to investigate this discrepancy.

The second-highest percentage of recharge is 31%, at Tamale, in which a semi-arid climate prevails and groundwater is the main source of water (John Aduakye, personal communication, 1 August 2009). The results of this study are promising and show that the aquifer system is sufficiently recharged in semi-arid northern Ghana (see Figure 6) compared to the estimated groundwater abstraction rates documented by Carrier (2008) and John Aduakye (personal communication, 1 August 2009) and the groundwater development maps of Gumma and Pavelic (2012) and Forkuor et al. (2013).

Despite the high rates of rainfall, Forkuor et al. (2013) found that lower rates of recharge are expected from the Oti sub-basin (Yendi station), which has a "steep topography" and hence generates 30–40% of the Volta River basin's annual, perennial flow (Barry et al., 2005). At Kete-Krachi station, the low recharge rate contradicts the findings of Gumma and Pavelic (2012). It is worth mention that the low recharge rates in the south, namely at Koforidua and Akim Oda, perfectly match with those predicted (15–22% of rainfall) by Alfa et al. (2011) and agree with the findings of Gumma and Pavelic (2012). The annual recharge rates for different soil types range from 120 mm to 153 mm in the Densu River basin (Alfa et al., 2011), and these data are very close to the average recharge rate at Koforidua.

According to Figure 6, high rates of recharge are expected in the south-west of Ghana, where a humid climate prevails, the rainfall is bimodal, and the land cover consists of intensive tropical forests. The potential for developing groundwater resources in the northern region is promising, as supported by Gumma and Pavelic (2012) and Forkuor et al. (2013). It is important to note that the total volume of recharge is approximately 63 km^3, or 21% of the rainfall volume, and produces a runoff volume of about 13%, or 38 km^3. Hence, the water excess comprises 34% of the rainfall volume across Ghana. The percentages of the water budget components from Figure 6 are a little different from those depicted in Table 5. This is because the figures from the former are corrected by the area of the corresponding polygon for each station.

Given the minimum recharge for the four principal sub-basins of the Volta River basin (see Barry et al., 2005), the groundwater resources of Ghana are adequately replenished and recharged. Across the country, the average annual recharge rate is 23% of the rainfall, suggesting a high potential for groundwater availability to satisfy the growing future demands.

Comparison to prior recharge estimates

Recharge rates computed by the water balance method cannot be directly compared to those estimated by water table fluctuation methods in which the water flows to the water table and directly contributes to the aquifer system. In other words, recharge rates determined by the water table fluctuation method and chloride mass balance represent "actual or net recharge", i.e. recharge minus ET from groundwater (see Martin, 2006). However, recharge rates computed by water balance represent "potential recharge" that may not fully reach the water table, given the presence of the soil moisture zone and loss due to ET.

Table 6. Comparison of estimated annual recharge between this study and previous studies in absolute and percentage terms.

Reference	Method	Rainfall range (mm/y)	Recharge (% of rainfall) Min.	Avg.	Max.	Study area
This study	Water balance	963–1432	11	23	41	Ghana
Houston (1982)	Penman-Grindley	640–1470	0.0	30.8	84.6	Kabwe in Zambia
Friesen et al. (2005)	Thornthwaite-Mather		0.0		10.0	Volta River basin
Martin (2006)	Chloride mass balance	910–1138	3.0	5.9	6.2	Upper East Region
	Water table fluctuation		1.4		12.5	
	Water balance		4.0		13.0	
	WaSiM-ETH model		7.2		14.3	
Nyagwambo (2006)	Chloride mass balance	507–962	4.0	12.0	25.0	Central Zimbabwe
	Water balance		6.0	8.0	9.0	
	Water table fluctuation		9.0	13.0	14.0	
Sandwidi (2007)	Water table fluctuation	573–1197	5.3	17.0	29.4	Southern Burkina Faso
Carrier (2008)	Chloride mass balance	800–1250	1.5	4.4	10.6	North of Ghana
	Water balance		1.8	6.8	15.9	
Obuobie (2008)	Chloride mass balance	870–1294	3.4	8.3	18.5	Upper East Region
	Water table fluctuation		2.5		16.0	White Volta of Ghana
	SWAT model			7.0		White Volta Basin
Alfa et al. (2011)	MIKE SHE	850–1650	15.0		22.0	Densu River basin

A modified Thornthwaite-Mather model developed by Friesen et al. (2005) showed that groundwater recharge could approach 10% of average annual rainfall in a wet year, while recharge is negligible in a dry year (see Table 6). In northern Ghana, recharge from water balance (as depicted in Table 6) was found to be lower than the estimates of chloride mass balance conducted by Carrier (2008).

Martin (2006) and Obuobie (2008) estimated groundwater recharge rates using chloride mass balance, water table fluctuation, and modeling approaches. In the Upper East Region of Ghana, chloride mass balance produced recharge rates from 3.4% to 18.5% and from 3% to 6.2% of the rainfall, according to Obuobie (2008) and Martin (2006), respectively. Applying the same method in the same Upper East Region resulted in recharge estimates that are three times the others (Table 6); this gives an indication of the high uncertainty of the methods despite the variation in rainfall rates. Using the water table fluctuation method, however, the results of the two studies matched, as both used the same specific yield range while the study areas varied. It should be noted that the Upper East Region is much wetter than the remaining parts of the White Volta River basin.

In the White Volta River, Martin (2006) and Wagner (2008) used the Water Balance Simulation Model, or WaSiM. Such models provide good simulations but require large amounts of data that are typically not available and may have to be assumed or obtained from calibration. It is important to mention that most prior studies were conducted by the GLOWA Project (http://www.glowa-volta.de/), and therefore these studies (e.g. Andreini et al., 2000; Friesen et al., 2005) used common approaches. Although the previous estimates are mostly lower than the estimates from this study, there are many unanswered questions about the methods and corresponding assumptions, given the significantly different trends and findings from these studies. In addition, Alfa et al. (2011) applied MIKE SHE to the Densu River basin in southern Ghana and found that annual recharge may approach 22% of rainfall.

Houston (1982) estimated recharge in a semiarid climate at Kabwe in Zambia using a modified version of the Penman-Grindley recharge model. The mean annual precipitation was 911 mm from 1965 to 1980, while the annual recharge varied from 26 mm to 771 mm and 0 to 534 mm under bare soil and open forest, respectively. The mean annual recharge values were 281 mm and 80 mm for bare soil and open forest, respectively. The water table fluctuation method was applied to the Kompienga Dam basin in Burkuna Faso, in which the annual rainfall is 830 mm (Sandwidi, 2007). The recharge rate was estimated to be 5.3% to 29.4% of the annual rainfall. In another study in Zimbabwe, Nyagwambo (2006) applied the chloride mass balance method in a semi-arid region and estimated the recharge rate to be between 4% and 25% of rainfall. Groundwater studies across many tropical countries of Africa have shown that actual recharge may exceed 20% of the annual rainfall (as estimated by the water table fluctuation method; see Nyagwambo, 2006), indicating high rates of recharge. All abovementioned independent studies suggest that recharge rates are likely to be higher than a few per cent of rainfall in similar and neighbouring environments of Ghana.

Conclusions

This study provides water balance information for all of Ghana using ET estimates from a modified complementary method that relies on meteorological data only. Ghana has a complex hydrology, varying from humid regions in the south to semi-arid and arid regions in the north. Given the fact that ET is an influential driver of water balance, estimating ET with available data is a prerequisite to any accurate estimation of groundwater recharge. Given the limited data on land cover and land use available in Ghana, the use of simple but reliable methods to estimate ET from readily available data is an important step. In this work, the proposed modified GG model of Anayah (2012) is used to estimate country-wide ET for Ghana, followed by estimating groundwater recharge.

In general, the estimates from the modified GG model that represents the complementary methods are in good agreement with crop ET calculated using the classical methods. In selected stations such as Kete-Krachi and Sunyani, where other types of land uses such as water bodies and urban areas are present, the modified GG model captured total water loss from all sources, confirming the argument that the complementary methods are capable of predicting ET independent of land use/cover information. More importantly, the modified GG model that uses the complementary relationships requires meteorological data only, making the analysis simple and straightforward.

The results indicate that the actual ET ranges from 51% to 74% of the total annual rainfall, with an average of 65%. The spatial distribution shows that actual ET decreases from south-east to north-west; the trend switches to north-east in northern Ghana, following the rainfall distribution. The P-T equation used to compute ETW from the modified GG model is close to the P-M equation used in the ASCE grass reference (GR) method, and the relative absolute difference of annual values is around 9%. Comparing the ET estimates from the GG model and the GR method, the average RMSE and R^2 over the 10 stations are 22 mm/month and 0.65, respectively. Most importantly, this comparison shows that the modified GG model provides regional-scale estimates of ET from point observations of meteorological data only.

Although many recharge studies have been conducted in this region, none have been rigorously tested and evaluated using direct recharge measurements. There are a few exceptions, as cited in this study, that are of limited spatial and temporal scales and cannot be used at a national level. Therefore, prior studies cannot be directly followed. The water balance equation was used on a country-wide scale from 2000 to 2005. The results show that Ghana has on average 1211 mm of annual rainfall, of which 65% accounts for ET, 12% for surface runoff, and the remaining 23% for groundwater recharge. The natural groundwater recharge rates range from 11% to 41% of the average annual rainfall rates.

An obvious trend is found, that groundwater recharge generally increases with latitude in Ghana. These results are encouraging in northern Ghana, where a semi-arid climate prevails and groundwater is a major source of water. These findings are supported by recent studies conducted in Ghana (e.g. Forkuor et al., 2013; Gumma & Pavelic, 2012). When compared to the minimum recharge rates for the four principal sub-basins of the Volta River basin provided by Barry et al. (2005), this study indicates that the aquifer systems of Ghana are adequately recharged. This suggests a high potential for developing these groundwater resources to accommodate the increasing demands for water. Still, water quality is a concern recently, because of pollution produced by the disposal of residential and agricultural wastes (see Carrier, 2008). The other concern is the drying climate trend experienced in Ghana over the past 50 years (Lacombe, McCartney, & Forkuor, 2012), which suggests the necessity of better management of the available water resources to minimize the impacts of climate change and climate variability in the next century.

Acknowledgements

This study was financially supported by the Utah Water Research Laboratory at Utah State University. The authors would like to acknowledge the IWMI for research collaboration and providing travel funds from the project on Groundwater in Sub-Saharan Africa, funded by the Rockefeller Foundation. More specifically, the assistance of Dr Boubacar Barry, Dr Emmanuel Obuobie and Mr Gerald Forkuor of IWMI is appreciated. The authors would like to acknowledge Dr Enoch Asare of the Water Resources Commission of Ghana and personnel of the Ministry of Food and Agriculture and Meteorological Services of Ghana for the valuable data provided for this study. The assistance and comments of Dr Marc-Andre Carrier and Mr James Racicot, who were with CIDA during the period of this study, are appreciated very much. The authors are grateful to the anonymous reviewer for comments and suggestions that greatly improved this study.

Note

1. These data from the Ghanaian Ministry of Food and Agriculture website (http://mofa.gov.gh/site/) are no longer available, but a copy can be obtained on request from the first author.

References

Acheampong, P. K. (1986). Evaluation of potential evapotranspiration methods for Ghana. *JeoJournal*, *12*(4), 409–415. doi:10.1007/BF00262364

Alfa, B., Hasholt, B., Jørgensen, N. O., & Banoeng-Yakubo, B. (2011). Rainfall and water resources of a coastal basin of Ghana. *Journal of Hydrologic Engineering*, *16*(4), 316–323. doi:10.1061/(ASCE)HE.1943-5584.0000314

Allen, R. G., Pereira, L. S., Raes, D., & Smith, M. (1998). *Crop evapotranspiration: Guidelines for computing crop water requirements – FAO irrigation and drainage* (Paper No. 56). Retrieved from FAO website: http://www.fao.org/docrep/X0490E/X0490E00.htm

Allen, R. G., Walter, I. A., Elliot, R., Howell, T., Itenfisu, D., & Jensen M. (Eds.). (2005). *The ASCE standardized reference evapotranspiration equation*. American Society of Civil

Engineers Environmental and Water Resource Institute (ASCE-EWRI). Retrieved from http://www.kimberly.uidaho.edu/water/asceewri/ascestzdetmain2005.pdf

Anayah, F. M. (2012). *Improving complementary methods to predict evapotranspiration for data deficit conditions and global applications under climate change* (Unpublished doctoral dissertation). Utah State University.

Andreini, M., van de Giesen, N., van Edig, A., Fosu, M., & Andah, W. (2000). *Volta basin water balance*. ZEF – Discussion Papers on Development Policy (Report No. 21). Bonn. Retrieved from ZEF website: http://www.zef.de/fileadmin/webfiles/downloads/zef_dp/zef-dp21-00.pdf

Asare, D. K., Banini, G. K., Ayeh, E. O., & Amenorpe, G. 2011. Estimation of potential evapotranspiration for a coastal Savannah environment: Comparison of methods. *International Journal of Sustainable Agriculture*, *3*(2), 65–70.

Bagayoko, F. (2006). *Impact of land-use intensity on evaporation and surface runoff: processes and parameters for Eastern Burkina Faso, West Africa* (Unpublished doctoral dissertation). University of Bonn, Germany.

Barry, B., Obuobie, E., Andreini, M., Andah, W., & Pluquet, M. (2005). *The Volta River Basin: Comprehensive assessment of water management in agriculture, comparative study of river basin development and management*. International Water Management Institute (IWMI). Retrieved from IWMI website http://www.iwmi.cgiar.org/assessment/files_new/research_projects/River_Basin_Development_and_Management/VoltaRiverBasin_Boubacar.pdf

Brutsaert, W. (1982). *Evaporation into the atmosphere: Theory, history, and applications*. Dordrecht, Holland: D. Reidel Publishing Company.

Brutsaert, W., & Stricker, H. (1979). An advection aridity approach to estimate actual regional evaporation. *Water Resources Research*, *15*(2), 443–450. doi:10.1029/WR015i002p00443

Carrier, M. (2008). *Hydrogeological synthesis of Northern Ghana* (Unpublished master's thesis). Avant-Garde University, Canada.

Chiew, F. H., & Leahy, C. P. 2003. Comparison of evapotranspiration variables in evapotranspiration maps for Australia with commonly used evapotranspiration variables. *Australian Journal of Water Resources*, *7*(1), 1–11.

Compaore, H., Hendrickx, J., Hong, S., Friesenc, F., van de Giesen, N. C., Rodgersd, C., & Vlek, P. (2008). Evaporation mapping at two scales using optical imagery in the White Volta Basin, Upper East Ghana. *Physics and Chemistry of the Earth*, *33*, 127–140. doi:10.1016/j.pce.2007.04.021

de Martonne, E. (1925). *Traité de géographie physique [Treaty of physical geography]*. Paris: A. Colin.

Doyle, P. (1990). Modelling catchment evaporation: An objective comparison of the Penman and Morton approaches. *Journal of Hydrology*, *121*, 257–276. doi:10.1016/0022-1694(90)90235-P

FAO. (n.d.). Crop calendar. Food and Agriculture Organization of the United Nations. Retrieved from: http://www.fao.org/agriculture/seed/cropcalendar/welcome.do

Forkuor, G., Pavelic, P., Asare, E., & Obuobie, E. (2013). Modelling potential areas of groundwater development for agriculture in northern Ghana using GIS/RS. *Hydrological Sciences Journal*, *58*(2), 437–451. doi:10.1080/02626667.2012.754101

Friesen, J., Andreini, M., Andah, W., Amisigo, B., & van de Giesen, N. (2005). *Storage capacity and long-term water balance of the Volta Basin, West Africa*. Proceedings of the Seventh IAHS Scientific Assembly, Brazil.

Granger, R. J., & Gray, D. M. (1989). Evaporation from natural nonsaturated surfaces. *Journal of Hydrology*, *111*, 21–29. doi:10.1016/0022-1694(89)90249-7

Granger, R. J., & Gray, D. M. (1990). Examination of Morton's CRAE model for estimating daily evaporation from field-sized areas. *Journal of Hydrology*, *120*, 309–325. doi:10.1016/0022-1694(90)90156-R

Gumma, M. K., & Pavelic, P. (2012). Mapping of groundwater potential zones across Ghana using remote sensing, geographic information systems, and spatial modeling. *Environmental Monitoring and Assessment*, *185*(4), 3561–3579. doi:10.1007/s10661-012-2810-y

Hargreaves, G. L., Hargreaves, G. H., & Riley, J. P. (1985). Agricultural benefits for Senegal River Basin. *ASCE Journal of Irrigation and Drainage Engineering*, *111*(2), 113–124. doi:10.1061/(ASCE)0733-9437(1985)111:2(113)

Hobbins, M. T., Ramirez, J. A., Brown, T. C., & Claessens, L. H. (2001). The complementary relationship in estimation of regional evapotranspiration: The Complementary Relationship Areal Evapotranspiration and Advection-Aridity models. *Water Resources*

Research, 37(5), 1367–1387. Retrieved from http://wolfweb.unr.edu/homepage/tylers/index.html/pdfpapers/hobbinsa.pdf
Houston, J. F. T. (1982). Rainfall and recharge to dolomite aquifer in a semi-arid climate at Kabwe, Zambia. *Journal of Hydrology, 59*, 173–187. doi:10.1016/0022-1694(82)90010-5
Jensen, M. E., Burman, R. D., & Allen, R. G. (1990). *Evapotranspiration and irrigation water requirements* (Report No. 70). New York: ASCE.
Jensen, M. E., & Haise, H. R. (1963). Estimating evapotranspiration from solar radiation. *ASCE Journal of the Irrigation and Drainage Division, 89,* 15–41. Retrieved from http://eprints.nwisrl.ars.usda.gov/1248/1/23.pdf
Lacombe, G., McCartney, M., & Forkuor, G. (2012). Drying climate in Ghana over the period 1960-2005: Evidence from the resampling-based Mann-Kendall test at local and regional levels. *Hydrological Sciences Journal, 57*(8), 1594–1609. doi:10.1080/02626667.2012.728291
Martin, N. (2006). *Development of a water balance for the Atankwidi catchment, West Africa – A case study of groundwater recharge in a semi-arid climate* (Unpublished doctoral dissertation). University of Gottingen, Germany.
Morton, F. I. (1983). Operational estimates of areal evapotranspiration and their significance to the science and practice of hydrology. *Journal of Hydrology, 66*, 1–76.
Nyagwambo, N. L. (2006). *Groundwater recharge estimation and water resources assessment in a tropical crystalline basement aquifer*. Unpublished doctoral dissertation, Delft University of Technology, Netherlands.
Obuobie, E. (2008). *Estimation of groundwater recharge in the context of future climate change in the White Volta River Basin, West Africa* (Unpublished doctoral dissertation). University of Bonn, Germany.
O'Driscoll, M., Clinton, S., Jefferson, A., Manda, A., & McMillan, S. (2010). Urbanization effects on watershed hydrology and in-stream processes in the southern United States. *Water, 2*(3), 605–648. doi:10.3390/w2030605
Oguntunde, P. G., van de Giesen, N., & Andreini, M. (2005, October). *Modeling regional evapotranspiration in the Volta Basin, West Africa: Testing the complementary relationship hypothesis*. Paper presented at the International Agricultural Research for Development Conference, Stuttgart-Hohenheim, Germany.
Penman, H. L. (1948). Natural evaporation from open water, bare soil, and grass. *Proceedings of the Royal Society of London,* A193(1032), 120–146. Retrieved from http://www.jstor.org/discover/10.2307/98151?uid=3738872&uid=2129&uid=2&uid=70&uid=4&sid=21101563615891
Priestley, C. H. B., & Taylor, R. J. (1972). On the assessment of surface heat flux and evaporation using large-scale parameters. *Monthly Weather Review, 100*(2), 81–92. Retrieved from http://docs.lib.noaa.gov/rescue/mwr/100/mwr-100-02-0081.pdf
Rietveld, M. R. (1978). A new method for estimating the regression coefficients in the formula relating solar radiation to sunshine. *Agricultural Meteorology, 19*, 243–252. doi:10.1016/0002-1571(78)90014-6
Sandwidi, W. J. P. (2007). *Groundwater potential to supply population demand within the Kompienga dam basin in Burkina Faso* (Unpublished doctoral dissertation). University of Bonn, Germany.
Schuttemeyer, D., Schillings, C. H., Moene, A. F., & de Bruin, H. A. R. (2007). Satellite-based actual evapotranspiration over drying semiarid terrain in West Africa. *Journal of Applied Meteorology and Climatology, 46*(1), 97–111. doi:10.1175/JAM2444.1
Shuttleworth, W. J., & Wallace, J. S. (2009). Calculating the water requirements of irrigated crops in Australia using the Matt-Shuttleworth approach. *Transactions of the ASABE, 52*(6), 1895–1906. Retrieved from https://elibrary.asabe.org/abstract.asp?aid=29217&t=2&redir=&redirType=
Statistics, Research and Information Directorate. (2010). *Agriculture in Ghana: Facts and figures (2009)*. Ghana: Ministry of Food and Agriculture. Retrieved from http://mofa.gov.gh/site/
Turc, L. (1961). Estimation of irrigation water requirements, potential evapotranspiration: A simple climatic formula evolved up to date. *Annals of Agronomy, 12*, 13–49.
Wagner, S. (2008). *Water balance in a poorly gauged basin in West Africa using atmospheric modelling and remote sensing information* (Unpublished doctoral dissertation). University of Stuttgart, Germany.

Appendix 1. Classical methods of calculating evapotranspiration

Potential ET (ETP) is widely used by hydrologists, meteorologists and agronomists. Three concepts are common in the scientific community and sometimes used interchangeably: ETP, reference ET (ET_0) and open water evaporation (E_p). In water resources studies, ETP is widely used to indicate the climate-driven water demand that is governed only by atmospheric conditions and not restricted by availability of water on the surface. However, ETP and ET_0 are different, because ETP does not depend on a reference crop while ET_0 is specified for a given reference crop. Moreover, ET_0 is extensively used in irrigation practices. In contrast, open water evaporation, introduced by Penman (1948), differs from ETP since ETP is influenced by the soil-plant system. In effect, the ambiguity in using ETP is still present, and it largely depends on the application. Brutsaert (1982) mentioned that ETP is often estimated by meteorological data that are not necessarily measured under potential conditions. In comparison, water resources studies focus on seasonal or annual estimates at regional scales, as opposed to local estimates.

Worldwide, the common classical methods for estimating ETP are the temperature methods (Hargreaves et al., 1985), radiation methods (Priestley & Taylor, 1972), and combination methods (Penman, 1948). Among all the methods of estimating ETP, the Food and Agriculture Organization (FAO) version of the Penman-Monteith equation (Allen et al., 1998), similar to the American Society of Civil Engineers (ASCE) version (i.e. Allen et al., 2005), is currently considered to be the best method.

As described by Allen et al. (1998) and Allen et al. (2005), ETP is computed for a reference crop such as grass or alfalfa and then multiplied by a crop coefficient for a given crop to determine crop ET. These crop coefficients are computed using prevailing climatic conditions for water-unstressed plant communities, mostly under humid environments. Extrapolation of these coefficients to arid and semi-arid climates is therefore questionable (Shuttleworth & Wallace, 2009). Another limitation is the lack of information on crop coefficients and growing cycles for different regions or countries. Examples include crops such as yam, cocoyam and plantain that are common in Ghana. For instance, maize is grown in Ghana but not addressed by Allen et al. (1998). Maize grown in Nigeria has published information, but climatic differences are significant between that country and Ghana, affecting the growing cycle.

Groundwater potential for dry-season irrigation in north-eastern Ghana

Emmanuel Obuobie, Deborah Ofori, Sampson Kwaku Agodzo and Collins Okrah

This paper assesses the groundwater resource potential for dry-season vegetable irrigation in two areas of north-eastern Ghana. It uses multiple methods, including geophysical surveying, recharge estimation, and water quality analysis. Results indicate that groundwater abstractions for all purposes are small compared to recharge. The quality of groundwater in both study areas is suitable for irrigation, but a few wells had high nitrate levels for drinking water. There is a potential to expand dry-season irrigation with groundwater 14- to 18-fold in the study areas.

Introduction

Agriculture, particularly at a small scale, is a major contributor to the Ghanaian economy. It accounts for about 28% of the GDP and employs over 56% of the work-force, who are mostly vulnerable rural dwellers (CIA, 2012). Therefore, growth in the agricultural sector will aid the growth of the economy as well as ensuring employment to reduce rural–urban migration. Ultimately, it will ensure food security and contribute immensely to the health and well-being of the Ghanaian population.

Agriculture in Ghana is largely rain-fed. However, given the unpredictable nature of the rainfall, irrigation development is one of few strategies available for increasing agricultural production. Globally, there is a strong positive correlation between wider use of irrigation and lower poverty rates (Lipton, Litchfield, & Faures, 2003). In Africa, only 3% of cropland is irrigated; thus, the region has experienced only marginal reduction in poverty since the 1990s. In contrast, regions that have large proportions of irrigated area (East Asia, the Pacific, North Africa and the Middle East) have experienced the greatest poverty reduction (Lipton et al., 2003). On the other hand, the large-scale irrigation boom experienced in parts of Asia was made possible through infrastructure development, access to cheap energy, credit and market integration – factors that catalyzed massive private investment. These facilities are not readily available in Sub-Saharan Africa, which includes Ghana.

The irrigation potential in Ghana is estimated to be in the range of 120,000–500,000 ha (Agodzo, Huibers, Chenini, van Lier, & Duran, 2003). The total area under irrigation in 1996 was estimated at 11,000 ha (Kyei-Baffour & Ofori, 2006). This represents only 0.26% of the total land area under cultivation. The irrigated area since 1996 has largely remained the same (Memuna & Cofie, 2005). Irrigation of some arable land in the country could not happen due to the prohibitive capital investment required for channelling surface water over long distances to the irrigable land. Widespread availability of groundwater is therefore a major asset that can greatly influence agricultural production. Developing groundwater for smallholder irrigation holds promise for strengthening livelihoods and improving food security (Molden, 2007). However, the use of groundwater for irrigation in Ghana is not totally new. Large-scale production of shallot and other vegetables using shallow groundwater in the Keta Strip has provided enormous income to the indigenous inhabitants (Kortatsi, Young, & Mensah-Bonsu, 2005).

Similarly, shallow groundwater irrigation using hand-dug wells and dugouts has been spreading in northern Ghana and is mostly found in low-lying areas. It is an important source of revenue for supplementing household income. However, its sustainability in the near future cannot be guaranteed because little is known about the quantity and quality of groundwater in many of the areas where it is practised. The hydrogeology in most parts of Ghana is highly variable. Over 90% of the land area is underlain by crystalline or consolidated rocks, which are characterized by low-permeability aquifers with limited storage and extent. While the groundwater seems to be adequate for domestic water supply and small-scale supplemental irrigation, it may not be able to support the intensive irrigation development potential that may be envisaged to enhance Ghana's agricultural development agenda. In this regard, refined quantitative answers are needed to understand the availability of groundwater in areas where shallow groundwater irrigation is, or could be, practised and subsequently the potential to expand groundwater irrigation to improve food security and livelihoods.

This study, therefore, sought to investigate the groundwater resource potential for supporting groundwater-based irrigation in two selected areas within two administrative districts in north-eastern Ghana and the implications for food security and livelihoods.

Description of study areas

This study was conducted in the Zalerigu and Sapeliga areas in the Upper East Region (UER) of Ghana. The UER is one of the 10 administrative regions of Ghana and located in the extreme north-east of the country. The Zalerigu area (118 km^2) encompasses the Zalerigu community and is situated in the Talensi-Nabdam District (912 km^2) of the UER (Figure 1). The Sapeliga area (86 km^2) is located in the Bawku West District (979 km^2) of the UER and covers the Sapeliga community. Both districts were selected for this study because they are well known for irrigation activities in the region, especially irrigation with groundwater (Dittoh & Awuni, 2012).

Talensi-Nabdam District

The Talensi-Nabdam District lies between latitude 10°15′ and 10°60′ N and longitude 0°31′ and 0°05′ W. The climate is classified as tropical, with two distinct seasons: a rainy season (May–October) and a long dry season (October–April). The annual rainfall ranges between 880 and 1100 mm. The area experiences a maximum temperature of 45 °C in April and a minimum of 12 °C in December. The vegetation is guinea savannah woodland consisting of short deciduous trees and a ground flora of grass. The district is underlain

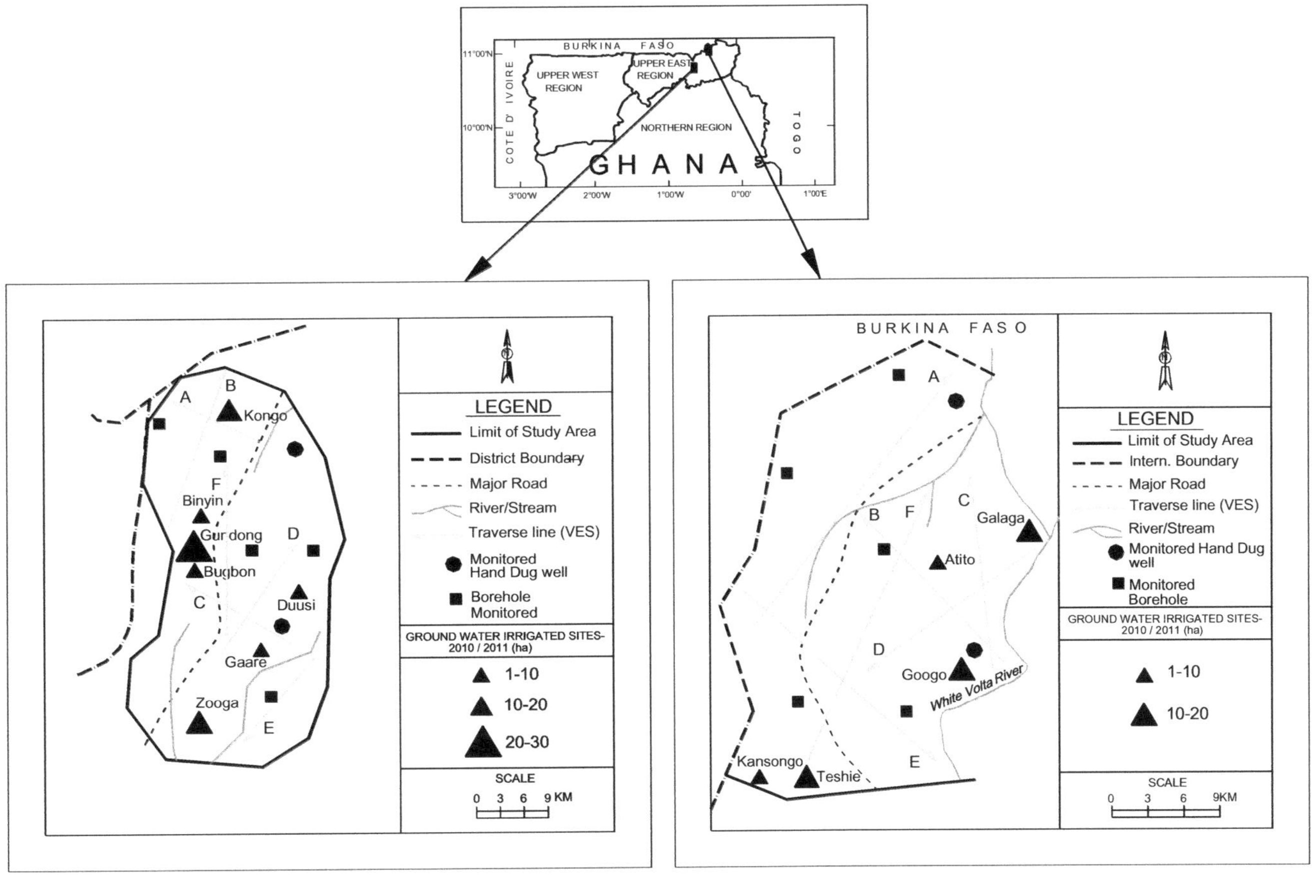

Figure 1. Map of the study areas, Sapeliga (top) and Zalerigu (bottom), showing the locations of groundwater-irrigated sites (2010/2011 irrigation season), boreholes/dug wells sampled, and VES traverse lines.

by the Dahomeyan formation, Birimian sediments, Birimian volcanic, basal sandstone and Tarkwaian formation. The Zalerigu area is largely underlain by Birimian volcanic, which consists of metamorphosed lava, pyroclastic rocks, hypabyssal basic intrusive, phyllite and greywacke (Agyekum & Dapaah-Siakwan, 2008; Obuobie & Barry, 2012).

The total population of the district is 115,020, consisting of 57,702 males and 57,318 females (GSS, 2012). The population is mainly rural, with about 90% uneducated. Agriculture, mainly crop farming, is the major economic activity in the district. It is the main source of employment and accounts for about 90% of local GDP. The district has about 49,200 ha of cultivable land, with about 15,465 ha currently under cultivation. Agriculture is largely rain-fed, and cultivation of vegetables in the dry season depends on irrigation.

Bawku West District

The Bawku West District is located between latitudes 10°30′N and 11°10′ N, and between longitudes 0°20′ and 0°35′ E. The land area is about 979 km^2. Similar to the Telensi-Nabdam district, the Bawku West district has a tropical climate, which consists of a rainy season of about five months (May–September) and a dry season of seven months (October–April). The rainfall pattern is unimodal, with mean annual values in the range of 800–1000 mm (Obuobie, 2008). Temperature varies from a minimum of 18 °C in December to a maximum of 38 °C in April, with a mean value of 28 °C. The geology of the district comprises Dahomeyan super group (61%), basal sandstone (18%), Birimian sediments (16%), and Birimian volcanic (5%). The Sapeliga area within the district is underlain by Dahomeyan (mostly undifferentiated granitoid) (Agyekum & Dapaah-Siakwan, 2008; Obuobie & Barry, 2012).

The district has a total population of about 94,034, comprising 45,114 males (48%) and 48,920 females (52%) (GSS, 2012). Settlements are mainly rural (91% rural, 9% urban). Small-scale agriculture is the main source of livelihood in the district and employs over 80% of the work-force. The district has a total cultivable area of 58,406 ha, but less than half of this amount is presently cultivated. Vegetables are the main crops grown in the dry season, using runoff harvested in small reservoirs and shallow groundwater.

Methods and data

Geophysical investigation

Geophysical investigations were conducted in the Zalerigu and Sapeliga areas to determine the thickness of the groundwater aquifers. The investigation was done using electromagnetic (EM) profiling and vertical electrical sounding (VES), as described briefly below.

Electromagnetic profiling

Electromagnetic profiling measurements were made along 8 traverse lines (233 m on average) in the Zalerigu area and 6 traverse lines (216 m on average) in Sapeliga, using a Geonics EM 34–3 ground conductivity meter. This equipment uses the principle of electromagnetic induction to directly measure apparent conductivity. Details of the principle of electromagnetic induction can be obtained from McNeil (1980).

A 40 m inter-coil separation cable was used for the apparent conductivity measurements in the two study areas. For that length of cable, the maximum depth below the ground that could be investigated was 30 m when operated in the horizontal dipole (HD) mode and 60 m when operated in the vertical dipole (VD) mode. Measurements were made in both HD and VD modes at intervals of 10 m along each traverse line.

Vertical electrical sounding

VES was done at 150 points at Sapeliga and 120 points at Zalerigu, at intervals of 20 m. This was done to determine the depth and thickness of the shallow groundwater aquifers. Generally, the points were selected at random along well defined traverses. In a few instances, however, the selection of VES points was influenced by the availability of space for setting up the survey equipment. This was the case for portions of traverse lines that were heavily cropped at the time of the survey. The measurements were done with an SAS 1000C Terrameter, adopting the Schlumberger electrode expansion procedure (Loke, 2002). The resulting data from the VES were analyzed using RESIST software (Veplen, 1988). The depth of investigation varied from zero to 60 m below ground level. The quality of the VES data was controlled by plotting the measured values in the field while the sounding was in progress. Unrealistic values were rejected and the sounding repeated at the same point when necessary.

Estimation of groundwater volume

The volume of groundwater (V_{GW}) that can be stored on an annual basis (potential groundwater storage) in the regolith (residual soil and saprolite) in the study areas was computed as the product of the average porosity of the aquifer material (n_{av}), the average cross-sectional area (A_{av}), and the length of the study area (L):

$$V_{GW} = n^*_{av} A^*_{av} L \tag{1}$$

Estimation of porosity

Porosity was computed based on the work of Vukovic and Soro (1992), who derived porosity from its empirical relationship with the coefficient of grain uniformity:

$$n = 0.255(1 + 0.83^U) \tag{2}$$

where n is the porosity and U is the coefficient of grain uniformity, given as:

$$U = \frac{d_{60}}{d_{10}} \tag{3}$$

where d_{60} and d_{10} are the grain diameters in mm for which 60% and 10% of the soil samples, respectively, are finer than the remainder of the sample.

Soil samples were taken from 2 layers (0–50 cm and 50–100 cm) of selected hand-dug wells used by farmers for irrigation (3 wells per study area) and analyzed at the laboratory of the Soil Research Institute, Kumasi, Ghana, to determine d_{60} and d_{10}. The laboratory results were used to compute porosity per Equation (2). The computed porosity values were compared with estimates made from the relationship between porosity and dry bulk density, using soil dry bulk density data from previous studies in the Upper East Region (Amegashie, 2009).

Estimation of cross-sectional area

The cross-sectional area for each study area was computed using Simpson's rule (Matthews, 2004). The extent of the aquifer in each study area was divided into six sections. Each section was divided into strips to estimate the cross-sectional area. The depth between the upper part of the topsoil and the near lower part of the saprolite was taken to contain the shallow groundwater that is of interest to this study (Figure 2). The estimated areas of the strips were summed to obtain the cross-sectional areas for each section. The areas of the various cross-sections were averaged to obtain the average cross-sectional area for each study area.

Mapping groundwater-irrigated plots

Groundwater-irrigated plots were mapped in a detailed field survey at Sapeliga and Zalerigu from December 2010 to February 2011 as part of estimating the total land area irrigated with groundwater. This was achieved with the aid of a hand-held GPS receiver. In each area, the coordinates (longitude and latitude) and the elevation of the vertices of the plots were captured with the GPS. The coordinates were later retrieved and processed in the Esri ArcView geographic information system to estimate the size of plots cultivated. The individual plot sizes were then aggregated to obtain the total land area cultivated with groundwater in each study area.

Estimation of groundwater uses

Household interviews were used to determine the proportion of groundwater abstracted from boreholes that is used for domestic water supplies, livestock watering and industry.

TYPICAL LITHOLOGIES	THICKNESS
Reddish-brown silty quartz sand with	0.6 up to 1m
lateritic concretions	up to 5 m
Reddish-brown to cholate medium-grained sand completely weathered	0.8 – 10m
Highly weathered multi-coloured granite, medium-grained	up to 5.1 m
Deeply weathered coloured granite, semi-hard and medium-grained with feldspar, fractures filled with secondary clay	up to 21m
Largely unweathered bedrock with some stairing on fractures	~

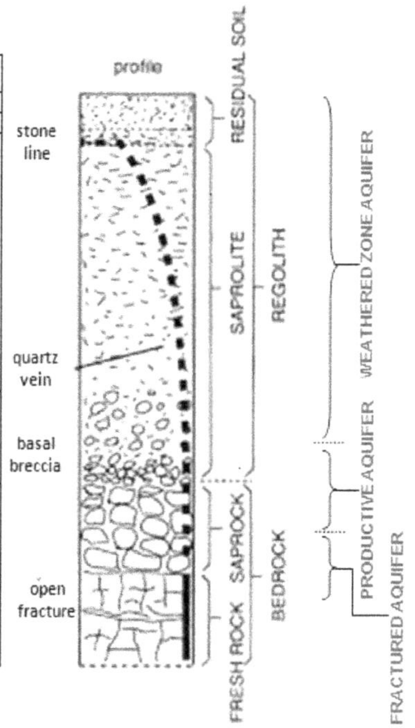

Figure 2. Schematic representation of the hydrogeological profile in the Zalerigu area, north-eastern Ghana (modified from Chilton & Foster, 1995; HAP, 2006).

A total of 55 households were interviewed, 30 in Zalerigu and 25 in Sapeliga. The rationale for using the household was to allow for discussions among members of the same household to agree on a common use pattern for each household, thereby eliminating divergent views of individual members of the same household.

Estimation of groundwater abstraction

The total volume of groundwater abstracted in each of the study areas was computed based on groundwater abstracted from boreholes for domestic use, livestock watering and industrial use, and groundwater abstracted from hand-dug wells for dry-season irrigation. Groundwater abstracted for livestock watering was considered to be included in domestic groundwater use because at all the borehole sites in the study areas, water troughs were provided as part of the borehole design to collect overflows of water from which cattle and other livestock are watered. In both study areas, groundwater for domestic use, industry and livestock watering is almost entirely abstracted from drilled boreholes and was therefore calculated by multiplying the average yield of boreholes by the number of functional boreholes. Hand-dug wells are seldom used for domestic purposes. The yields of boreholes in the study areas were obtained from borehole data provided by the Community Water and Sanitation Agency (CWSA), Upper East Regional Office, Bolgatanga, Ghana. The volume of water abstracted was based on the assumption of an average daily pumping duration of 8 hours. A similar approach has been used by Agyekum and Dapaah-Siakwan (2008) and Martin and van de Giesen (2005) to estimate the volume of groundwater abstracted in the UER of Ghana and the Volta Basin, respectively. During field surveys, a total of 31 and 34 functional boreholes were found in the Sapeliga and Zalerigu areas, respectively. Based on the available yield information on the functional boreholes, the average groundwater yields computed were 2.6 m^3/h for the Sapeliga area and 3.4 m^3/h for the Zalerigu area (Table 1).

Groundwater abstracted for irrigation was computed based on the number of hand-dug wells used in each study area in the 2010/2011 irrigation season (172 wells in Zalerigu and 125 wells in Sapeliga) and the average yield of hand-dug wells. There is hardly any water use data for hand-dug wells and dugouts in the study areas and in Ghana in general. This is possibly because hand-dug wells (excluding modernized hand-dug wells) and dugouts are constructed by individuals for their personal use and therefore do not get any support from NGOs or the government. Kortatsi (1994) estimated the yield of hand-dug wells to range from 0–26 m^3/day, with an average of 6 m^3/day. In the absence of other information, this average yield value was used in this study.

Table 1. Characteristics of functional boreholes in the Zalerigu and Sapeliga areas in north-eastern Ghana.

Area	Geology	No. of functional boreholes	Borehole success rate* (%)	Average borehole depth (m)	Transmissivity (m^2/d)	Yield (m^3/h)
Zalerigu	Birimian	34	75 (58–83)	41 (35–62)	6.4 (0.8–62)	3.4 (0.5–26)
Sapeliga	Dahomeyan	31	39 (28–47)	52 (45–70)	3.9 (0.5–32)	2.6 (0.7–15)

Note: *Successful boreholes are generally defined as those with a minimum yield of about 0.78 m^3/h.
Data source: Community Water and Sanitation Agency, Ghana.

Groundwater quality monitoring

The quality of groundwater in the 2 study areas was monitored monthly for a period of 11 months (January to November 2010) to determine the suitability of the groundwater for domestic and agricultural uses. Water samples were taken from five boreholes across each study area for analysis. In addition to the boreholes, two hand-dug wells in each area were monitored for four months (January to April 2010). A total of 65 samples were taken from each of the study areas. Water quality parameters analyzed included pH, temperature, electrical conductivity, sodium, calcium, magnesium, sulfate, nitrate, carbonate, bicarbonate, chloride and total dissolved solids. Electrical conductivity, pH and temperature were measured *in situ* using an Eijkelkamp 18.21 multi-parameter analyzer field kit; the remaining parameters were analyzed in the laboratory. Sampling and analysis were done in accordance with the protocols described by Claasen (1982) and by Barcelona, Gibb, Helfrich, and Garske (1985).

Results and discussion

Aquifer and borehole characteristics

The occurrence of groundwater in the entire north-east of Ghana, including the two study areas, is controlled by the extent of weathering of the overburden material, the presence of fractures and the degree of their interconnectivity (Agyekum & Dapaah-Siakwan, 2008). Analysis of well logs and aquifer characteristics of functional boreholes in the study areas (obtained from the CWSA, Bolgatanga, and the Water Research Institute, Accra) reveals that there are, generally, three main groundwater aquifers located within the overburden and in the fresh rocks. These are: the weathered zone aquifer; the bedrock–weathered zone interface aquifer; and the fractured zone aquifer. The weathered zone aquifer can be found within the regolith, which consists of the topsoil and the highly weathered mantle or saprolite (Figure 2). This aquifer occurs as either phreatic or semiconfined, with high static water levels (1.5–4.0 m below ground level). Its yield depends strongly on the thickness of the weathered mantle and the topography of the area. The thickness of the weathered mantle, from the upper part of the topsoil, varies between 8 and 36 m in Zalerigu and between 11 and 34 m in Sapeliga. In most parts of the study areas, the topsoil is made up of sandy clay material, which has allowed the development of discontinuous shallow perched aquifers at some locations.

The bedrock–weathered zone interface aquifer is the most productive groundwater aquifer in the study areas. This is also the case for most areas in Ghana and the Volta Basin underlain by Precambrian crystalline rocks (Martin, 2006; Obuobie & Barry, 2012). It comprises the lower part of the regolith and the upper part of the saprock. The depth to this aquifer ranges between 32 and 49 m in the Sapeliga area and between 35 and 58 m in the Zalerigu area. The thickness of this aquifer varies from 3 to 5 m in the Sapeliga area. For the Zalerigu area, the thickness ranges between 2 and 7 m. Many of the boreholes in the study areas and in northern Ghana tap water from this aquifer. The fractured zone aquifer is the deepest of the three main aquifers and is located in the saprock. It occurs under semiconfined to confined conditions and is relatively high yielding.

Table 1 presents a summary of some characteristics of boreholes in the two study areas. Comparatively, the chance of drilling a successful borehole in the Zalerigu area is far higher than in Sapeliga (Table 1). Also, transmissivity and borehole yields are higher in Zalerigu than in Sapeliga. Generally, boreholes in the Sapeliga area are deeper than those in the Zalerigu area.

Groundwater storage

Measurements from the EM profiling in the study areas were plotted to obtain dipole response curves. The vertical and horizontal dipole responses were compared to determine the existence of structures associated with the occurrence of groundwater. Samples of VES measurements are presented in Table 2. The VES results suggest that both study areas are underlain by four distinct subsurface geological layers, as depicted by available lithological logs (Figure 2).

The first layer is the residual soil, which consists of reddish-brown sandy clay material at the top (topsoil) and lateritic concretions at the bottom. The thickness of the topsoil ranges between 0.6 and 1 m in the Sapeliga area and between 0.5 and 0.9 m in the Zalerigu area. The topsoil at Sapeliga has a higher clay content compared to Zalerigu, and therefore permeability is lower. The thickness of the residual soil is generally less than 5 m in both study areas (Figure 2). The second layer, saprolite, consists of highly weathered and conductive materials of variable resistivity and thickness. The upper to the middle parts of this layer contain the shallow groundwater or the weathered zone aquifer, which is tapped by small-scale farmers for dry-season irrigation in the two study areas. The lower part of this layer interfaces with the upper part of the third layer, saprock, to house the most productive aquifer in north-eastern Ghana. The productive aquifer is located at a greater depth than hand-dug wells reach and is therefore not regarded as a shallow aquifer from a groundwater-irrigation point of view. The thickness of the saprolite ranges from 0.8 to 15.1 m in the Sapeliga area, depending on the extent of weathering. The mean thickness and standard deviation were computed to be 5.8 m and 3.2 m, respectively. In the Zalerigu area, the thickness of this layer ranges from 1.4 m to 18.7 m, with mean and standard deviation of 8.4 m and 2.7 m, respectively. The third and fourth layers are the fractured bedrock (saprock) and fresh rock.

The total volume of groundwater that can be stored in the shallow aquifer in the Zalerigu and Sapeliga areas was computed using Equation (1). The lengths of the Zalerigu and Sapeliga study areas were estimated (using the XTools extension in ArcView) to be 16,385 m and 13,565 m, respectively. The average porosity values computed over a depth of a metre and used in this study, based on Equations (2) and (3), were 0.258 for Zalerigu and 0.273 for Sapeliga. Compared to the porosity values of 0.311–0.509 estimated based on the relationship between porosity and dry bulk density and data from Amegashie (2009), the porosity values used in this study appear to be on the low side. However, the values obtained in this study could be considered average values, accounting for variations in porosity over the entire depth of the shallow aquifers in the study areas. It is generally

Table 2. Sample VES data at two locations in the Zalerigu and Sapeliga areas, north-eastern Ghana.

VES point	Layer	Apparent resistivity (Ω m)	Thickness (m)	Depth (m)
Zalerigu (10°47′24″N 0°40′55″W)	1	275	0.9	0.9
	2	84	1.6	1.9
	3	1901	11.5	12.7
	4	161	–	–
Sapeliga (11°05′24″N 0°27′19″W)	1	96	0.7	0.7
	2	37	2.3	3.1
	3	2053	14.9	16.2
	4	237	–	–

Table 3. Estimated annual groundwater storage in the Zalerigu and Sapeliga areas, north-eastern Ghana.

Parameter	Zalerigu	Sapeliga
Average porosity	0.283	0.275
Average cross-sectional area (m^2)	18,830	16,500
Length of study area (m)	16,385	13,565
Potential groundwater storage (10^6 m^3)	87.4	61.6

known that the porosity of surface soils is higher than in subsurface soils due to compaction of subsurface layers. Such variation was explicitly considered within a metre depth in both study areas and not over the entire depths of the shallow aquifers.

The groundwater storage in the shallow aquifer was estimated to be 87.4×10^6 m^3 for the Zalerigu area and 61.6×10^6 m^3 for the Sapeliga area (Table 3).

Groundwater recharge and abstraction

Recharge to the groundwater aquifers in the study areas was determined from past recharge studies in areas of north-eastern Ghana that encompassed the two study areas (Martin, 2006; Obuobie, 2008). The two past studies employed multiple recharge estimation techniques, including water-table fluctuation, chloride mass balance and modelling to estimate the recharge. The recharge values obtained at locations in and around the study areas with the different techniques in the past studies were averaged to obtain average recharge values.

The annual recharge obtained for the Zalerigu area ranged from 33.7 mm to 183.2 mm (3.4% – 18.5% of the mean annual rainfall of 990 mm), with an average of 63.4 mm (6.4% of mean annual rainfall). In the Sapeliga area, the annual recharge was estimated to be between 37.4 mm and 84.8 mm (4.1% – 9.3% of the annual rainfall of 912 mm). The average recharge overall was estimated to be 50.2 mm (5.5% of annual rainfall). The recharge values obtained for the study areas largely compare favourably with recharge estimates obtained by independent studies in Northern Ghana, including the Upper East Region where this study was done (Acheampong, 1988; Friesen, Andreini, Andah, Amisigo, & van de Giesen, 2005; Carrier, Lefebvre, Racicot & Asare, 2008) but somewhat differs from values (20%–31% of annual rainfall) obtained by Anayah, Kaluarachchi, Pavelic & Smakhtin (2013) for their study sites in northern Ghana (Bole, Yendi, Tamale,Wa and Navrongo). The differences could be attributed to differences in the methods used in the two studies as well as the scale at which the estimates were made. While the recharge values used in this study could be described as actual recharge (since they determine the water flux to the watertable), the work of Anayah et al. (2013) was based on a water balance approach that estimates potential recharge below the root zone of vegetation (Scalon, Healy, & Cook, 2002). However, it is important to note that the work of Anayah et al. (2013) is a valuable first step in the estimation of recharge at a larger scale. In volume terms, recharge to the shallow groundwater was computed to be 7.6×10^6 m^3 and 4.3×10^6 m^3, respectively, in the Zalerigu and Sapeliga areas.

The annual total volume of groundwater abstracted from boreholes and hand-dug wells in the study areas was estimated to be 0.49×10^6 m^3 for the Zalerigu area and 0.36×10^6 m^3 for the Sapeliga area (Table 4).

Table 4. Estimated annual groundwater abstraction in Zalerigu and Sapeliga areas, north-eastern Ghana.

Parameter	Zalerigu	Sapeliga
Volume of groundwater abstracted from boreholes (10^6 m^3)	0.36	0.24
Volume of groundwater abstracted from hand-dug wells (10^6 m^3)	0.13	0.12
Total annual groundwater abstraction (10^6 m^3)	0.49	0.36

Table 5. Quality of groundwater in Zalerigu and Sapeliga, north-eastern Ghana.

	Sapeliga ($N = 65$)	Zalerigu ($N = 65$)	WHO drinking water guideline (2004)	Irrigation water quality guideline*
pH	6.9 – 7.5	6.9 – 7.6	6.5–8.5	6.5 – 8.4
Temperature (°C)	31 – 33	31 – 32	–	–
Electrical conductivity (μS/cm)	210 – 583	260 – 748	–	0 – 3000
Chloride Cl (mg/l)	0.1 – 4.7	0.3 – 3.9	250	0 – 1065
Sodium Na (mg/l)	38.1 – 53.1	24.5 – 30.6	200	0 – 920
Fluoride F (mg/l)	0.7 – 1.2	0.5 – 0.9	1.5	–
Manganese Mn (mg/l)	0.1 – 0.4	0.1 – 0.7	–	0 – 120
Calcium Ca (mg/l)	23.7 – 49.4	23.3 – 42.5	200	0 – 400
Nitrate NO_3-N (mg/l)	1.9 – 25.9	0.3 – 31.7	10	0 – 620
Bicarbonate HCO_3 (mg/l)	276 – 283	283 – 312	–	0 – 610
Total dissolved solids (mg/l)	128.5 – 372.8	165.4 – 478.3	1000	0 – 2000

*From Ayers and Westcot (1994). According to the authors, these guidelines are intended to cover the wide range of conditions encountered in irrigated agriculture. Several basic assumptions have been used to define their range of usability. The guidelines may be adjusted to fit the local situation, depending on the conditions of use and availability of sufficient experience, field trials, research or observations.

Groundwater quality

The quality of groundwater in both study areas can generally be described as good for multiple uses, including drinking and irrigation (Table 5). All the analyzed water quality parameters were largely within the WHO guideline values for drinking water and fully within the recommended guideline values for irrigation (Ayers & Westcot, 1994). Ghana does not have national guidelines for irrigation water quality. There were a few cases of nitrate values above the recommended levels for drinking water supply (10 mg/l), which gives reason for concern (when used for domestic water supply), though the levels were far below the upper threshold for irrigation (620 mg/l). These included a borehole and two hand-dug wells in Zalerigu, with nitrate levels in the range of 14.3–25.9 mg/l, and 2 boreholes and a hand-dug well in Sapeliga, with nitrate levels ranging from 11.2 to 31.7 mg/l. The excess nitrate levels may be attributed to the use of nitrate-based fertilizers for agricultural purposes as well as indiscriminate disposal of human waste.

Groundwater uses

Groundwater is used for multiple purposes in the study areas. The uses are domestic water supply, which includes livestock watering and industrial use, and irrigation of crops. Groundwater for domestic water supply is abstracted from boreholes, while

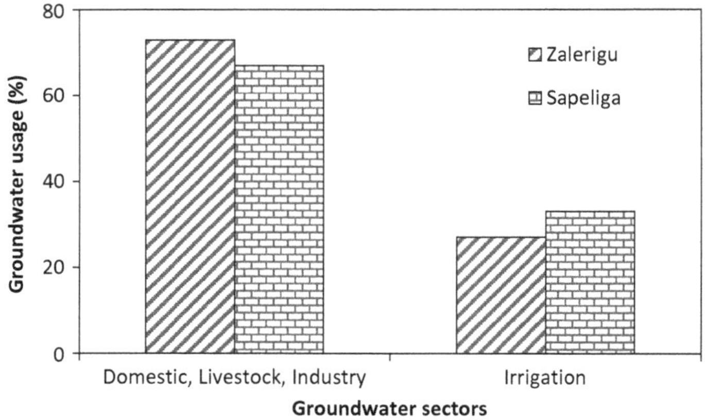

Figure 3. Groundwater uses in the Zalerigu and Sapeliga areas, north-eastern Ghana.

irrigation groundwater is abstracted through hand-dug wells. In both study areas, domestic water supply (abstracted from boreholes) is the dominant user of groundwater (Table 4). About 73% and 67% of the groundwater abstracted in the Zalerigu and Sapeliga areas, respectively, is used for domestic water supply, including livestock watering and industry (Figure 3). Irrigation accounts for 27% and 33% of usage, respectively, in Zalerigu and Sapeliga. The domestic water supply in Zalerigu consists of domestic usage (67%), livestock watering (4%) and industry (2%). In Sapeliga, domestic water supply comprises domestic usage (60%), livestock watering (5%) and industry (2%).

Characteristics of groundwater irrigation

Irrigation of crops, mostly vegetables, is an important use of groundwater in both the Zalerigu and Sapeliga areas. The practice takes place solely in the dry season, from November to March. In both areas, irrigators use hand-dug wells and dugouts to extract groundwater from low-lying areas, like floodplains, valley bottoms and alluvial channels along the courses of ephemeral streams, to produce vegetables on plot sizes between 0.03 and 0.21 ha in the Zalerigu area and between 0.01 and 0.16 ha in the Sapeliga area. Results from field mapping showed that a total of 51.8 ha and 40.3 ha were put under groundwater irrigation in the Zalerigu and Sapeliga areas, respectively, in the 2010/2011 irrigation season. The area cultivated varies from one year to another, and it is influenced by the profitability of the previous season, access to credit facilities, availability of land (willingness of landowners to rent or give out land for free), and the rainfall situation of the preceding rainfall season, among other factors. Many of the irrigators do not own the land they cultivate. Lands may be rented or cultivated for a "token" (i.e. a gift in the form of farm produce) in the case where the irrigator has family relations with the landowner. Many of the landowners engage in rain-fed agriculture and do not cultivate in the dry season. The main vegetables grown in the study areas are tomato, onion and pepper.

Watering of crops is done mostly by buckets and cans, though a few of the irrigators have invested in small-capacity motorized pumps (mostly petrol-driven) to lift water from wells to irrigate their crops. As with irrigation, all other farm operations (e.g. field preparation, planting and weeding) are done manually. This makes the practice tedious and labour intensive. Unfortunately, the period of groundwater irrigation coincides with the period of labour shortage in the study areas and northern Ghana at large. Many young people migrate to the south of Ghana to do menial jobs during the dry season. Irrigators mostly rely on traditional knowledge and local experience of water availability, acquired from their parents and from many years of practice, to identify suitable locations for wells. Drilling of wells is done manually, using small implements such as hoes, pick-axes and shovels. Generally, the average depth of a hand-dug well is about 1.4 m in the Zalerigu area and 1.8 m in the Sapeliga area. The digging of the wells is done in phases during the cultivation period. There is generally an initial drilling done to a depth of about 1 m to collect water for irrigation at the initial stages of the crop life cycle. The wells are later deepened two to three times, as the water-table sinks, to collect more water for irrigation at the flowering and maturity stages of the crop.

Benefits and constraints of groundwater irrigation

Many of the irrigators in the study areas mentioned that groundwater-irrigated vegetable production is profitable and an important source of income to the household. However, they do take losses in years of poor rainfall or low recharge, which results in less water being available in the shallow aquifers for crop water needs. This study did not include in-depth data collection and analysis to determine the level of revenue irrigators make from groundwater irrigation. However, an attempt was made at quantifying the profits through an extensive informal discussion with an irrigator at Zalerigu who is a landowner and cultivated about 1.2 ha of tomato, onion and pepper. The net revenue, based on the 2008/2009 irrigation season, was GHC650 (equivalent to USD458, based on the April 2010 exchange rate). Similar profit levels were reported by Dittoh and Awuni (2012) and Laube, Awo, and Schraven (2008) in a survey of groundwater irrigators in north-eastern Ghana.

Some of the irrigators are full-time farmers in the rainy season, but many of them are employed in other sectors and therefore profits from groundwater irrigation are considered additional income. This additional income is substantial, considering that the annual average household income in north-eastern Ghana (Upper East Region) in 2006 was reported to be less than GHC130 (USD141.30 at an exchange rate of USD1 = GHC0.92) (GSS, 2008). Groundwater irrigation, therefore, can be an effective strategy for overcoming poverty in north-eastern Ghana.

Notwithstanding that groundwater irrigation is profitable, it is constrained by many challenges, including lack of access to credit facilities, land access limitations, lack of appropriate drilling technology, market access and extension services.

Irrigation expansion, food security and livelihoods

The potential for using groundwater to expand irrigated agriculture in the study areas was assessed based on the recharge to the groundwater aquifers and the crop water requirements of the main vegetable grown in the areas (tomato). Currently, the total volume of groundwater abstracted from the Zalerigu area represents 6.4% of the annual recharge. In the Sapeliga area, the abstracted volume represents 8.4% of the recharge. Barring other influencing factors, the quantities of groundwater available in both study areas are capable of sustaining groundwater-irrigated vegetable production for now and in the near future. Comparing the recharge volumes and the crop water requirement for the main vegetable grown in both study areas, there exists the potential to increase groundwater-irrigated land from the present 52 ha to 930 ha in the Zalerigu area (an 18-fold increase), based on 50% of the recharge (using the assumptions of Pavelic, Villholth, Shu, Rebelo, & Smakhtin, 2013) and subject to the availability of sufficient suitable lands in low-lying areas. Similarly, there is a potential to increase groundwater-irrigated land from the current 40 ha to 548 ha in the Sapeliga area (a 14-fold increase).

The expansion of groundwater irrigation could offer many more households the opportunity to increase their income and alleviate poverty. Higher income may result in an improved standard of living and quality of life. It will also mean improved food security for the local market, as well as markets in the southern parts of Ghana, because the Upper East region is one of the food baskets of the nation.

Groundwater irrigation policies

Presently, groundwater irrigation is recognized by the Government of Ghana and considered as part of the informal irrigation subsector in the current irrigation policy of Ghana (MOFA, 2010). Prior to this recognition in 2010, groundwater irrigation, together with other individual and community initiated irrigation schemes, were unrecognized. Yet, the informal irrigation subsector was and still is bigger than the formal irrigation subsector and produces the bulk of irrigated output in the country (MOFA, 2010). As a result of the non-recognition until very recently, the subsector is faced with many constraints described by Dittoh and Awuni (2012).

The current irrigation policy proposes four major objectives to address the constraints of irrigation in Ghana, including and in some cases with special emphasis on informal irrigation, namely: performance and growth, socio-economic inclusion, responsible production and enhanced services. The need to realize and support the productive capacity of the informal irrigation sector to increase productivity and ensure growth in the irrigation sector is clearly identified as one of the key objectives of the current policy. This is to be achieved partly through: (i) raising productivity of agricultural water for irrigation, livestock watering and aquaculture; (ii) enhancing production potential of ongoing irrigation activities; and (iii) developing new irrigation areas according to demand and feasibility (MOFA, 2010). It is understood that it may take several years for groundwater irrigators to see the benefits of the current irrigation policy, partly because the policy advocates re-structuring of the very institution responsible for implementing the policy (Ghana Irrigation Development Authority), which takes time to do. However, the recognition of the informal irrigation subsector, including groundwater irrigation, can be seen as an important step forward for the subsector.

Conclusions

The potential for using groundwater to expand dry-season irrigation to improve food security and livelihoods in the Zalerigu and Sapeliga areas in north-eastern Ghana has been assessed using multiple methods.

A total of 92 ha of irrigable land (52 ha in the Zalerigu area and 40 ha in the Sapeliga area) was irrigated with groundwater in the 2010/2011 irrigation season. Total groundwater storage in Zanlerigu and Sapeliga was estimated to be 87.4×10^6 m^3 and 61.6×10^6 m^3, respectively. The annual recharge to the groundwater aquifers was estimated to be 7.6×10^6 m^3 (6.4% of the mean annual rainfall) in the Zalerigu area and 4.3×10^6 m^3 (5.5% of the mean annual rainfall) in the Sapeliga area. The abstracted volume in Zalerigu represents 6.4% of the recharge, while in Sapeliga the abstracted volume represents 8.4% of the recharge. The quality of groundwater in both study areas is of good quality for irrigation, but a few of the wells have nitrate levels beyond the WHO recommended levels for drinking water.

There is a large potential to expand groundwater irrigation in both study areas to other low-lying areas and beyond, depending on the availability of sufficient suitable lands. The current recharge volumes are far more than what is abstracted for all uses. If appropriate and affordable technologies for siting and drilling wells are made available and irrigators are supported with access to credit facilities, land, stable market and extension services, groundwater irrigation could be expanded to increase food production and create employment, contributing to improved food security and livelihoods in north-eastern Ghana and beyond. Recent changes in national irrigation policies, formally recognizing the role of the informal irrigation subsector, are a promising development.

Acknowledgements

This study was done as part of the Groundwater in Sub-Saharan Africa: Implications for Food Security and Livelihoods project led by the International Water Management Institute (IWMI) and funded by the Rockefeller Foundation.

References

Acheampong, P. K. (1988). Water balance analysis for Ghana. *Geography*, 73(2), 125–131.

Agodzo, S. K., Huibers, F. P., Chenini, F., van Lier, J. B., & Duran, A. (2003). *Use of wastewater in irrigated agriculture. Country studies from Bolivia, Ghana and Tunisia. Volume 2: Ghana.* Wageningen: WUR-(W4F-Wastewater). ISBN 90-6754-704–2.

Agyekum, W. A., & Dapaah-Siakwan, S. (2008). The occurrence of ground water in the North Eastern Ghana. In: S. M. A. Adelana & A. M. MacDonald (Eds.), *Applied ground water studies in Africa*. New York, NY: CRC Press.

Amegashie, K. B. (2009). *Assessment of catchment erosion, sedimentation and nutrient export into small reservoirs from their catchment in the Upper East Region of Ghana* (MSc dissertation). Kwame Nkrumah University of Science and Technology, Kumasi, Ghana.

Anayah, F. M., Kaluarachchi, J. J., Pavelic, P., & Smakhtin, V. (2013). Predicting regional evapotranspiration and groundwater recharge: A country-wide study of Ghana. *Water International*, doi: 10.1080/02508060.2013.821642

Ayers, R. S., & Westcot, D. W. (1994). *Water quality for agriculture*. (FAO Irrigation and drainage paper 29.Rev.1). Food and Agriculture Organization of the United Nations, Rome, Italy. Retrieved from http://www.fao.org/docrep/003/T0234E/T0234E00.htm

Barcelona, M. J., Gibb, J. P., Helfrich, J. A., & Garske, E. E. (1985). *Practical guide for groundwater sampling*. EPA/600/2-/85/104.

Carrier, M. A., Lefebvre, R., Racicot, J., & Asare, E. B. (2008). Access to sanitation and safe water: Global partnership and local actions. Northern Ghana Hydrogeological Assessment Project. 33rd WEDC International conference, Accra, Ghana. pp. 353–361. Available from [http://wedc.lboro.ac.uk/resources/conference/33/Carrier_MA_GHA.pdf]

Chilton, P. J., & Foster, S. S. D. (1995). Hydrogeological characterisation and water-supply potential of basement aquifers in tropical Africa. *Hydrogeology Journal*, 3(1), 36–49.

CIA. (2012). World Factbook. Retrieved from https://www.cia.gov/library/publications/the-world-factbook/geos/gh.html

Claasen, H. C. (1982). *Guidelines and techniques for obtaining water samples that accurately represent the water chemistry of an aquifer*. U.S. Geological Survey. Open file report 82–1024.

Dittoh, S., & Awuni, A. J. (2012). Groundwater use for food security and livelihoods in the Upper East Region. Final project report submitted to the International Water Management Institute as part of the Rockefeller Foundation funded project "Groundwater in sub-Saharan Africa: Implications for food security and livelihoods".

Friesen J., Andreini, M., Andah, W., Amisigo, B., & van de Giesen, N. (2005). *Storage capacity and long-term water balance of the Volta Basin, West Africa.* IAHS Publication no. 296, 138–145.

GSS (Ghana Statistical Service). (2008). Ghana living standard survey. Report of the fifth round (GLSS 5). Retrieved from http://www.statsghana.gov.gh/docfiles/glss5_report.pdf

GSS (Ghana Statistical Services). (2012). 2010 Population and housing census. Summary report of final results. Accra: Author.

HAP (Hydrogeological Assessment Project). (2006). Hydrological assessment of the Northern Regions of Ghana: A bibliographical review of selected papers. CIDA, WRC, SNC-LAVALIN International.

Kortatsi, B. K. (1994). Groundwater utilization in Ghana. In: *Future groundwater resources at risk (proceedings of the Helsinki conference)*, Wallingford: IAHS Press, IAHS Publ. No. 222, pp. 149–156.

Kortatsi, B. K., Young, E., & Mensah-Bonsu, A. (2005). Potential impact of large scale abstraction on the quality of shallow groundwater for irrigation in the Keta Strip, Ghana. *West African Journal of Applied Ecology*, 8, 1–12.

Kyei-Baffour, N., & Ofori, E. (2006). Irrigation development and management in Ghana: Prospects and challenges. *Journal of Science and Technology*, 26, 148–159.

Laube, W., Awo, M., & Schraven, B. (2008). *Erratic rains and erratic markets: Environmental change, economic globalization and the expansion of shallow groundwater irrigation in West Africa. ZEF* Working Paper Series 30. Bonn, Germany: Centre for Development Research (ZEF), University of Bonn.

Lipton, M., Litchfield, J., & Faures, J.-M. (2003). The effects of irrigation on poverty: A framework for analysis. *Water Policy*, 5(5/6), 413–427.

Loke M. H. (1999). *Electrical imaging surveys for environmental and engineering studies – A practical guide to 2D and 3D surveys*. Austin, TX: Advanced Geosciences Inc.

Martin, N. (2006). *Development of a water balance for the Atankwidi catchment, West Africa – a case study of groundwater recharge in a semi-arid climate*, Ecology and Development Series, No. 41, Cuvillier Verlag Goättingen.

Martin, N., & van de Giesen, N. (2005). Spatial distribution of groundwater use and groundwater potential in the Volta River basin of Ghana and Burkina Faso. *IWRA Water International*, 30(2), 239–249.

Matthews, J. H. (2004). Simpson's 3/8 rule for numerical integration. *Numerical analysis – numerical methods project*. California State University, Fullerton. Retrieved from http://math.fullerton.edu/mathews/n2003/Simpson38RuleMod.html

McNeil, J. D. (1980). *General theory of terrain conductivity mapping using inductive electromagnetic techniques*. Technical Note TN-6, Geonics Ltd., Ontario.

Memuna, M. M., & Cofie, O. O. (2005). Effects of farming practices on the performance of rice (Oryza sativa) in selected farms of Ashaiman Irrigation Project. In: E. H. Y. Bobobee & A. Bart-Plange (Eds.), *Hunger without frontiers* (pp. 244–254). Kumasi: GSAE/WASAE.

MOFA (Ministry of Food and Agriculture, Ghana). (2010). *National irrigation policies, strategies and regulatory measures*. MOFA/FAO.

Molden, D. (Ed.). (2007). *Water for food, water for life. A comprehensive assessment of water management in agriculture*. London: Earthscan.

Obuobie, E. (2008). *Estimation of groundwater recharge in the context of future climate change in the White Volta River Basin, West Africa*. Ecology and Development Series No. 62. Retrieved from http://www.glowa-volta.de/publ_theses.html

Obuobie, E., & Barry, B. (2012). Ghana. In: P. Pavelic, M. Giordano, B. Keraita, T. Rao, & V. Ramesh (Eds.), *Groundwater availability and use in Sub-Saharan Africa: A review of 15 countries: Chapter 4* (pp.43–64). Colombo: International Water Management Institute (IWMI). DOI: 10.5337/2012.213.

Pavelic, P., Villholth, K. G., Shu, Y., Rebelo, L.-M., & Smakhtin, V. (2013). Smallholder groundwater irrigation in sub-Saharan Africa: Country-level estimates of development potential. *Water International*, *38*(4), 392–407. doi: 10.1080/02508060.2013.819601

Scanlon, B. R., Healy, R. W., & Cook, P. G. (2002). Choosing appropriate techniques for quantifying groundwater recharge. *Hydrogeology Journal*, 10, 18–39.

Veplen, V. (1988). RESIST: A computer program to process resistivity sounding data on PC compatibles. *Computer and Geosciences*, 19(5), 691–703.

Vukovic, M., & Soro, A. (1992). *Determination of hydraulic conductivity of porous media from grain-size composition*. Littleton, CO: Water Resource Publications.

WHO (World Health Organization. (2004). *Guidelines for drinking water quality, Vol. 1. Recommendation* (3rd ed.). Geneva: Author.

Constraints and opportunities for groundwater irrigation arising from hydrologic shifts in the Iullemmeden Basin, south-western Niger

Bio Mohamadou Torou, Guillaume Favreau, Bruno Barbier, Paul Pavelic, Mahamadou Illou and Fatoumata Sidibé

Land-use-change-induced increases in shallow groundwater levels across parts of the Sahel in recent years have coincided with expanded use of groundwater for irrigation. This study was conducted to assess the potential linkages and livelihood implications based on a field survey of nine villages building on previous hydrological studies. The results show that irrigators lack effective means of production and mostly rely on manual methods. Borehole usage is more profitable and reliable than shallower wells. Overall incomes from irrigation are relatively small and severely constrained by the limited field scale due to high establishment and operating costs.

Introduction

Groundwater irrigation (GWI) is booming worldwide. In India much of the recent gain of the green revolution was fueled by GWI (Dorin & Landy, 2002). In North Africa and the Middle East, pumping from aquifers with limited or no replenishment capacity has become widespread, allowing farmers to adopt drip irrigation, to save water, to expand the irrigated area and to diversify production (Kuper et al., 2009). In the West African Sahel, GWI is in its infancy (Siebert et al., 2010) and scientific knowledge is still insufficient (Barbier et al., 2011; Sally & Abernethy, 2002). GWI has developed in the inland valleys, along large river valleys, in some fossil paleochannels and around cities (Payen & Gillet, 2007). GWI emerges in rural areas through shallow wells and with boreholes, which are linked to deeper aquifers. The World Bank has supported GWI in Nigeria for over 20 years through the Fadama project, which showed remarkable returns on investment (Nkonya, Philip, & Mogues, 2008).

Small-scale private irrigation plays an increasingly important role in national strategies and policies, because results are often considered better than large-scale or village-level irrigation schemes (Sonou & Abric, 2010). Evidence of its efficacy in reducing poverty is scarce in Africa but probably important, as indicated in semi-arid northern Mali (Dillon, 2008).

GWI is often seen as an obvious solution for Sahelian farmers in areas remote from major rivers and dams, especially as an adaptation strategy to address climate variability and/or soil fertility reduction. As such, it is thought to have a vital role in reducing food insecurity and poverty. Groundwater reserves are increasingly viewed as an important alternative to surface water, which is scarce during the dry season because reservoirs with large volumes are expensive to build in the flat landscape and the sandy soils of western Niger. Also, villages remote from the river valleys can only rely on smaller, more distributed sources such as deep or phreatic aquifers and on small seasonal ponds.

The development of GWI at any appreciable scale raises concerns regarding its sustainability and the potential for groundwater depletion. In India the aquifers of the Haryana and Punjab regions have fallen by several tens of metres in some places (Dorin & Landy, 2002). In northern Africa, similar trends were observed in semi-arid regions with limited control over pumping (Leduc et al., 2007). In several parts of the Sahel, the aquifers have been depleted by the droughts of the 1970s and 1980s and struggle to recover, even though the rainfall has since improved. One can question the merits of developing GWI in the Sahel while the population's domestic water needs are not yet fully satisfied and where large herds and flocks of ruminants struggle to get permanent access to boreholes and wells during the dry season. The real question, however, is not so much whether to establish GWI but under what conditions, and how to continue to address domestic and livestock needs.

In some areas of the Sahel, the groundwater level has risen over the last decades, even during the drought years. In the Iullemmeden basin in south-western Niger, it has risen by about 4 metres over the past 50 years while rainfall patterns were approximately at average levels or even lower (Favreau, Nazoumou, Leblanc, Guéro, & Goni, 2012). Hydrologic studies have revealed that this is driven by land clearing and rain-fed crop expansion at the expense of natural vegetation; the increased runoff has collected in hundreds of natural ponds, where it infiltrates and percolates down to the saturated zone (Favreau et al., 2012; Massuel et al., 2011). When the landscape was not as highly cultivated (prior to the 1960s), the deeper-rooted natural vegetation would efficiently capture rainfall and evapotranspiration would return this moisture back to the atmosphere. While similar hydrological shifts in arid parts in Australia, the south-western USA, Pakistan and elsewhere have seen the emergence of major secondary salinization problems (Ghassemi, Jakeman, & Nix, 1995), in the Sahel, where there is not the vast store of salt at shallow depths present elsewhere, and where recharge in concentrated in discrete parts of the landscape, a more optimistic situation has emerged.

While groundwater levels were rising, small-scale irrigated agriculture also expanded in this region (Adamou, 2008), raising a number of issues regarding the opportunities and impacts of the enhanced water availability on irrigation development. This study seeks to answer specific questions concerning the sources of water used for irrigation, the socio-economic costs and benefits of GWI, and the major constraints faced by smallholder farmers who practice GWI in the Sahel region.

Setting and approach

Study area

The basin of Kori de Dantiandou (KD, also known as the Dantiandou Valley), situated within the Iullemmeden Basin in south-western Niger, was selected for study due to the availability of a number of previous hydrological and hydrogeological studies (e.g. Favreau et al., 2009).

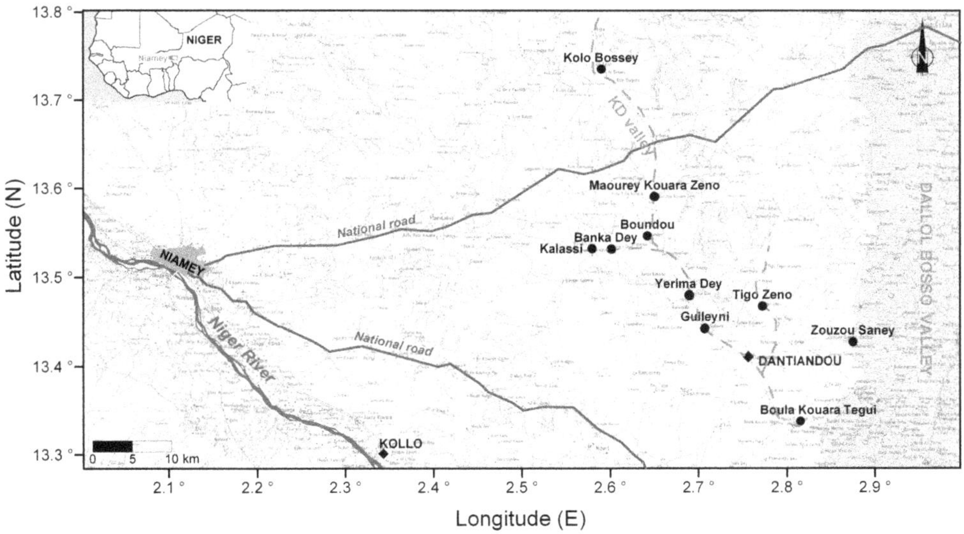

Figure 1. Location of the study area in southwest Niger. The dotted line represents the Dantiandou Valley, where irrigation using groundwater occurs. Most of the valley downstream of the village of Dantiandou at the time of the field surveys (December 2011) was experiencing outcropping groundwater and permanent ponding, creating new opportunities for expanding irrigation using readily available, shallow groundwater.

The vast sedimentary aquifer, underlying an area of about 10,000 km^2 with a mean annual rainfall of about 500 mm, is located roughly between Niamey, the capital of Niger, and the Dallol Bosso, a large (10 km) fossil valley connecting the Aïr Mountains (in northern Niger) to the Niger River (Figure 1). The water table is within 5 m of the surface in the Dallol Bosso Valley and the downstream of the KD. Water infiltrates into the aquifer from the surface (Massuel et al., 2011). Hydrological and hydrogeological investigations show that the system of endorheic lowlands here provides favourable conditions for deep infiltration and groundwater recharge (Favreau et al., 2009).

Ponds (partially clogged by clay deposits) and wetlands collect most of the runoff after intense rain events; aquifer recharge mainly occurs by concentrated infiltration below seasonal ponds. The water budget of ponds shows that 90% of the accumulated water infiltrates to the water table (Massuel et al., 2011). Naturally, the increase in recharge within the basin is accompanied by a long-term rise in the water table. Groundwater recharge is estimated at about 25 mm/y; pre-clearing values (before 1950–1960) were estimated as only a few mm/y, an order of magnitude less (Favreau et al., 2009). At a larger scale, within the Iullemmeden Basin, direct and indirect evidence confirms that this phenomenon of land clearing and rising water tables has been widespread over more than 100,000 km^2 at various time periods during the twentieth century (Favreau et al., 2012). Because groundwater recharge is focused on point sources and does not incur large evapotranspiration losses, groundwater is mainly fresh and highly suitable for irrigation, with about 45 mg/L mean total dissolved solids (Favreau et al., 2002).

Social setting

Niger's population was estimated in 2010 at just over 15 million people, with a natural growth rate of 3.3%, a total fertility of 7.1 children per woman and life expectancy at birth of 58 years (Institut National de Statistique du Niger, 2010, pers. comm.). Mainly rural, population is distributed unevenly on the national territory. The southern part of the country, along the border with Nigeria, is home to over 90% of the population. The western part belongs to the Niger River basin and has one of the highest population densities. In this semi-arid region, rain-fed millet cropping and extensive cattle rearing are the main rural livelihood activities (Hiernaux & Turner, 2002). The KD basin mainly lies within the department of Kollo, east of the peri-urban community of Niamey. The rural community of Dantiandou belongs entirely to the KD valley.

Dantiandou, with about 35,000 inhabitants (INS, 2010) is located in the Department of Kollo, which supports about 431,000 inhabitants. Its population is composed of various ethnic groups, mainly Zarmas and Fulanis. Zarmas are more specialized in cropping, while Fulanis are traditionally more oriented towards animal husbandry; but both groups undertake cultivation and breed cattle. Dwellings are centred about well-defined villages for Zarmas, but are more extended and dispersed for Fulanis.

The main resources in these communities are related to agriculture, with millet the dominant crop and staple food (Barbier & Hazell, 2000). In the more humid lowlands, some farmers plant mango trees and annual crops such as rice, peanuts, cassava and sweet potatoes. Sorghum is sometimes grown alongside receding ponds on clay soils close to villages. Because of the lack of perennial flow in the KD throughout the year, the only permanent water resource is groundwater. Village wells lie at the centre of social life and are the source of fresh water for people and animals; water lifting is almost always performed manually (Figure 2). The total amount collected by the villagers for all uses (e.g. domestic and pastoral) is estimated as between 10 L and 30 L per person per day (Favreau et al., 2009). This quantity depends on the accessibility to the water, with depths to standing water levels varying between villages, from negligible for groundwater-fed ponds to ~70 m on plateaus. The populations of the villages surveyed ranged from 200 to 800 inhabitants (2001 census).

With a human population density of 30 inhabitants per km^2, the total volume abstracted annually within the KD area of 5000 km^2 is no more than 1.6×10^6 m^3 (0.3 mm). This value increases with population growth (3.3% per year, INS 2010). Seasonal variability of groundwater consumption per capita was not reported to be significant, even though

Figure 2. Examples of small-scale irrigation using shallow (about 6 m) groundwater from cemented wells screened within the phreatic aquifer at (a) Yerima Dey and (b) Guileyni (February 2009, Monique Oï, ©IRD). At Yerima Dey, irrigation was promoted by NGOs and was in effect by the early 2000s. At Guileyni, irrigation during the dry season using the old (1981) domestic, cemented well started in December 2001 and has been continuous thereafter in November to April of each year.

Source: Field survey, 2011.

during the rainy season the village population and livestock are spread over the countryside, reducing the pressure on village wells.

Farm sampling and interviews

A sample of 90 irrigators was selected from 9 villages in the Dantiandou Valley in July 2011. The questionnaire was administered to a man or woman in charge of a garden to obtain information about the technical and socio-economic aspects of the household.

These irrigators derive their water from a range of sources. There are five types of sources of water, as given in Table 1 – with one qualification. Underlying the shallow unconfined aquifer, deeper confined aquifers are tapped by a few artesian boreholes. This water is also used for irrigation, although it is warmer (34 °C) and more saline (about 1 g/L). Environmental isotope geochemistry indicates that this is fossil water, dating back to more humid and/or cooler climates during the Quaternary period (Le Gal La Salle et al., 2001). In the analysis, the farmers using artesian (fossil/non-renewable aquifer) boreholes were grouped with the other users.

In 10 villages (nine study villages and one additional village) 118 water sources were counted (Table 1). Two-thirds of the sources are cemented wells, only one of which was broken. Fifteen boreholes were drilled, but 7 were broken and 6 were free-flowing artesian under the pressure of basin-scale deep aquifers. The survey identified 16 ponds, of which 10 were intermittently filled by surface water. All villages (except one) have access to at least one pond.

Table 1. Sources of water in the 10 selected communities.

	Kalassi	Bankadey	Boundou	Guileyni	Yerimadey	Boula Kouara Tegui	Tigo Zeno	Kolo Bossey	Maourey Kouara Zeno	Zouzou Saney	Total
1. Boreholes											
– Artesian			1	1	1		1	1	1		6
– Water tower										1	1
– Foot pump		1									1
– Foot pump (broken)						2			1		3
– Hand pump						1				1	2
– Hand pump (broken)	2					1				1	4
2. Cemented wells											
– Within villages	3		1	27	12	1	2	1			47
– Outside villages		2		17	3					7	29
– Broken									1		1
3. Traditional wells											
– Broken	1	1		1	1		1			1	6
– In use						1		1			2
4. Ponds											
– Temporary		1	1		3		2	1		2	10
– Permanent artesian				1	1						2
– Permanent						2			2		4
Total water sources	6	5	3	47	21	8	6	4	5	13	118

Table 2. Gender-wise and source-wise sample of farmers in each village.

	Gender		Water source				
Village	Female	Male	Cemented wells	Boreholes	Ponds	Traditional wells	Total
Banka Dey	10	0	8	2	0	0	10
	100%	*0%*	*80%*	*20%*	*0%*	*0%*	*100%*
Yerimadey	10	0	10	0	0	0	10
	100%	*0%*	*100%*	*0%*	*0%*	*0%*	*100%*
Boundou	5	4	0	9	0	0	9
	56%	*44%*	*0%*	*100%*	*0%*	*0%*	*100%*
Guileyni	7	3	5	5	0	0	10
	70%	*30%*	*50%*	*50%*	*0%*	*0%*	*100%*
Maourey K.Z.	6	4	0	10	0	0	10
	60%	*40%*	*0%*	*100%*	*0%*	*0%*	*100%*
Boula Kouara T.	5	6	1	0	6	4	11
	45%	*55%*	*9%*	*0%*	*55%*	*36%*	*100%*
Kolo Bossey	2	8	0	10	0	0	10
	20%	*80%*	*0%*	*100%*	*0%*	*0%*	*100%*
Zouzou Saney	5	5	0	0	1	9	10
	50%	*50%*	*0%*	*0%*	*10%*	*90%*	*100%*
Kalassi	5	5	10	0	0	0	10
	50%	*50%*	*100%*	*0%*	*0%*	*0%*	*100%*
Total	55	35	34	36	7	13	90
	61%	*39%*	*38%*	*40%*	*8%*	*14%*	*100%*

Source: Field survey, 2011.

Ten farmers were selected in each of the 9 villages. In each village a water source was identified where gardeners (both men and women) could be found. Though the majority of irrigators are women, they do not necessarily own or manage the land. Many women work on household plots where the men are considered the heads (Table 2).

Most of the 90 interviewed farmers take water from cemented wells in 5 communities, but a large number also take water from boreholes in 6 communities. A small number take water from traditional wells, but only in two communities, in a valley where the water table is particularly close to the surface (<4 m). Four villages have temporary ponds but only one or two interviewed farmers per village use these ponds. It would appear therefore that most gardeners tend to concentrate around cemented wells and boreholes (Figure 3); ponds were usually too temporary or fluctuating in levels. Unlined shallow wells can be also be used in some rare instances.

Results

Irrigator characteristics

All interviewed irrigators are from the Zarma ethnic group, all married, between early 20s and 70s in age. Two-thirds of the interviewed irrigators were women. The first irrigators started gardening from the early 1980s. The development of gardens was regular, without boom and bust during the drought years, suggesting that the decision to start a garden is not linked to shocks or to particular events – except for artesian bore drilling, which could have created new opportunities.

Three-quarters of the farmers produce onions. The second crop is tomato, then cabbage and a large variety of secondary crops. The method of water lifting is by hand and foot

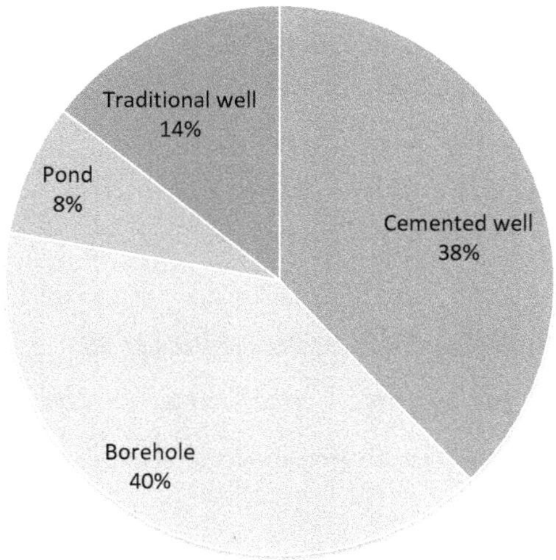

Figure 3. Distribution of irrigators according to the four main types of irrigation water source.
Source: Field survey, 2011.

pumps on boreholes and various very simple devices on wells and ponds. There are no *shadoof* or motor pumps in the surveyed area. Almost half the farmers use watering cans to irrigate; the remainder use various improvised devices with small holes to distribute water to the plants.

Groundwater infrastructure and water management

The financing of infrastructure varies among the types of water source (Figure 4). Boreholes are funded mainly by NGOs and the government, while some cemented wells are funded by some wealthier individuals. Irrigators also dig traditional wells themselves, at their own cost.

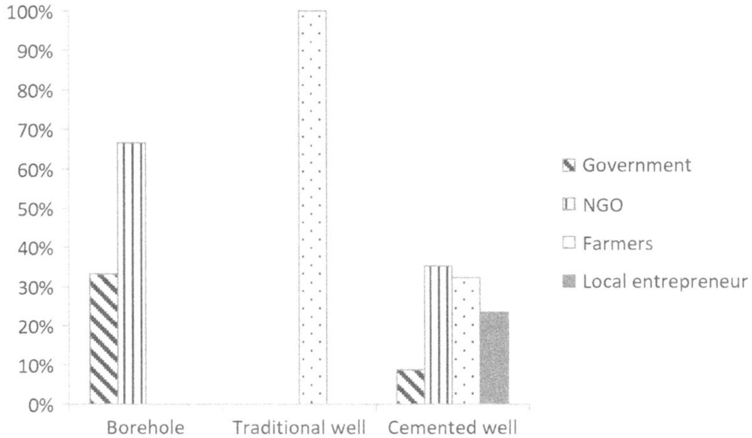

Figure 4. Means of funding the different water sources.
Source: Field survey, 2011.

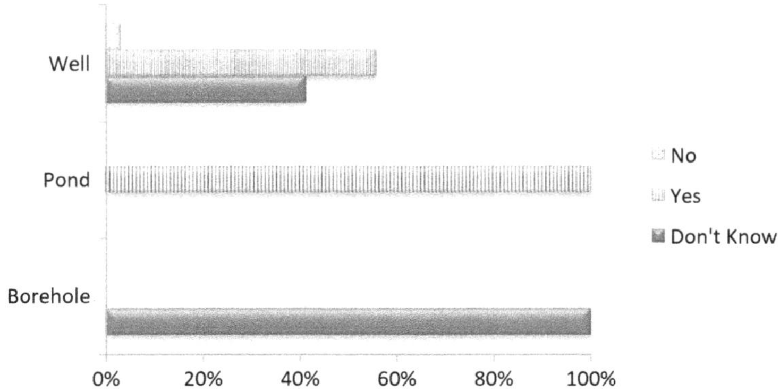

Figure 5. Answer to the question "has the groundwater level increased over the last years?". Source: Field survey, 2011.

The link between water source and technical mobilization of water is very important and significant (p-value = .000, Cramér's $V = 0.768$), given that all boreholes are topped with a hand- or foot-propelled pump while wells and cemented wells are connected to manual lifting.

The average distance between the ground surface and the standing water level was 18 m in cemented wells and less than 5 m in traditional wells. Borehole users do not know the depth of the groundwater. Those who use cemented wells cannot provide numbers for changes in the water depth (Figure 5). The average water use for irrigation is 6 months of the year for all water sources (for boreholes, 5 months). Irrigation is not continuous: GWI was reported to last seven months. In cemented wells, GWI was shorter (less than 5 months), because there is competition between irrigation and domestic use. Nonetheless, 90% of the farmers do not see any competition between water users.

The average garden is 1274 m^2 (0.127 ha) but size differs greatly between gardens. The gardeners using wells have larger areas (3362 m^2) compared to borehole users (559 m^2). The correlation between the type of water source and the size of the irrigated plot is strong (86%).

Land ownership is either individual or collective, meaning that the land has been allocated within a common plot, often distributed according to the participants' manpower. The perceived land tenure is quite complex. One-third consider the plot their own; these are overwhelmingly men. A large fraction belong to an extended family. One collective garden belongs to the village chief. And one fraction of the sample use a plot belonging to a parent. Irrigators who use ponds and traditional wells tend to manage individual gardens, while those using cemented wells and boreholes tend to manage more collective gardens (Figure 6).

Interviewed women cultivate much smaller areas than men (Table 3). Women often cultivate in groups within a small fenced garden. Interviewed men have much larger plots, but usually cultivate with their family. When a woman controls the plot, it often means that the man has migrated temporarily.

Incomes and food security

Farm income and agricultural expenditures are highly variable (Table 4). The standard deviation is very large for each class of irrigators. Average income is quite small given

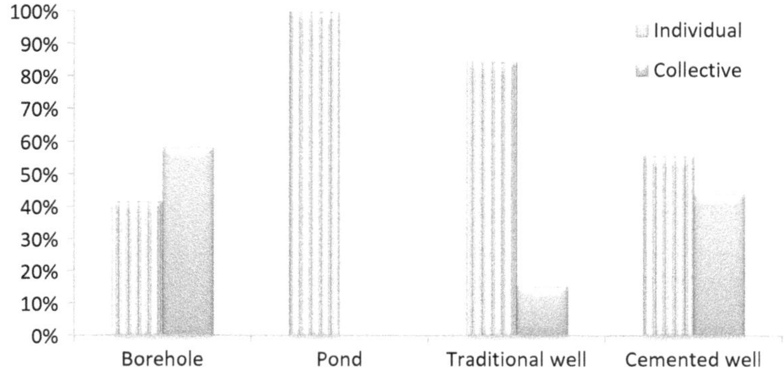

Figure 6. Type of water source and the control of the irrigated area. This control is different from ownership. Farmers consider that control is individual when the farmer is alone within the boundary of the garden, while collective usually means that there is a common fence around several gardens.
Source: Field survey, 2011.

Table 3. Sizes of the irrigated plot by gender.

		Area (m^2)	
	Person	Average	Std-dev
Women	55	550	1384
Men	35	2411	3282

Source: Field survey, 2011.

Table 4. Garden areas and economic parameters by type of water sources.

Water source	Garden Number	Average Garden Area (m^2)	Farm workers	Farm gross income (FCFA)	Farm expenses (FCFA)	Net income (FCFA)	Net income per worker (FCFA)
Borehole	36	559	3.9	68 079	4 593	63 486	16 095
Pond	7	1207	4.9	67 857	14 393	53 464	11 007
Traditional well	13	3362	4.5	58 673	25 565	33 108	7 421
Cemented well	34	1245	4.0	31 132	7 819	23 313	5 871
Total sample	90	1273	4.1	52 746	9 603	43 142	10 522

Source: Field survey, 2011.

the labour invested. A typical farm has around 4 family workers and gains a net income of XOF40,000 per season, which means around XOF10,000 per worker. Recognizing that the average daily wage for labour in the area is between XOF1000 and XOF1500 (the monthly SMIG [Salaire Minimum Interprofessionnel Garanti] was about €46 in August, 2012, in Niger, based on an exchange rate of €1 = XOF655.957) leads one to question the economic viability of the practice. Irrigation is also a very labour-intensive activity, which suggests that the opportunity cost of labour is very low in the area. Indeed, the customary employments of traditional gold mining and temporary migration have been hampered by recent crises in the Ivory Coast (2010–2011) and northern Nigeria (since 2011).

Table 5. Months of food availability and expenses for food by type of water source.

Water source	Food availability in months	Food expenses (FCFA)
Borehole	10.1	110 917
Pond	9.0	174 429
Traditional well	9.1	174 519
Cemented well	9.1	90 419
Sample	9.5	117 300

Note: Food expenses include the purchase of grain to cover the deficit in the household grain production.
Source: Field survey, 2011.

Irrigation is most profitable for those who use boreholes, followed by ponds, traditional wells and cemented wells. The last are probably the least profitable because the aquifer they draw from is the deepest. While incomes vary depending on the type of water source, the link is weak. Other factors explain differences in income and agricultural expenditure. The link between agricultural expenditure and type of irrigator is highly significant.

Food availability differs little between the farms in the sample. More than half of the farmers do not consider themselves self-sufficient. Most interviewed think they can cover 9 months of their needs, which is not that far from self-sufficiency (Table 5). On the other hand, the relations between irrigation water source and the number of months of food availability (96%), and between source and food expenses (71%), are strong. However, only the link between water sources and food expenditure is significant. Those who manage boreholes produce slightly more food, but irrigators with cemented wells buy less food. Those who use ponds and traditional wells spend a little more on food purchased on the local market. This could be because they have less water control.

System constraints

Around 41% of the households obtain some credit from NGOs to buy inputs. Over 60% of the irrigators who use cemented wells have access to credit, against an average of 20% for the others (Figure 7). Since NGOs often built the cemented wells they also provided microcredit programmes to the well users. Just a few farmers are able to obtain credit from traders or the government.

Two-thirds of the farmers apply inorganic fertilizers to their vegetables, and all use organic fertilizers in various quantities, but on average the equivalent of several small carts per farm. Almost half the farmers benefit from fencing, usually constructed from acacia shrub; one-quarter of the gardens use barbed wire or *grillage*; and one-quarter use millet stems from the previous rainy season (see e.g. Figure 2).

Even those who use boreholes with water all year round stop cultivating in April, probably partly because it becomes too hot for the crops and too hot to work, but mostly because the rain-fed cropping labour is about to start. Only two male farmers reportedly irrigate all year round.

The major problems reported were lack of fences to protect crops from ruminants, lack of modern water-lifting devices and lack of easy access to seeds and fertilizers. Only a few farmers mentioned a lack of water, attack by pests, or lack of land or markets.

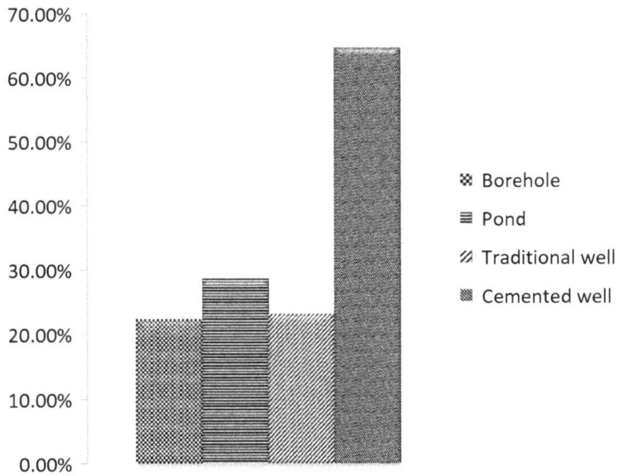

Figure 7. Access to credit for different water sources.
Source: Field survey, 2011.

Discussion

Three questions posed earlier are discussed here: (1) the link between groundwater infrastructure and water resources management; (2) impacts on incomes and food security; and (3) the main constraints to expanding GWI.

Effect of groundwater level rise on irrigation

All of the traditional-well users observed that groundwater levels have been rising over the past decades, and more than half of the users of cemented wells also think the same. Since cemented wells usually do not require annual servicing or repair, most farmers cannot really observe a change in the depth to the water table. Moreover, cemented wells are usually deeper and the relative change in depth is less (e.g. a change of 1 or 2 m for a well 18 m deep is less apparent than a 1 m change for a 5 m well). Only farmers with traditional wells could provide an estimate of the increase in water level (between 1 and 4 m, and around 2 m on average). Among the farmers using cemented wells, those who were able to sense the depth change owned shallower wells. Our piezometric surveys indicate that when the phreatic water table is getting shallower (e.g. about 7 m in depth in Guileyni, Figure 8) it can be used successfully for irrigation, whereas for deeper water tables, only free-flowing artesian bores get used for irrigation.

Groundwater levels may continue to rise in future because the physical causes of the rise, i.e. land clearing on the plateaus and changing land uses in the cultivated valley, are still going on. The vegetative cover in the region is still being degraded, and unless farmers adopt stronger land conservation practices, runoff generation into ponds that feed the water table will continue to increase. Though efforts have been made by farmers, NGOs and the government to control runoff and erosion, it will be difficult to reverse the trend on a large scale in the near future.

Though most wells are permanent throughout the year, most farmers start to irrigate in December and end in March (or up to April or May in some instances). Artesian boreholes

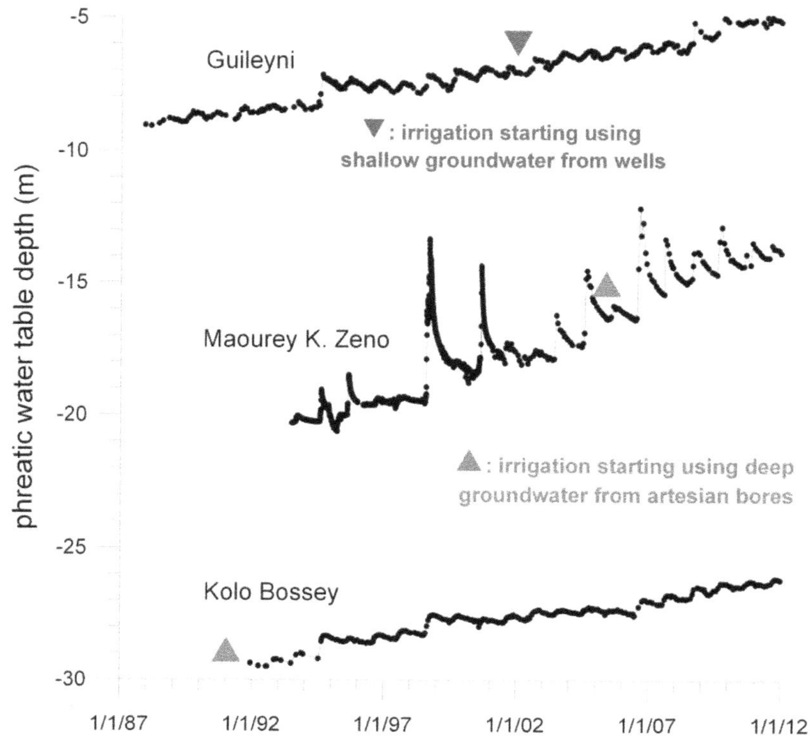

Figure 8. Examples of long-term rises in the water table for three of the study sites.

Note: The water table has been rising for the whole unconfined aquifer in SW Niger since the 1950s (Favreau et al., 2012). Triangle symbols indicate starts in small scale irrigation using groundwater for each of the three villages.

that bring water to the surface without requiring energy inputs can, from one viewpoint, be seen as an excellent opportunity for villagers. However, they typically tap fossil water that is non-renewable on a human time scale (Le Gal La Salle et al., 2001). Issues of sustainable levels of exploitation become more pressing under these circumstances.

Water for domestic uses or irrigation is closely related to the water source. Wells are typically used for irrigation, and boreholes for drinking water, mainly for humans but also for domestic animals. The more sophisticated the means of water lifting, the more people come together to produce vegetables, but it is hard to separate out cause and effect. Since the boreholes and cemented wells are usually financed by NGOs, government, or community leaders, collective management by the community is a donor requirement, not necessarily an endogenous effect.

Impact of GWI on incomes and food security

As noted, results from the sample suggest that garden-scale irrigation is not greatly improving farmers' overall income. Gardens are very small, and farmers irrigate mainly with manual buckets and watering cans, rather than pumps. Levels of production are very small – for example only a few bags of onions after four months of intense labour. Incomes are very low when compared to daily wages in the area and even when compared to the value of rain-fed crops. A typical household of 10 persons produces millet on an average of 6 ha of land in this part of Niger (de Rouw, 2004; Hiernaux & Turner, 2002). A hectare of rainfed millet produces around 150–300 kg per year without fertilization and up to 600 kg per year when manure is used (Hiernaux, pers. com.). Millet prices typically range from XOF100 to XOF150 per kilogram at harvest time and input use is usually nil. This provides a household net yearly income of between XOF240,000 and XOF360,000, or around XOF30,000 per person. These returns are well below the poverty-line value of about XOF50,000 per person in rural areas in Niger (PNUD/INS, 2008). The additional income from vegetables therefore represents only 5–15% of the income from rainfed crops. However, households need this activity to reach food security. There are few other economic activities worthy of pursuit in the region other than small-animal fattening. Many men migrate to cities or to neighbouring countries during the dry season to supplement the poor rain-fed production. Although the impact of GWI on poverty and food security seems marginal on average, it could be decisive during food-crisis years. The impact on nutritional enhancement was not estimated in this study.

Farmers with boreholes generate higher incomes than other irrigators – typically double. This can be explained by easier access to groundwater all year long and by better water control. Water levels in artesian boreholes are quite stable and secure, while wells tend to be less secure and typically lower by a few decimetres to metres by the end of the dry season (March–April; Figure 8), creating uncertainty for vegetable planters as to when the well might run dry. Borehole users have higher food expenses. However, food expenses are probably not a good indicator of self-sufficiency; this may only indicate that wealthier farmers buy new and more diversified types of food. The link between food purchase and household food security in terms of months of home-grown grain availability is not statistically strong.

Constraint to GWI

For the majority of the interviewed gardeners, the development of GWI is not hampered by a lack of water availability or water salinity, a potential constraint for irrigation in semi-arid regions (see e.g. Leduc et al., 2007; Pavelic, Smakhtin, Favreau, & Villholth, 2012). Those who use boreholes have easy access to water. Nor is lack of land a strong limiting factor, because land near water sources can be rented to other farmers during the dry season. But not all villages and villagers have access to irrigated plots.

Despite the significant numbers of wells and boreholes in the area, only a small fraction of farmers invest in gardening. First, gardeners lack production inputs such as fences, tools, fertilizers, and pesticides. Credit is available through NGOs in the area, but the expressed problems and low incomes suggest that irrigators are severely constrained by a lack of upfront and working capital.

It is not certain that using assets such as land titles to provide collateral to banks would improve the access to capital through credit. Plots are far too small and not necessarily privately owned. The market for land is limited or non-existent anyway. Also, it is not yet always easy to access land titles, especially for women.

New fencing options such as electrified fences, potentially solar-powered ones, need to be explored. Given the high price of fences, communities themselves should find alternative local solutions for keeping small and large ruminants under control in the vicinity of the gardens.

Conclusions

The initial research questions of this study were related to the functioning of small private groundwater irrigation in rural areas of south-western Niger. We analyzed incomes and food security from various water sources. The survey of 90 irrigators in 10 communities where irrigation was known to have started during the past decades provides a fair picture of the current situation. The first major finding is that that most irrigators are aware of the increase in groundwater level over the last few decades, and this phenomenon has steadily reduced the effort required in pumping and withdrawal. However, it is not clear, except where the water table is now outcropping permanently, that it has triggered an expansion of irrigation, even though very few irrigators think that water availability is a major problem.

Second, current water resources are under-exploited because of a severe lack of means of production. Though credit is available to a few farmers through NGOs, it is usually not aimed at farm inputs. NGOs and the government should favour access to credit to help farmers purchase small materials, fences, seed and fertilizers. Public support can be made available under the umbrella of a programme of adaptation to climate change.

The third finding is that labour productivity is still very low. Farm income from irrigation is very small because of the size of the plots. Irrigators using deeper cemented wells have the lowest income, suggesting that water lifting by hand is a limiting factor. Small-scale irrigation is not very profitable at this stage but remains widespread because there is little alternative. The opportunity cost of labour is very low in the area, meaning that there is little alternative to small-scale irrigation during the six-to-seven-month dry season. GWI involves an increasing number of small farmers, especially women. It probably reduces outmigration of men or contributes to improved food security for their families during their absence. It also seems to be a sustainable recourse against drought and poverty.

Acknowledgements

We are grateful to the farmers who made their time available to answer the questionnaires during field surveys. Piezometric data were made available by the Niger Ministry of Hydraulics in Niamey, and through the AMMA-Catch Hydrological Observatory in West Africa, funded by the French Ministry of Research (http://www.amma-catch.org). This study contributes to a larger research project financially supported by the Rockefeller Foundation through Project Number 2008-AGR-305, Groundwater in Sub-Saharan Africa: Implications for Food Security and Livelihoods. Pierre Hiernaux (CNRS, Toulouse, France) is also thanked for constructive comments on an early version of the manuscript.

References

Adamou, M. M. (2008). La production maraîchère dans la ville de Niamey (Niger): Évolutions et impacts sur les populations. In FAO (ed.) *Séminaire régional: "Irrigation informelle : Importance et perspective en Afrique de l'Ouest et du centre*. Ouagadougou: FAO.

Barbier, B., & Hazell, P. B. R. (2000). *Implications of population growth and declining access to transhumant areas and sustainability of agro-pastoral systems in the semi-arid areas of Niger in property right and collective action*. Washington, DC: IFPRI.

Barbier, B., Ouedrago, H., Dembélé, Y., Yacouba, H., Barry, B., & Jamin, J. Y. (2011). L'agriculture irriguée dans le Sahel ouest-africain. *Cahiers Agricultures*, *20*(1–2), 24–33. [20110526]. doi:http://dx.doi.org/10.1684/agr.2011.0475

de Rouw, A. (2004). Improving yields and reducing risks in pearl millet farming in the African Sahel. *Agricultural Systems*, *81*, 73–93.

Dillon, A. (2008). Access to irrigation and the escape from poverty evidence from Northern Mali. IFPRI Discussion Paper 00782, July 2008.

Dorin, B., & Landy, F. (2002). *Agriculture et alimentation de l'Inde: Les vertes années (1947–2000)*. Paris: INRA.

Favreau, G., Cappelaere, B., Massuel, S., Leblanc, M., Boucher, M., Boulain, N., & Leduc, C. (2009). Land clearing, climate variability, and water resources increase in semiarid southwest Niger: A review. *Water Resources Research*, *45*, W00A16. doi:10.1029/2007WR006785

Favreau, G., Leduc, C., Marlin, C., Dray, M., Taupin, J.-D., Massault, M., . . . Babic, M. (2002). Estimate of recharge of a rising water table in semiarid Niger from ^{3}H and ^{14}C modeling. *Ground Water*, *40*(2), 144–151.

Favreau, G., Nazoumou, Y., Leblanc, M., Guéro, A., & Goni, I. B. (2012). Groundwater resources increase in the Iullemmeden Basin, West Africa. In H. Treidel, J. L. Martin-Bordes, Gurdak, J. J. (Eds.), *Climate change effects on groundwater resources: A global synthesis of findings and recommendations* (International Contributions to Hydrogeology (27), Chap. 7, pp. 113–128). Leiden, The Netherlands: CRC Press Inc.

Ghassemi, F., Jakeman, A. J., & Nix, H. A. (1995). *Salinisation of land and water resources – human causes, extent, management and case studies* (1st ed., 526pp). Sydney: University of New South Wales Press Ltd.

Hiernaux, P., & Turner, M. D. (2002). The influence of farmer and pastoralist management practices on desertification processes in the Sahel. In J. F. Reynolds & D. M. Stattford Smith (Eds.), *Global desertification: Do human cause deserts?* (Chap. 8, pp. 135–148). Berlin: Dahlem University Press.

Kuper, M., Bouarfa, S., Errahj, M., Faysse, N., Hammani, A., Marlet, S., . . . Vincent, B. (2009). A crop needs more than a drop: Towards a new praxis in irrigation management in North Africa. *Irrigation and Drainage*, *58*(3), S231–S239.

Leduc, C., Ben Ammar, S., Favreau, G., Beji, R., Virrion, R., Lacombe, G., . . . Zouari, K. (2007). Impacts of hydrological changes in the Mediterranean zone: Environmental modifications and rural development in the Merguellil catchment, central Tunisia. *Hydrological Sciences Journal*, *52*(6), 1162–1178.

Le Gal La Salle, C., Marlin, C., Leduc, C., Taupin, J. -D., Massault, M., & Favreau, G. (2001). Renewal rate estimation of groundwater based on radioactive tracers (^{3}H, ^{14}C) in an unconfined aquifer in a semi-arid area, Iullemeden Basin, Niger. *Journal of Hydrology*, *254*, 145–156. doi:10.1016/S0022-1694(01)00491-7

Massuel, S., Cappelaere, B., Favreau, G., Leduc, C., Lebel, T., & Vischel, T. (2011). Integrated surface water -- groundwater modelling in the context of increasing water reserves of a regional Sahelian aquifer. *Hydrological Sciences Journal*, *56*(7), 1242–1264.

Nkonya, E., Philip, D., & Mogues, T. (2008). *From the ground up: Impacts of a pro – poor community-driven development project in Nigeria* (Discussion Paper 756). Washington, DC: IFPRI.

Pavelic, P., Smakhtin, V., Favreau, G., & Villholth, K. G. (2012). Water-balance approach for assessing potential for smallholder groundwater irrigation in Sub-Saharan Africa. *Water SA*, *38*(3), 399–405.

Payen, J., & Gillet, V. (2007). L'irrigation informelle en Afrique de l'Ouest. Une solution ou un problème? Issue paper, Document thématique no. 6. FAO, Rome, IPTRID. Retrieved from ftp://ftp.fao.org/docrep/fao/010/a1429f/a1429f00.pdf

PNUD/INS. (2008). Rapport sur l'état de la pauvreté. Niger 2008. Retrieved from http://www.pnud.ne/rap_eval/Rapport_etat_pauvrete2008.pdf

Sally, H., & Abernethy, C. L. (2002). *Private irrigation in sub-Saharan Africa: Proceedings of regional seminar on private sector participation and irrigation expansion in sub-Saharan Africa, 22–26 October 2001, Accra, Ghana*. Colombo: IWMI. Retrieved from http://publications.iwmi.org/pdf/H030864_TOCOA.pdf

Siebert, S., Burke, J., Faures, J. M., Frenken, K., Hoogeveen, J., Döll, P., & Portmann, F. T. (2010). Groundwater use for irrigation – a global inventory. *Hydrology and Earth System Sciences Discussion*, *7*, 3977–4021.

Sonou, M., & Abric, S. (2010). Capitalisation d'expériences sur le développement de la petite irrigation privée pour des productions à haute valeur ajoutée en Afrique de l'Ouest. Technical Report ARID, FAO, IWMI.

Afterword

This volume is dedicated to better understanding why groundwater is so underutilized in Sub-Saharan Africa (SSA). The opportunities offered by SSA's vast shallow aquifers remain largely underexploited, and only 5% of the area equipped for irrigation uses groundwater.[1] The Volume provides evidence and offers policy propositions for helping smallholder farmers in SSA realize the benefits of groundwater irrigation. We argue that doing this can catalyze climate-resilient agricultural growth, aid poverty reduction and improve food security across the region.

The chapters were originally published in 2013 as papers in a Special Issue of *Water International* journal. While much of what is reported holds true even today, there have been two significant developments relevant to the prospects for small-scale irrigation in Africa. One, there is a renewed interest and investment in 'farmer-led irrigation development' as opposed to state-driven/public irrigation.[2] This aligns well with the key policy proposition discussed in this volume. The second stems from the rapid decline in cost and growing popularity of solar pumps in Asia and Africa. We believe that by delivering reliable and affordable green energy, solar pumps can address a key barrier to expansion of groundwater irrigation in SSA.

Given the centrality of energy access to smallholder irrigation, it is not surprising that solar pumps have become critical to the irrigation discourse in SSA. Much of the early solar irrigation research in SSA has focused on small (sub-kW) affordable solar pumps and the development of new business models of delivering 'solar irrigation as a service'.[3] The technology has matured to the point, where we need more work in the space of sustainable financing options and smallholder-friendly business and institutional models and approaches to solar irrigation promotion. Towards this, there are also tremendous opportunities for Africa and Asia to learn from each other's experiments and experiences.

A lot of the barriers to solar irrigation expansion in SSA are directly inherited from existing barriers to smallholder irrigation expansion. A key barrier is poorly developed irrigation equipment markets and value chains. Trade and market regulations have been relaxed across borders within Africa, particularly through the recently established agreement on the African Continental Free Trade Area (AfCFTA).[4] This will help penetration of irrigation and energy infrastructure and equipment and support markets for agricultural inputs and outputs for smallholder farmers. The International Solar Alliance (ISA) is another key stakeholder that is trying to break these barriers.

Finally, the African Ministers' Council on Water (AMCOW), which is the African apex organization for sustainable and equitable water development and service delivery, is strongly embracing groundwater for Africa's resilience and socioeconomic transformation, highlighting and calling for attention among member states to enhance regional integration through better cooperation on shared groundwater resources, while also enhancing the focus on groundwater for social upliftment through better water access for multiple uses and in particular for agricultural livelihoods of smallholder farmers.[5]

To achieve the potential, groundwater resources need to be developed effectively and managed sustainably. For groundwater use to be sustainable, however, efforts must not impinge on the quantity or quality available for domestic uses and environmental processes and services. Local management of the resource, with appropriate national and external support, will be central to avoid depletion and intersectoral conflicts. At a comparatively early stage of groundwater development, it is crucial to put in place effective systems of governance. SSA's untapped groundwater resources represent a major opportunity to enable more widespread groundwater irrigation, but it needs to go hand in hand with greater emphasis on groundwater governance.[6]

Paul Pavelic, Karen G. Villholth and Shilp Verma

Notes

1. United Nations (2022). *The United Nations World Water Development Report 2022: Groundwater: Making the invisible visible.* UNESCO, Paris.
2. Wiggins, S. and B. Lankford (2019) Farmer-led irrigation in sub-Saharan Africa: synthesis of current understandings. London: Evidence & Policy Group of the DFID-ESRC Growth Research Programme, Overseas Development Institute.
3. Efficiency for Access Coalition (2019). Solar Water Pump Outlook 2019: Global trends and market opportunities. Report prepared by CLASP/Energy Saving Trust/Dalberg, September 2019, 32p. https://efficiencyforaccess.org/publications/solar-water-pump-outlook-2019-global-trends-and-market-opportunities
4. World Bank (2022). Making the Most of the African Continental Free Trade Area - Leveraging Trade and Foreign Direct Investment to Boost Growth and Reduce Poverty. 158 p. https://openknowledge.worldbank.org/handle/10986/37623
5. African Ministers' Council on Water (AMCOW) (2022). Groundwater for Africa's Resilience and Socioeconomic Transformation. A Call for greater attention to investing in groundwater to sustainably meet growing water demands under global environmental change. High-level Summary of White Paper. ISBN 978-978-56367-5-2. https://knowledgehub.amcow-online.org/resource/groundwater-for-africas-resilience-and-socioeconomic-transformation
6. Villholth, K.G., J. van der Gun, E. López-Gunn, K. Conti, and A. Garrido (Eds.) (2018). *Advances in Groundwater Governance.* Taylor & Francis Group, London, UK. ISBN: 1-138-02980-4. https://gripp.iwmi.org/wp-content/uploads/sites/2/2019/01/advances-in-groundwater-governance.pdf.

Index

For Product Safety Concerns and Information please contact our EU
representative GPSR@taylorandfrancis.com
Taylor & Francis Verlag GmbH, Kaufingerstraße 24, 80331 München, Germany

www.ingramcontent.com/pod-product-compliance
Ingram Content Group UK Ltd.
Pitfield, Milton Keynes, MK11 3LW, UK
UKHW052049230926
475702UK00025B/983
* 9 7 8 1 0 3 2 3 9 1 9 1 5 *